Oral Cancer: Prophylaxis, Etiopathogenesis and Treatment

Oral Cancer: Prophylaxis, Etiopathogenesis and Treatment

Guest Editors

Violeta Popovici
Emma Adriana Ozon

MDPI • Basel • Beijing • Wuhan • Barcelona • Belgrade • Novi Sad • Cluj • Manchester

Guest Editors
Violeta Popovici
INCE-CEMONT
Romanian Academy
Vatra-Dornei
Romania

Emma Adriana Ozon
Pharmacy Department
Carol Davila University of
Medicine and Pharmacy
Bucharest
Romania

Editorial Office
MDPI AG
Grosspeteranlage 5
4052 Basel, Switzerland

This is a reprint of the Special Issue, published open access by the journal *Current Issues in Molecular Biology* (ISSN 1467-3045), freely accessible at: https://www.mdpi.com/journal/cimb/special_issues/U2TX90NIK1.

For citation purposes, cite each article independently as indicated on the article page online and as indicated below:

Lastname, A.A.; Lastname, B.B. Article Title. *Journal Name* **Year**, *Volume Number*, Page Range.

ISBN 978-3-7258-3015-2 (Hbk)
ISBN 978-3-7258-3016-9 (PDF)
https://doi.org/10.3390/books978-3-7258-3016-9

Contents

current issues in molecular biology

MDPI

Editorial

Oral Cancer: Prophylaxis, Etiopathogenesis and Treatment

Violeta Popovici [1,*] and Emma Adriana Ozon [2]

1 Center for Mountain Economics, "Costin C. Kirițescu" National Institute of Economic Research (INCE-CEMONT), Romanian Academy, 725700 Vatra-Dornei, Romania
2 Department of Pharmaceutical Technology and Biopharmacy, Faculty of Pharmacy, "Carol Davila" University of Medicine and Pharmacy, 020956 Bucharest, Romania; emma.budura@umfcd.ro
* Correspondence: violeta.popovici@ce-mont.ro

Oral cancer contributes to approximately 3–10% of all cancer mortality worldwide, and its incidence is continuously increasing due to environmental conditions and harmful habits of the modern lifestyle [1]. In association with hereditary predisposition, chronic inflammation, and infectious diseases, these factors raise the frequency of oral cavity malignancies. Major intervention involves extensive surgery associated with radio and chemotherapy [1]. Moreover, advancements in cancer immunotherapy have not yet established the specific mechanisms by which immune cells influence tumor progression and immune evasion, and oral cancers continue to inadequately respond to the current treatment protocols. The abovementioned aspects underline this Special Issue, which has brought together the most recent research on molecular mechanisms implied in prophylaxis, etiopathogenesis, and the treatment of oral cancer. New and valuable data have been added to this field by the authors of the 10 articles contained in this publications: nine original papers and one review.

Oral squamous cell carcinoma (OSCC) represents 90% of oral neoplasias; it is the sixth most frequent malignancy in the world, with an overall 5-year survival rate below 50% due to its modest outcomes, tardive diagnosis, aggressive local invasiveness, recurrences, and metastases. OSCC belongs to the comprehensive group of head and neck squamous cell carcinomas (HNSCCs), which affects the cell lining of the oral cavity, pharynx, and larynx. The local–regional recurrence rate for advanced carcinomas is over 50% despite administration of multimodal therapy, and understanding the biology of HNSCCs is essential to ameliorate their prognosis [1].

Thus, the authors of the first contribution integrated the cells isolated from a hypopharyngeal tumor of a squamous cell carcinoma patient (FaDu) in an oral mucosa model (OMM) obtained by seeding primary human fibroblasts and keratinocytes onto a porcine small intestinal submucosa with preserved mucosa (SIS/MUC). They generated a 3D model that mimics the crosstalk at the tumor front of human HNSCC by enabling cellular and stromal interactions and revealing the features of tumor cell invasiveness [1]. Their work created the premise of integrating further tumor microenvironment components to establish the molecular mechanisms of the most effective anticancer therapy.

One of the most important components of stromal cells is tumor-associated macrophages, which are related to poor prognoses. The second contribution suggests that M2 macrophage-derived exosomes could induce OSCC cell proliferation, invasion, and migration and inhibit tumor cell apoptosis by transferring miRNA-23a-3p into tumor cells [2]. Additionally, the authors revealed that phosphatase and tensin homolog (PTEN), a well-known tumor-suppressor gene, could be a potential cellular target for miRNA-23a-3p to promote OSCC development.

In a retrospective study, Takasaki et al. examined the surgically resected tissues of 70 patients with oral potentially malignant disorders (OPMDs) via immunochemistry and statistically analyzed the association of the target proteins' (p53, p62, Ki67, and XPO1) expressions with intracellular distribution, malignant transformation, and clinicopathological characteristics [3]. They found that Ki67, a well-known marker of cell proliferation shows a

Citation: Popovici, V.; Ozon, E.A. Oral Cancer: Prophylaxis, Etiopathogenesis and Treatment. *Curr. Issues Mol. Biol.* **2024**, *46*, 12911–12913. https://doi.org/10.3390/cimb46110768

Received: 29 October 2024
Accepted: 11 November 2024
Published: 13 November 2024

significant positive correlation with p62 expression in the cell cytoplasm and aggregation expression and a negative one with p62 expression in the nucleus. This third contribution suggests that the autophagy-related multidomain protein p62 could be a potential biomarker for the risk of the neoplastic transformation of OPMDs [3].

Using human in vitro and ex vivo models of oral dysplasia from the tongue, Peña-Oyarzún et al. found that 1,25-(OH)2D3 increased nuclear vitamin D receptors (VDR) and membranous expression of E-cadherin and diminished the Ki67 expression and nuclear localization of β-catenin [4]. Thus, this fourth contribution shows that 0.1 μM Calcitriol treatment could diminish the risk of malignant transformation of OPMDs by reducing cell proliferation, migration, and β-catenin signaling [4].

Many types of cancer alter the energetic metabolism, thus decreasing oxidative processes and raising glucose uptake and glycolysis. Glycolytic activity significantly increases the development of HNSCCs, and this particularity could be considered a potential therapeutic target. Therefore, in the fifth contribution, Kleszcz et al. show that the expression of glycolytic enzymes in tongue squamous cell carcinoma could be modulated by the Wnt signaling pathway inhibitors PRI-724 and IWP-O1 [5].

MicroRNAs (miRNAs) are endogenous small RNA molecules that are single-stranded and non-coding; they circulate in a stable form and display aberrant expression in various malignancies. Differentially expressed miRNAs could be potential biomarkers for cancer screening. Moreover, next-generation cancer therapy could be based on miRNA modulation. In the sixth contribution, the authors identified a five-miRNA diagnostic model associated with tongue squamous cell carcinoma patients. They highlighted the remarkable diagnostic potential of miR-196b and selected five hub genes from the target ones of miR-196b [6].

CD44 is a transmembrane protein with important roles in cell proliferation, adhesion, migration, and lymphocyte activation. Various types of CD44 are expressed in cancer cells, especially in the advanced stages, and their expression is detected using monoclonal antibodies. In the seventh contribution, Ishikawa et al. analyzed C44Mab-18 (a novel anti-CD44 variant with 10 monoclonal antibodies) for immunohistochemical analysis of oral squamous cell carcinomas [7].

Recent studies investigating genome alteration in HNSCC patients have attracted wide attention [8]. The eighth contribution of this Special Issue describes novel mutations in MUC6 and MUC16, providing new insight into the genetic alternation in mucin genes among oropharyngeal squamous cell carcinoma (OPSCC) patients. This research could initiate further studies, including larger cohorts, to recognize the pattern in which the mutations affect oropharyngeal carcinogenesis.

Understanding the molecular mechanisms implied in the development and progression of OSCC is essential for improving diagnostic and therapeutic strategies. Considering oral squamous cell carcinomas in dogs as an excellent model for studying human counterparts, Files et al. investigated the significance of two key molecular components, Cox-2 and EGFR, in canine OSCC. Their findings revealed that Cox-2 was highly expressed in 70.6% of cases, while EGFR overexpression was observed in 44.1%; Cox-2 overexpression was associated with the histological grade of malignancy (HGM) and EGFR with vascular invasion. Therefore, the results of this ninth contribution suggest that Cox-2 and EGFR could be promising biomarkers and potential therapeutic targets, leading to the development of novel treatment strategies for OSCC therapy.

Analyzing the major risk factors for HNSCC, the tenth contribution highlights that human papillomavirus (HPV) is correlated with a high incidence of oropharyngeal cancers. Moreover, the accessed literature data show that HPV vaccines approved for cervical cancer prevention in females had a notable impact on HNSCC incidence [9]. In addition, this comprehensive review investigates various mechanisms of inducing immunogenicity against HNSCC cells, including traditional approaches (cell-mediated cytotoxicity induced by antigens), as well as innovative strategies (to counteract tumor immune escape mechanisms or stimulate the immune system's cytotoxic activity against neoplastic cells). The last three contributions are listed below.

Funding: This editorial received no external funding.

Acknowledgments: The Academic Editors of this Special Issue are grateful to all Contributors and Editors of *CIMB*, an MDPI journal, for their excellent collaboration and support.

Conflicts of Interest: The authors declare no conflicts of interest.

List of Contributions

1. Gaździcka, J.; Biernacki, K.; Salatino, S.; Gołąbek, K.; Hudy, D.; Świętek, A.; Miśkiewicz-Orczyk, K.; Koniewska, A.; Misiołek, M.; Strzelczyk, J.K. Sequencing Analysis of MUC6 and MUC16 Gene Fragments in Patients with Oropharyngeal Squamous Cell Carcinoma Reveals Novel Mutations: A Preliminary Study. *Curr. Issues Mol. Biol.* **2023**, *45*, 5645–5661. https://doi.org/10.3390/cimb45070356
2. Files, R.; Santos, C.; Queiroga, F.L.; Silva, F.; Delgado, L.; Pires, I.; Prada, J. Investigating Cox-2 and EGFR as Biomarkers in Canine Oral Squamous Cell Carcinoma: Implications for Diagnosis and Therapy. *Curr. Issues Mol. Biol.* **2024**, *46*, 485–497. https://doi.org/10.3390/cimb46010031
3. Meliante, P.G.; Petrella, C.; Fiore, M.; Minni, A.; Barbato, C. Head and Neck Squamous Cell Carcinoma Vaccine: Current Landscape and Perspectives. *Curr. Issues Mol. Biol.* **2023**, *45*, 9215–9233. https://doi.org/10.3390/cimb45110577

References

1. Stöth, M.; Mineif, A.T.; Sauer, F.; Meyer, T.J.; Mueller-Diesing, F.; Haug, L.; Scherzad, A.; Steinke, M.; Rossi, A.; Hackenberg, S. A Tissue Engineered 3D Model of Cancer Cell Invasion for Human Head and Neck Squamous-Cell Carcinoma. *Curr. Issues Mol. Biol.* **2024**, *46*, 4049–4062. [CrossRef] [PubMed]
2. Li, J.; Bao, Y.; Peng, S.; Jiang, C.; Zhu, L.; Zou, S.; Xu, J.; Li, Y. M2 Macrophages-Derived Exosomal miRNA-23a-3p Promotes the Progression of Oral Squamous Cell Carcinoma by Targeting PTEN. *Curr. Issues Mol. Biol.* **2023**, *45*, 4936–4947. [CrossRef] [PubMed]
3. Takasaki, R.; Uchida, F.; Takaoka, S.; Ishii, R.; Fukuzawa, S.; Warabi, E.; Ishibashi-Kanno, N.; Yamagata, K.; Bukawa, H.; Yanagawa, T. p62 Is a Potential Biomarker for Risk of Malignant Transformation of Oral Potentially Malignant Disorders (OPMDs). *Curr. Issues Mol. Biol.* **2023**, *45*, 7630–7641. [CrossRef] [PubMed]
4. Peña-Oyarzún, D.; Guzmán, C.; Kretschmar, C.; Torres, V.A.; Maturana-Ramirez, A.; Aitken, J.; Reyes, M. Calcitriol Treatment Decreases Cell Migration, Viability and β-Catenin Signaling in Oral Dysplasia. *Curr. Issues Mol. Biol.* **2024**, *46*, 3050–3062. [CrossRef] [PubMed]
5. Kleszcz, R.; Paluszczak, J.; Belka, M.; Krajka-Kuźniak, V. PRI-724 and IWP-O1 Wnt Signaling Pathway Inhibitors Modulate the Expression of Glycolytic Enzymes in Tongue Cancer Cell Lines. *Curr. Issues Mol. Biol.* **2023**, *45*, 9579–9592. [CrossRef] [PubMed]
6. Park, Y.-N.; Ryu, J.-K.; Ju, Y. The Potential MicroRNA Diagnostic Biomarkers in Oral Squamous Cell Carcinoma of the Tongue. *Curr. Issues Mol. Biol.* **2024**, *46*, 6746–6756. [CrossRef] [PubMed]
7. Ishikawa, K.; Suzuki, H.; Kaneko, M.K.; Kato, Y. Establishment of a Novel Anti-CD44 Variant 10 Monoclonal Antibody C44Mab-18 for Immunohistochemical Analysis against Oral Squamous Cell Carcinomas. *Curr. Issues Mol. Biol.* **2023**, *45*, 5248–5262. [CrossRef] [PubMed]
8. Chen, X.; Zhao, W.; Chen, S.; Yu, D. Mutation Profiles of Oral Squamous Cell Carcinoma Cells. *Adv. Oral Maxillofac. Surg.* **2021**, *2*, 100026. [CrossRef]
9. Villa, A.; Patton, L.L.; Giuliano, A.R.; Estrich, C.G.; Pahlke, S.C.; O'Brien, K.K.; Lipman, R.D.; Araujo, M.W.B. Summary of the Evidence on the Safety, Efficacy, and Effectiveness of Human Papillomavirus Vaccines. *J. Am. Dent. Assoc.* **2020**, *151*, 245–254. [CrossRef] [PubMed]

current issues in molecular biology

Article

M2 Macrophages-Derived Exosomal miRNA-23a-3p Promotes the Progression of Oral Squamous Cell Carcinoma by Targeting PTEN

Jun Li [1,2,3,†], Yongjie Bao [1,2,3,†], Sisi Peng [1,2,3], Chao Jiang [1,2,3], Luying Zhu [1,2,3], Sihai Zou [1,2,3], Jie Xu [1,2,3,*] and Yong Li [1,2,3,*]

1 College of Stomatology, Chongqing Medical University, Chongqing 401147, China
2 Chongqing Key Laboratory for Oral Diseases and Biomedical Sciences, Chongqing 401147, China
3 Chongqing Municipal Key Laboratory of Oral Biomedical Engineering of Higher Education, Chongqing 401147, China
* Correspondence: xujie@hospital.cqmu.edu.cn (J.X.); 500081@hospital.cqmu.edu.cn (Y.L.)
† These authors contributed equally to this work.

Abstract: Exosomes from tumor cells and immune cells regulate the tumor microenvironment through the biomolecules or microRNAs (miRNAs) they carry. This research aims to investigate the role of miRNA in exosomes derived from tumor-associated macrophages (TAMs) in the progression of oral squamous cell carcinoma (OSCC). RT-qPCR and Western blotting assays were used to determine the expression of genes and proteins in OSCC cells. CCK-8, Scratch assay and invasion-related proteins were utilized to detect the malignant progression of tumor cells. High-throughput sequencing predicted differentially expressed miRNAs in exosomes secreted by M0 and M2 macrophages. Compared with exosomes from M0 macrophages, exosomes from M2 macrophages led to enhanced proliferation and invasion of OSCC cells and inhibited their apoptosis. High-throughput sequencing results show that miR-23a-3p is differentially expressed in exosomes from M0 and M2 macrophages. MiRNA target gene database predicts that phosphatase and tensin homolog (PTEN) are target genes of miR-23a-3p. Further studies revealed that transfection of miR-23a-3p mimics inhibited PTEN expression in vivo and in vitro and promoted the malignant progression of OSCC cells, which was reversed by miR-23a-3p inhibitors. MiR-23a-3p in exosomes derived from M2 macrophages promotes malignant progression of OSCC. PTEN is a potential intracellular target of miR-23a-3p. MiR-23a-3p, an M2 macrophage-associated exosome, is a promising target for the future treatment of OSCC.

Keywords: OSCC; M2 macrophages; exosome; MiR-23a-3p; PTEN

Citation: Li, J.; Bao, Y.; Peng, S.; Jiang, C.; Zhu, L.; Zou, S.; Xu, J.; Li, Y. M2 Macrophages-Derived Exosomal miRNA-23a-3p Promotes the Progression of Oral Squamous Cell Carcinoma by Targeting PTEN. *Curr. Issues Mol. Biol.* **2023**, *45*, 4936–4947. https://doi.org/10.3390/cimb45060314

Academic Editor: Violeta Popovici

Received: 16 May 2023
Revised: 31 May 2023
Accepted: 1 June 2023
Published: 7 June 2023

1. Introduction

More than 90% of oral cancers are reported to be squamous cell carcinomas [1]. Oral squamous cell carcinoma is considered the most common maxillofacial malignancy, with a high rate of lymph node metastasis and a low five-year survival rate [2]. Currently, the treatment strategy combines surgery, radiotherapy, chemotherapy, and other methods, but achieving satisfactory results takes time [3]. Thus, it is necessary to find a more effective systemic therapy.

One of the critical components of stromal cells is tumor-associated macrophages, which are associated with poor prognoses [4]. Macrophages adopt different polarization states depending on the type of stimulus. Two central polarization states of the macrophage phenotype have been described: M1 macrophages and M2 macrophages [5]. M1 macrophages tend to promote inflammation and counteract tumors, whereas M2 macrophages frequently resist inflammation and contribute to tumors [6]. The macrophages in TME are primarily M1 polarized and play the anti-tumor role in the early stage of the tumor. Meanwhile, in the late stage of cancer, the macrophages in TME can change from M1 to M2 macrophages and participate in the progression of the tumor [7]. The promoting effect of M2 macrophages on the development of tumors has been confirmed in many reports, including gastric cancer and OSCC [8,9]. A study also shows that when the proportion of M2/M1 macrophage increases, it is often associated with tumor proliferation and unfavorable prognosis [10]. However, the mechanisms underlying the promotion of OSCC progression by M2 macrophages are still in need of further exploration.

In the TME, cells secrete soluble factors and two types of extracellular vesicles: microvesicles and exosomes. Exosomes with a diameter of 30–100 nm are membranous vesicles that cells secrete into the extracellular environment, which contain many proteins, lipids and genetic materials from parental cells [11]. This is one of the main communication pathways in TME and it transmits a variety of proteins, mRNAs and miRNAs between cells. As small non-coding RNA, miRNA regulate gene expression at the post-translational stage [12]. They can be used as inhibitors or enhancers of crucial signaling pathways and proteins, affecting various aspects of cancer biology [13].

Based on the evidence described above, we collected exosomes from M0 and M2 macrophages and treated tumor cells with them in order to better explore the role played by M2 macrophage exosomes in tumors. This study shows that miRNA-23a-3p is abundantly expressed in M2 macrophage exosomes and promotes malignant progression of tumor cells. PTEN is a target potential for miRNA-23a-3p. This study offers a new target for the prevention and treatment of OSCC.

2. Materials and Methods

2.1. Cell Culture and Transfection

Cal-27 (ATCC, Manassas, VA, USA) cells were cultured at 37 °C in an incubator with 5% CO2 using high sugar BMEM medium (Biological Industries, Kibbutz, Israel) containing fetal bovine serum (Biological Industries, Israel). Cal-27 are seeded into a 6-well plate when they are cultured to a density of 70–80%. Cal-27 cells were transfected with miR-23a-3p mimic using EntransterTM-R4000 (Engreen, Beijing, China) following the protocols of Ribobio (Guangzhou, China).

THP-1 (ATCC, USA) were cultured with RPMI-1640 medium (Biological Industries, Israel). When the density of THP-1 reached 70–80%, 100 ng/mL PMA (Sigma-Aldrich, St Louis, MS, USA) was added for 24 h to induce M0 macrophages. The medium was then changed and the stimulation of M0 macrophages was continued with 20 ng/mL IL-4 (Sigma-Aldrich, USA) for 24 h to induce the formation of M2 macrophages [14].

2.2. Isolation and Characterization of Exosomes

M0 and M2 macrophages were cultured in serum-free RPMI-1640 medium for 24 h. Exosomes in the cell supernatant were obtained via differential centrifugation and stored at −80 °C. After the exosomes were diluted using 1× PBS, VivaCell Biosciences and ZetaView PMX 110 (Particle Metrix, Meerbusch, Germany) were used to measure the size and concentration of the exosomes. The system was calibrated using 110 nm polystyrene particles. The temperature was 26.62 °C.

2.3. Reverse Transcription Polymerase Chain Reaction (RT-PCR)

Total intracellular RNA was extracted by RNAiso Plus kit (TaKaRa, Nogihigashi, Japan). RNA concentration was identified by NanoDrop 2000 spectrophotometer (Thermo, Waltham, MA, USA). cDNA was generated using SYBR Premix Ex Taq II (TaKaRa, Japan). TB Green (TaKaRa, Japan) and cDNA were mixed, and qPCR was performed for 40 cycles in Bio-Rad's PCR system. The total reaction volume was 20 μL. GAPDH and U6 were used as internal references. Relative quantification of genes was performed using the $2^{-\Delta\Delta Ct}$ method.

2.4. Western Blot

The treated cells were lysed with RIPA (Beyotime, Shanghai, China) containing protease inhibitors. The protein concentration was determined with the BCA protein quantification kit (Beyotime, China). After that, the protein was separated by SDS-PAGE gel electrophoresis and transferred to a PVDF membrane. PVDF membranes were closed with 5% skim milk powder for 2 h at room temperature, and the primary antibodies CD206 (1:1000; CST), CD163 (1:1000; Abcam), CD68 (1:1000; Abcam), CD86 (1:1000; Bioss), CD63 (1:500; Abcam), CD9 (1:500; Abcam), Caspase-3 (1:750; Bioss), Caspase-9 (1:750; Bioss), MMP-2 (1:1000; HUABIO), MMP-9 (1:1000; HUABIO), E-cadherin (1:1000; Abcam), Vimentin (1:1000; Abcam), and PTEN (1:1000; Abcam) were incubated at 4 °C overnight. After incubation of the secondary antibody for 1 h at room temperature, ECL reagent (Beyotime, Shanghai, China) was added and the bands were observed by Bio-Rad GelDoc 2000.

2.5. Cell Counting Kit-8 Assay

Cell Counting Kit-8 was used to test cell viability. The cells were seeded into a 96-well plate (3×10^3/well) in the cell viability assay. After cell adhesion, the cells were stimulated with macrophage exosomes for 0, 24, 48, 72 h. Then, 10 μL CCK-8 solution was added to each well and the absorbance value was measured at 450 nm after incubation for 1 h.

2.6. Wound Healing Assay

The wound healing assay was performed to measure cell motility. Treated cells were inoculated into 6-well plates at a density of 5 × 105/mL and incubated for 24 h. Then, parallel lines were drawn in the plate, and the serum-free medium was used instead of the complete medium. Wound closure was observed under a microscope at 0, 24 and 48 h. Image J software version 1.44 (National Institutes of Health, Bethesda, MA, USA) was used to measure the wound area.

2.7. miRNA Sequencing and Bioinformatics

The supernatants of M0 and M2 macrophages were collected and centrifuged to remove cells and debris. miRNA extraction, library preparation, and sequencing were performed at a commercial facility (Ribobio, Guangzhou, China). The Illumina HiSeq 2500 platform was chosen for sequencing. Data were collected using Illumina analysis software 2.2.68 version. The target genes of miR-23a-3p were predicted in TargetScan, miRDB, and miRTarBase databases.

2.8. Animal Studies

BALB/c nude mice (female, 4 weeks old) were purchased from Slekkinda Laboratory Animal Co., Ltd. (Shanghai, China). The mice lived in an environment with sufficient light, air, food, and water, and no pathogens. Treated or untreated Cal-27 cells (2×10^6 cells in 0.1 mL PBS per mouse) were injected subcutaneously into the axillae of mice (n = 4). After one month, tumor-bearing mice were euthanized, and tumor tissue was removed and weighed. BALB/c nude mice were cultivated following the Guidelines for the Care and Use of Experimental Animals.

2.9. Tunel Staining

Paraffin sections were dewaxed with xylene and treated with 20 μg/mL proteinase K for 15–30 min. They were then washed 3 times with PBS. Immediately after, FITC-labeled TUNEL detection solution (Beyotime, Shanghai, China) was dropped on the sample, followed by incubation at 37 °C for 60 min in the dark. PBS wash was followed by dropwise addition of DAPI containing anti-fluorescence attenuation medium (Solarbio, Beijing, China), then the coverslip was fixed. FITC-labeled TUNEL-positive cells were observed at 488 nm excitation wavelength.

2.10. Immunohistochemistry

Immunohistochemical staining for Ki67 (1:200, Bioss) was performed. After conventional paraffin embedding and section, xenograft tumors were dewaxed with xylene, hydrated with ethanol, and sealed with 3% H_2O_2 for 10 min. After sections were closed with 10% goat serum, primary antibodies were incubated overnight at 4 °C. On the second day, a second antibody coupled with HRP was used for testing. Diaminobenzidine (DAB) was used as a chromogenic substrate. All slides were restained with hematoxylin. Images of tissue-stained sections were obtained using Olympus microscope (Olympus, Tokyo, Japan).

2.11. Statistical Analysis

All experiments were repeated three times. The data for each group were expressed as the mean ± standard deviation or the mean ± standard error of each group's mean (S.E.M.). Differences between groups were compared using *t*-test or one-way ANOVA. Statistical analyses and graphs were performed using SPSS 18.0 (Chicago, IL, USA) and GraphPad Prism 8 (La Jolla, CA, USA). $p < 0.05$ was considered to be statistically significant.

3. Results

3.1. Extraction and Identification of M0 and M2 Exosomes

After PMA and IL-4 stimulation, M0 and M2 macrophages were obtained. Under the microscope, M0 macrophages were observed to be round and M2 macrophages were observed to be spindle shaped (Figure 1A). Western blot and RT-PCR analysis were used to identify the surface markers of macrophages to demonstrate successful induction of M0 and M2 macrophages. The results showed that CD206 and CD163 were highly expressed in M2 macrophages, while CD86 (M1 macrophage marker) was not significantly different between M0 and M2 (Figure 1B,C). The exosomes in the supernatant were extracted via differential centrifugation (Figure 1D) and then identified. Transmission electron microscopy revealed that the vesicles isolated from the supernatant of macrophages have the characteristics of exosomes: double-layer membrane structure and disc shape (Figure 1E). As per the results displayed in the NTA detection, the diameter of the vesicles was mainly concentrated in 50–200 nm (Figure 1F). In addition, the exosome markers (CD9 and CD63) expression was observed via protein blot (Figure 1G). These findings illustrated that M0 and M2 macrophages were successfully induced, and their exosomes were extracted.

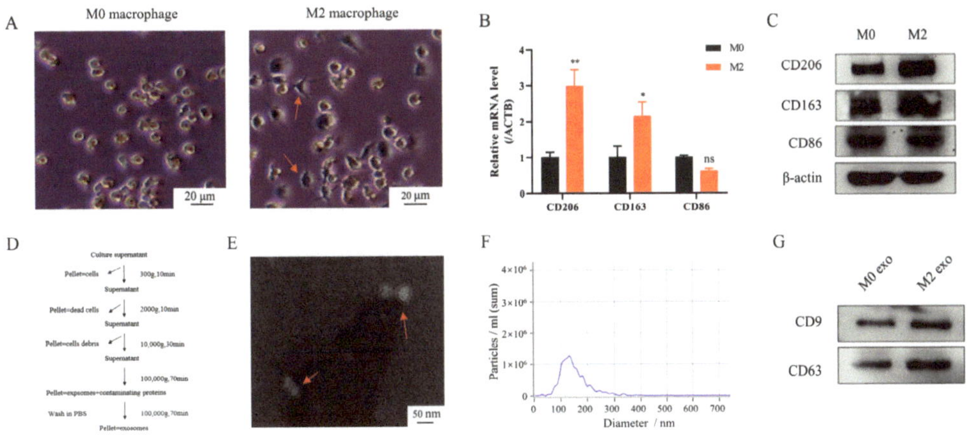

Figure 1. Extraction and identification of M0 and M2 exosomes. After 24 h of PMA (100 ng/mL) stimulation, THP-1 cell lines were treated with IL-4 (20 ng/mL) for 24 h. (**A**–**C**) Macrophage surface markers were detected. (**A**) Morphology of macrophages was observed under microscope.

(**B**) RT-PCR analysis, (**C**) Western blot analysis. (**D**) Exosomes were obtained from cell supernatant via differential centrifugation. (**E–G**) Identification of exosomes: (**E**) Identification of exosome structure under an electron microscope (20,000×). (**F**) Detection of the size and number of exosomes via nanoparticle tracking analysis (dilution factor: 1:1000). (**G**) Detection of exosome markers with Western blot analysis. Data represent the mean ± standard errors of three separate experiments. * $p < 0.05$, ** $p < 0.01$, ns: not significant according to one-way ANOVA (versus M0 group).

3.2. M2 Macrophage Promotes Malignant Progression of OSCC

Afterward, to probe the function of M2 macrophage exosomes on OSCC, CCK-8 was performed to detect the viability of tumor cells. As a result, compared with the control group, M0 sup group, and M0 exo group, the activity of Cal-27 cells treated with M2 macrophage supernatant and M2 macrophage exosomes increased significantly (Figure 2A). Additionally, Western blot and RT-PCR analysis observed that supernatants and exo of M2 macrophages significantly inhibited the expression of apoptosis-related proteins (Caspase-3 and Caspase-9) in tumor cells (Figure 2B,C). In contrast, there was no statistical difference between M0 sup, M0 exo, and the control group. These results suggest that exosomes of M2 macrophages inhibit the apoptosis of tumor cells and enhance their cell viability.

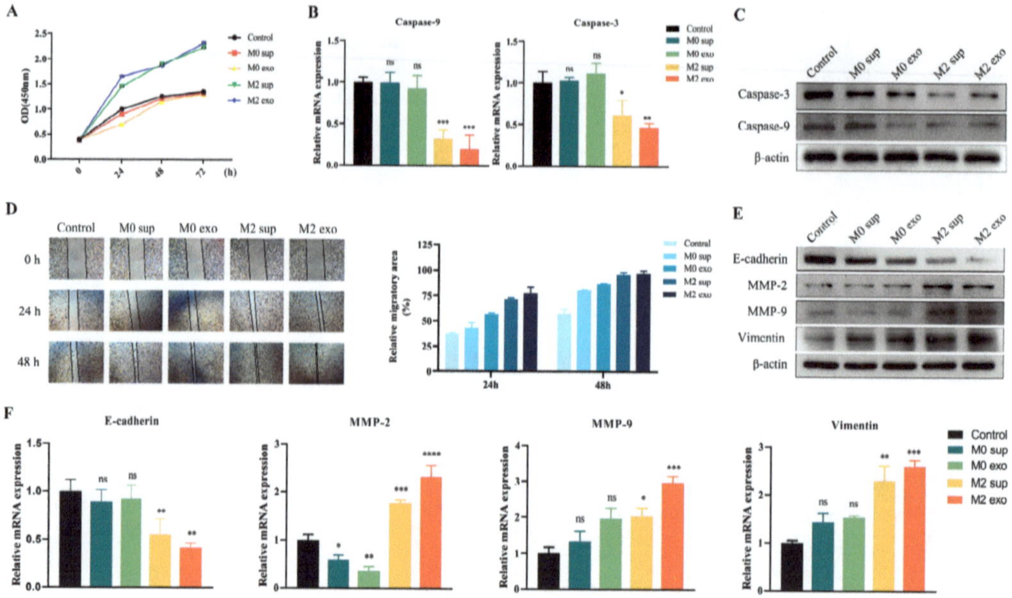

Figure 2. M2 macrophage exosomes promote the progression of OSCC cells (**A**) CCK-8 assays were performed to determine cell viability at 0, 24, 48, and 72 h after cell treatment. (**B,C**) Effect of M2 macrophage-derived exosomes on apoptosis-related cytokine in OSCC cells. (**B**) RT-PCR analysis. (**C**) Western blot analysis. (**D**) The migration activity of Cal-27 cells was measured via wound healing assay. The wound area was calculated at 0 h, 24 h, and 48 h after cell treatment. (**E,F**) The expression of EMT related protein in Cal-27 cells is affected by M2 macrophage-derived exosomes. (**E**) RT-PCR analysis. (**F**) Western blot analysis. * $p < 0.05$, ** $p < 0.01$, *** $p < 0.001$, **** $p < 0.0001$, ns: not significant by one-way ANOVA (versus control group).

Subsequently, the promoting effect of M2 macrophages on tumor migration and invasion was also observed. M2 macrophage exosome treatment enhanced the migration ability of tumor cells (Figure 2D). The increased MMP-2, MMP-9 and waveform proteins, as well as the decreased E-calmodulin in tumor cells, tend to promote the process of epithelial–mesenchymal transition (EMT), thus favoring tumor expansion into adjacent tissues. The results showed that the expression levels of MMP-2, MMP-9 and Vimentin increased, and E-cadherin decreased in the cells treated with M2 sup and M2 exo. In contrast, the expressions of MMP-2, MMP9 and E-cadherin in the M0-sup and M0-exo groups were not significantly different from those in the control group (Figure 2E,F). The above conclusions indicate that M2 macrophage exosomes can enhance the proliferation and migration ability of Cal-27 cells.

3.3. MiR-23a-3p Promotes Malignant Progression of OSCC Cells

To identify essential factors contributing to the malignant progression of OSCC cells in M2 exosomes, high-throughput sequencing was used to determine the differences in miRNA expression in M0 and M2 macrophage exosomes. Based on the high-throughput sequencing results, we obtained many miRNAs differentially expressed between M0 and M2 macrophages. Then, we screened the top ten miRNAs with statistical significance and the most considerable diversity. In the results, the content of miR-23a-3p in the exosomes of M2 macrophages was the highest. It was significantly higher than in the exosomes of M0 macrophages (Figure 3A). Next, we transfected Cal-27 cells with miR-23a-3p mimics (Figure 3B) to explore whether miR-23a-3p derived from M2 macrophage exosomes caused the malignant progression of Cal-27 cells described above. The results were consistent with our prediction that transfection of miR-23a-3p mimics caused Cal-27 cells to exhibit enhanced value-added capacity and inhibition of apoptosis (Figure 3C–E).

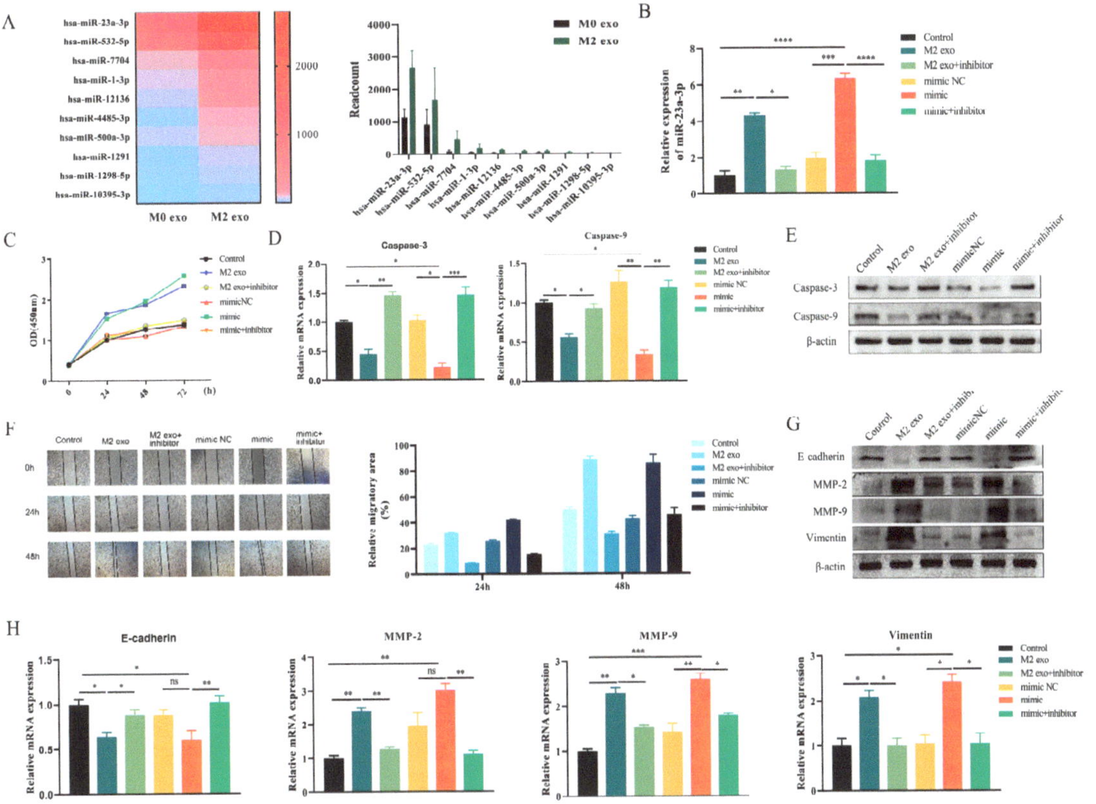

Figure 3. MiR-23a-3p promotes malignant progression of OSCC cells (**A**) In silico analysis of regulatory miRNAs related to M0 and M2 macrophage exosomes. Cal-27 was treated with M2 macrophage exosomes or miR-23a-3p mimics after stable transfection with miR-23a-3p inhibitor. (**B**) RT-PCR analysis of miR-23a-3p in Cal-27 cells after transfection. (**C**) CCK-8 assays were performed to determine

cell viability at 0, 24, 48, and 72 h after cell treatment. (**D**,**E**) RT-PCR and Western blot were used to detect the levels of apoptosis-related factors. (**F**) The motility of Cal-27 cells was determined via wound healing assays. EMT-associated proteins in Cal-27 cells were detected via Western blot (**G**) and RT-PCR (**H**). * $p < 0.05$, ** $p < 0.01$, *** $p < 0.001$, **** $p < 0.0001$, ns: not significant according to one-way ANOVA.

In line with this, high intracellular expression of miR-23a-3p also promoted the migration ability of Cal-27 cells (Figure 3F). Meanwhile, the expression of EMT-related proteins in tumor cells increased after transfection of miR-23a-3p, which facilitated its invasion (Figure 3G,H). These data indicate that MiR-23a-3p in exosomes secreted by M2-type macrophages can enter OSCC cells and enable OSCC cells to develop in a self-beneficial direction.

3.4. MiR-23a-3p Targets PTEN

The TargetScan, miRDB, and miRTarBase databases were applied to predict the target genes of miR-23a-3p to reveal the mechanism of malignant progression of OSCC cells caused by miR-23a-3p derived from exosomes of M2 macrophages. Then, 78 target genes of miR-23a-3p were screened from three MiRNA databases, and pathway enrichment of 78 target genes was obtained (Figure 4A,B). Among the 78 target genes we obtained, PTEN is a common tumor suppressor gene known to play an essential regulatory role in the occurrence and development of tumors [15–17]. In addition, we detected a match between the sequence of miR-23a-3p and PTEN, further suggesting that PTEN is a potential target for miR-23a-3p (Figure 4C). Therefore, we further studied the effect of miR-23a-3p on the expression of PTEN in tumor cells. RT-PCR and Western blot results showed that PTEN expression in Cal-27 was significantly inhibited when M2 macrophage exosomes were used to stimulate tumor cells. The same results were obtained through transfection of tumor cells with miR-23a-3p mimics, but this was reversed by inhibitors of miR-23a-3p (Figure 4D–F).

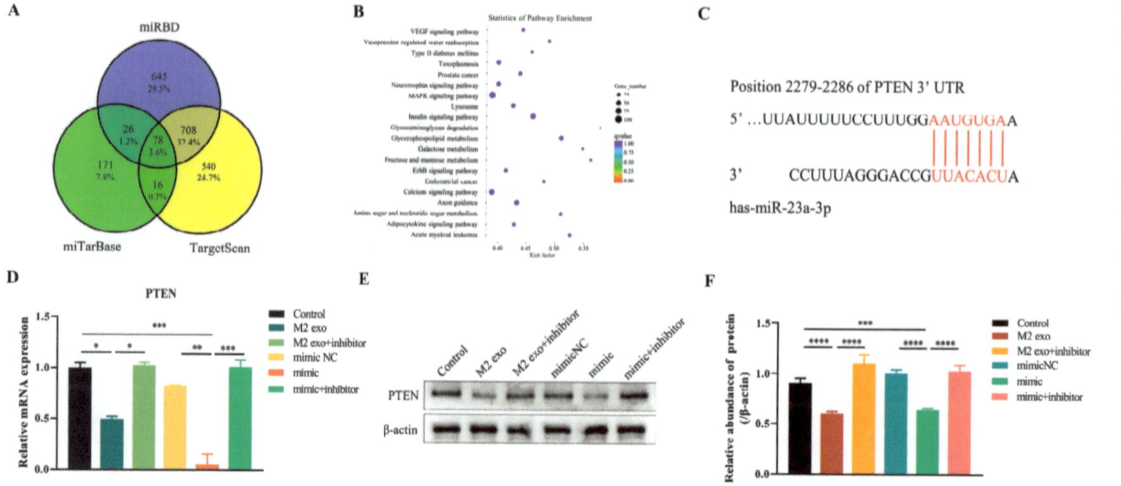

Figure 4. MiR-23a-3p targets PTEN (**A**) The three circles represent the target gene of miR-23a-3p in the three miRNA databases, and the middle part represents the intersection of the three datasets. (**B**) KEGG analysis of 78 target genes. (**C**) The predicted binding sites between miR-23a-3p and PTEN genes were predicted in the miRBD database. (**D**,**E**) RT-PCR and Western blot analysis of miR-23a-3p in Cal-27 cells. (**F**) Quantitative analysis. * $p < 0.05$, ** $p < 0.01$, *** $p < 0.001$, **** $p < 0.0001$ according to one-way ANOVA.

3.5. MiR-23a-3p Promotes Tumor Growth In Vivo

To further assess the effect of miR-23a-3p on tumors in vivo, we injected miR-23a-3p mimic-transfected Cal-27 cells subcutaneously into mice and observed the growth of tumors in vivo. The results showed a significant increase in tumor size and weight in the miR-23a-3p mimic group compared to the other groups (Figure 5A–C). In tumor specimens, the percentage of Tunel-positive cells was significantly lower in the miR-23a-3p transfected group, accompanied by an increase in Ki67-positive cells. This indicates that miR-23a-3p enables tumors to acquire greater proliferation ability (Figure 5D,E). In addition, consistent with the results of previous cellular experiments, miR-23a-3p also significantly inhibited the expression of PTEN in tumor tissues (Figure 5F,G).

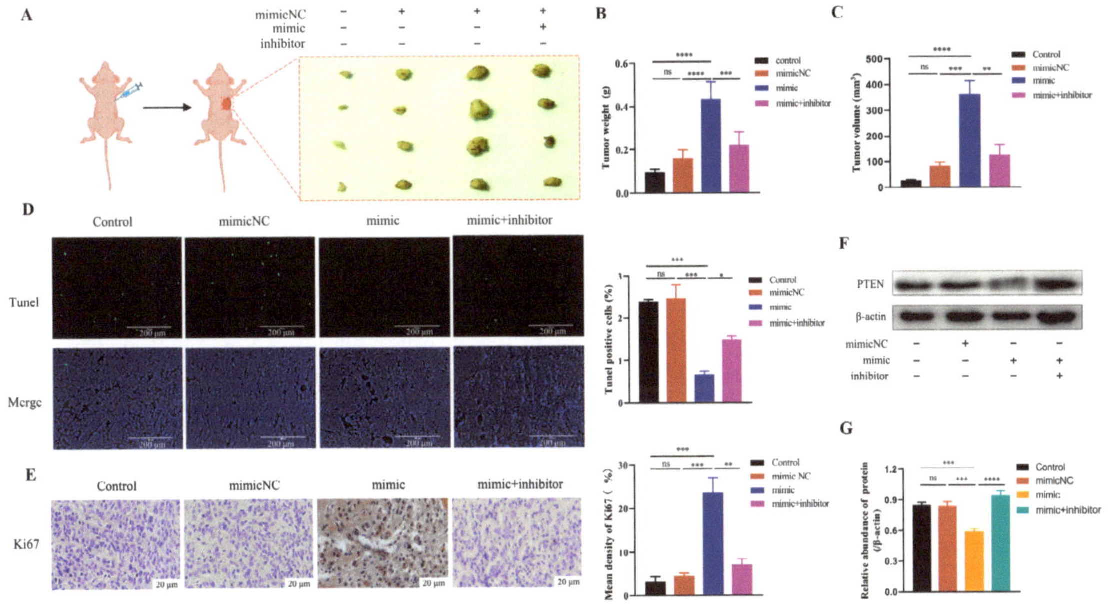

Figure 5. MiR-23a-3p promotes tumor growth in vivo. (**A**) The tumor. (**B**) Tumor weight. (**C**) Tumor volume. (**D**) Tunel-positive cells (green fluorescence) and their proportion. The nucleus is stained with DAPI (blue). (**E**) Immunohistochemical analyses of Ki67 in tumors. (**F**,**G**) Expression level of PTEN in tumor tissues. * $p < 0.05$, ** $p < 0.01$, *** $p < 0.001$, **** $p < 0.0001$, ns: not significant according to one-way ANOVA.

4. Discussion

Macrophages aggregated in the TME are commonly referred to as TAMs. Many studies have shown that TAMs are effective promoters of tumor development and metastasis [18]. In particular, the tumor microenvironment will be altered during the polarization of M0 macrophages to M2-type macrophages. M2 macrophages tend to exert immunosuppressive effects to control inflammation and frequently promote tumor progression [6]. Recently, many studies have shown that exosomes with immunosuppressive activity are released mainly by M2 macrophages, promoting cancer progression and treatment resistance [19,20].

Exosomes have been studied in different diseases, and they are both pathogenic and protective. They play multiple roles in the microenvironment, reshaping the extracellular matrix (ECM) and mediating the transmission of signals and molecules between cells [21]. As cell-secreted vesicles, exosomes are readily absorbed by cells. Therefore, they have potential applications in cancer immunotherapy [22]. More and more studies have shown that exosomes can change their cell origin and disease state by transporting biologically active substances [23,24].

When we extracted exosomes from M0 and M2 macrophages to stimulate OSCC cells, we found that M2 macrophage-derived exosomes positively modulated OSCC progression, which manifested as enhanced cell proliferation and migration and inhibition of apoptosis. EMT is a process of epithelial dysfunction and increased motility that is critical in in tumor progression, metastasis, and drug resistance [25]. The upregulation of MMP-2, MMP-9 and Vimentin and the inhibition of E-cadherin in tumor cells will promote EMT and contribute to tumor invasion [26–28]. In subsequent studies, we also found that the exosomes of M2 macrophages enhanced the migration ability of tumor cells and promoted the TMT process by regulating the expression of MMP-2, MMP-9, Vimentin, and E-cadherin. These pieces of evidence suggest that exosomes of M2 macrophages benefit tumor growth and metastasis.

MiRNA and other non-coding RNAs are important regulatory components of exosomes, and they are taken up by surrounding cells during the exosomal cycle [12]. They can be used as inhibitors or enhancers of key signaling pathways and proteins, thus affecting different aspects of cancer biology [13]. MiRNAs not only play a role in transcriptional activation, epigenetic regulation, and translation inhibition, but have also been found in the mitochondria and nucleus [29]. We speculate that a specific miRNA is also responsible for the malignant progression of Cal-27 cells in the exosomes of M2 macrophages. Fortunately, we identified a highly differentially expressed miRNA in the exosomes of M0 and M2 macrophages via high-throughput sequencing: miRNA-23a-3pin. Then, we transfected Cal-27 cells with mimics of miRNA-23a-3p to verify the effect of miRNA-23a-3p on OSCC. Not surprisingly, we obtained the same results in Cal-27 cells transfected with miRNA-23a-3p as when stimulated with M2 macrophage exosomes. This suggests that miRNA-23a-3p in M2 macrophage exosomes can lead to the development of OSCC.

Immediately, to explore the mechanism of mirNA-23A-3p's effect on OSCC cells, we analyzed the target genes of mirNA-23A-3p using the miRNA target gene database (targetscan, mirdb, and mirtarbase). PTEN mutation is one of the critical factors in human cancer development and is a known tumor suppressor gene [30,31]. Many studies have confirmed that the loss of PTEN can promote the development and metastasis of tumors, including breast cancer, testicular germ cell tumors, and cervical cancer [32,33]. In OSCC, the antitumor effect of PTEN has also been confirmed [34–36]. So, we boldly speculated that miRNA-23a-3p promoted the progress of OSCC by targeting the expression of PTEN in OSCC cells. We successfully observed a binding site between miRNA-23a-3p and PTEN and confirmed the targeting of PTEN by miRNA23a-3p at the cellular level. The tumor-promoting effect of miRNA-23a-3p achieved by inhibiting PTEN has also been observed in vivo.

5. Conclusions

In brief, the above findings suggest that M2 macrophage-derived exosomes promote OSCC cell proliferation, invasion, and migration and inhibit OSCC cell apoptosis by transferring miRNA-23a-3p into OSCC cells. Additionally, PTEN is a potential cellular target for miRNA-23a-3p to promote tumor development (Figure 6). This study may provide new biomarkers for the treatment of OSCC, but more work is still needed.

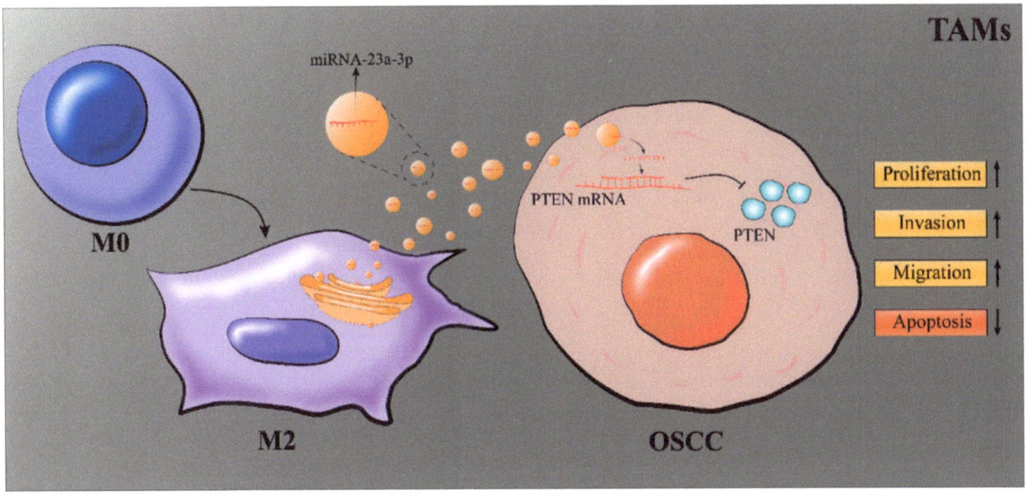

Figure 6. Graphical illustration of the effect of miRNA-23a-3p on OSCC cells.

Author Contributions: J.L.: Methodology (equal); visualization (equal); writing—original draft (equal); investigation (equal). Y.B.: Methodology (equal); writing—original draft (equal); formal analysis (equal). S.P.: Data curation (equal); methodology (equal); writing—review and editing (equal). C.J.: Data curation (equal); methodology (equal). L.Z.: Conceptualization (equal); methodology (equal). S.Z.: Data curation (equal); investigation (equal); methodology (equal). J.X.: Supervision (equal); writing—review and editing (equal). Y.L.: Conceptualization (equal); resources (equal); writing—review and editing (equal). All authors have read and agreed to the published version of the manuscript.

Funding: This research was supported by the National Natural Science Foundation of China (81800999), The Chongqing Technology Innovation and Application Development Special Project (cstc2019jscx-msxnX0173).

Institutional Review Board Statement: This study was approved by the Animal Ethics Committee of the Stomatological Hospital of Chongqing Medical University (CQHS-REC-2021/LSNo.21).

Data Availability Statement: Data supporting this study can be obtained according to the methods in this paper. Reasonable requests for data in the article can be obtained by contacting the corresponding author.

Conflicts of Interest: The authors declare that the research was conducted without any commercial or financial relationships that could be construed as a potential conflict of interest.

References

1. Khurshid, Z.; Zafar, M.S.; Khan, R.S.; Najeeb, S.; Slowey, P.D.; Rehman, I.U. Role of Salivary Biomarkers in Oral Cancer Detection. *Adv. Clin. Chem.* **2018**, *86*, 23–70. [CrossRef] [PubMed]
2. Gan, C.P.; Sam, K.K.; Yee, P.S.; Zainal, N.S.; Lee, B.K.B.; Rahman, Z.A.A.; Patel, V.; Tan, A.C.; Zain, R.B.; Cheong, S.C. IFITM3 knockdown reduces the expression of CCND1 and CDK4 and suppresses the growth of oral squamous cell carcinoma cells. *Cell. Oncol.* **2019**, *42*, 477–490. [CrossRef] [PubMed]
3. Khan, T.; Relitti, N.; Brindisi, M.; Magnano, S.; Zisterer, D.; Gemma, S.; Butini, S.; Campiani, G. Autophagy modulators for the treatment of oral and esophageal squamous cell carcinomas. *Med. Res. Rev.* **2019**, *40*, 1002–1060. [CrossRef] [PubMed]
4. Liu, M.; O'connor, R.S.; Trefely, S.; Graham, K.; Snyder, N.W.; Beatty, G.L. Metabolic rewiring of macrophages by CpG potentiates clearance of cancer cells and overcomes tumor-expressed CD47−mediated 'don't-eat-me' signal. *Nat. Immunol.* **2020**, *20*, 265–275. [CrossRef]
5. Lin, Y.; Xu, J.; Lan, H. Tumor-associated macrophages in tumor metastasis: Biological roles and clinical therapeutic applications. *J. Hematol. Oncol.* **2019**, *12*, 76. [CrossRef] [PubMed]
6. Baig, M.S.; Roy, A.; Rajpoot, S.; Liu, D.; Savai, R.; Banerjee, S.; Kawada, M.; Faisal, S.M.; Saluja, R.; Saqib, U.; et al. Tumor-derived exosomes in the regulation of macrophage polarization. *Inflamm. Res.* **2020**, *69*, 435–451. [CrossRef]

7. Zhang, B.; Miao, T.; Shen, X.; Bao, L.; Zhang, C.; Yan, C.; Wei, W.; Chen, J.; Xiao, L.; Sun, C.; et al. EB virus-induced ATR activation accelerates nasopharyngeal carcinoma growth via M2-type macrophages polarization. *Cell Death Dis.* **2020**, *11*, 742. [CrossRef]
8. Fu, X.-L.; Duan, W.; Su, C.-Y.; Mao, F.-Y.; Lv, Y.-P.; Teng, Y.-S.; Yu, P.-W.; Zhuang, Y.; Zhao, Y.-L. Interleukin 6 induces M2 macrophage differentiation by STAT3 activation that correlates with gastric cancer progression. *Cancer Immunol. Immunother.* **2017**, *66*, 1597–1608. [CrossRef]
9. Mori, K.; Hiroi, M.; Shimada, J.; Ohmori, Y. Infiltration of M2 Tumor-Associated Macrophages in Oral Squamous Cell Carcinoma Correlates with Tumor Malignancy. *Cancers* **2011**, *3*, 3726–3739. [CrossRef]
10. Dan, H.; Liu, S.; Liu, J.; Liu, D.; Yin, F.; Wei, Z.; Wang, J.; Zhou, Y.; Jiang, L.; Ji, N.; et al. RACK1 promotes cancer progression by increasing the M2/M1 macrophage ratio via the NF-κB pathway in oral squamous cell carcinoma. *Mol. Oncol.* **2020**, *14*, 795–807. [CrossRef]
11. Farran, B.; Nagaraju, G.P. Exosomes as therapeutic solutions for pancreatic cancer. *Drug Discov. Today* **2020**, *25*, 2245–2256. [CrossRef] [PubMed]
12. Ghosh, S.; Bhowmik, S.; Majumdar, S.; Goswami, A.; Chakraborty, J.; Gupta, S.; Aggarwal, S.; Ray, S.; Chatterjee, R.; Bhattacharyya, S.; et al. The exosome encapsulated microRNAs as circulating diagnostic marker for hepatocellular carcinoma with low alpha-fetoprotein. *Int. J. Cancer* **2020**, *147*, 2934–2947. [CrossRef] [PubMed]
13. Chatterjee, B.; Saha, P.; Bose, S.; Shukla, D.; Chatterjee, N.; Kumar, S.; Tripathi, P.P.; Srivastava, A.K. MicroRNAs: As Critical Regulators of Tumor- Associated Macrophages. *Int. J. Mol. Sci.* **2020**, *21*, 7117. [CrossRef]
14. Binenbaum, Y.; Fridman, E.; Yaari, Z.; Milman, N.; Schroeder, A.; Ben David, G.; Shlomi, T.; Gil, Z. Transfer of miRNA in Macrophage-Derived Exosomes Induces Drug Resistance in Pancreatic Adenocarcinoma. *Cancer Res.* **2018**, *78*, 5287–5299. [CrossRef]
15. Worby, C.A.; Dixon, J.E. PTEN. *Annu. Rev. Biochem.* **2014**, *83*, 641–669. [CrossRef] [PubMed]
16. Robinson, D.; Van Allen, E.M.; Wu, Y.-M.; Schultz, N.; Lonigro, R.J.; Mosquera, J.-M.; Montgomery, B.; Taplin, M.-E.; Pritchard, C.C.; Attard, G.; et al. Integrative Clinical Genomics of Advanced Prostate Cancer. *Cell* **2015**, *161*, 1215–1228. [CrossRef]
17. Zhao, J.; Chen, A.; Gartrell, R.D.; Silverman, A.M.; Aparicio, L.; Chu, T.; Bordbar, D.; Shan, D.; Samanamud, J.; Mahajan, A.; et al. Immune and genomic correlates of response to anti-PD-1 immunotherapy in glioblastoma. *Nat. Med.* **2019**, *25*, 462–469. [CrossRef]
18. Petty, A.J.; Li, A.; Wang, X.; Dai, R.; Heyman, B.; Hsu, D.; Huang, X.; Yang, Y. Hedgehog signaling promotes tumor-associated macrophage polarization to suppress intratumoral CD8+ T cell recruitment. *J. Clin. Investig.* **2019**, *129*, 5151–5162. [CrossRef]
19. Xiao, W.-H.; Yao, L.-P.; Li, M.; Wang, M.; Wu, L.; Jiang, M.-F.; Ma, H.-F.; Li, J.-Q.; Chen, G.-R. The Tumor-Associated Macrophage-M2-Cancer Cell Complex and the Observation of Heterogeneous Modification of the Morphological Structure of Lung Adenocarcinoma. *OncoTargets Ther.* **2020**, *13*, 11139–11149. [CrossRef]
20. Wu, A.T.; Srivastava, P.; Yadav, V.K.; Tzeng, D.T.; Iamsaard, S.; Su, E.C.-Y.; Hsiao, M.; Liu, M.-C. Ovatodiolide, isolated from Anisomeles indica, suppresses bladder carcinogenesis through suppression of mTOR/β-catenin/CDK6 and exosomal miR-21 derived from M2 tumor-associated macrophages. *Toxicol. Appl. Pharmacol.* **2020**, *401*, 115109. [CrossRef]
21. Mashouri, L.; Yousefi, H.; Aref, A.R.; Ahadi, A.M.; Molaei, F.; Alahari, S.K. Exosomes: Composition, biogenesis, and mechanisms in cancer metastasis and drug resistance. *Mol. Cancer* **2019**, *18*, 75. [CrossRef] [PubMed]
22. Lukic, A.; Wahlund, C.J.; Gómez, C.; Brodin, D.; Samuelsson, B.; Wheelock, C.E.; Gabrielsson, S.; Rådmark, O. Exosomes and cells from lung cancer pleural exudates transform LTC4 to LTD4, promoting cell migration and survival via CysLT1. *Cancer Lett.* **2018**, *444*, 1–8. [CrossRef] [PubMed]
23. Rayamajhi, S.; Nguyen, T.D.T.; Marasini, R.; Aryal, S. Macrophage-derived exosome-mimetic hybrid vesicles for tumor targeted drug delivery. *Acta Biomater.* **2019**, *94*, 482–494. [CrossRef] [PubMed]
24. Baassiri, A.; Nassar, F.; Mukherji, D.; Shamseddine, A.; Nasr, R.; Temraz, S. Exosomal Non Coding RNA in LIQUID Biopsies as a Promising Biomarker for Colorectal Cancer. *Int. J. Mol. Sci.* **2020**, *21*, 1398. [CrossRef]
25. Dongre, A.; Weinberg, R.A. New insights into the mechanisms of epithelial-mesenchymal transition and implications for cancer. *Nat. Rev. Mol. Cell Biol.* **2019**, *20*, 69–84. [CrossRef]
26. Yee, D.S.; Tang, Y.; Li, X.; Liu, Z.; Guo, Y.; Ghaffar, S.; McQueen, P.; Atreya, D.; Xie, J.; Simoneau, A.R.; et al. The Wnt inhibitory factor 1 restoration in prostate cancer cells was associated with reduced tumor growth, decreased capacity of cell migration and invasion and a reversal of epithelial to mesenchymal transition. *Mol. Cancer* **2010**, *9*, 162. [CrossRef]
27. Sobral, L.M.; Sousa, L.O.; Coletta, R.D.; Cabral, H.; Greene, L.J.; Tajara, E.H.; Gutkind, J.S.; Curti, C.; Leopoldino, A.M. Stable SET knockdown in head and neck squamous cell carcinoma promotes cell invasion and the mesenchymal-like phenotype in vitro, as well as necrosis, cisplatin sensitivity and lymph node metastasis in xenograft tumor models. *Mol. Cancer* **2014**, *13*, 32. [CrossRef]
28. Li, S.; Lu, J.; Chen, Y.; Xiong, N.; Li, L.; Zhang, J.; Yang, H.; Wu, C.; Zeng, H.; Liu, Y. MCP-1-induced ERK/GSK-3β/Snail signaling facilitates the epithelial–mesenchymal transition and promotes the migration of MCF-7 human breast carcinoma cells. *Cell. Mol. Immunol.* **2017**, *14*, 621–630. [CrossRef]
29. Dragomir, M.P.; Knutsen, E.; Calin, G.A. Classical and noncanonical functions of miRNAs in cancers. *Trends Genet.* **2022**, *38*, 379–394. [CrossRef]
30. Yehia, L.; Keel, E.; Eng, C. The Clinical Spectrum of *PTEN* Mutations. *Annu. Rev. Med.* **2020**, *71*, 103–116. [CrossRef]
31. Brewer, T.; Yehia, L.; Bazeley, P.; Eng, C. Exome sequencing reveals a distinct somatic genomic landscape in breast cancer from women with germline PTEN variants. *Am. J. Hum. Genet.* **2022**, *109*, 1520–1533. [CrossRef] [PubMed]

32. Turnham, D.J.; Bullock, N.; Dass, M.S.; Staffurth, J.N.; Pearson, H.B. The PTEN Conundrum: How to Target PTEN-Deficient Prostate Cancer. *Cells* **2020**, *9*, 2342. [CrossRef] [PubMed]
33. Parsons, R. Discovery of the PTEN Tumor Suppressor and Its Connection to the PI3K and AKT Oncogenes. *Cold Spring Harb. Perspect. Med.* **2020**, *10*, a036129. [CrossRef] [PubMed]
34. Miyahara, L.A.N.; Pontes, F.S.C.; Burbano, R.M.R.; Neto, N.C.; Guimarães, D.M.; Fonseca, F.P.; Pontes, H.A.R. *PTEN* allelic loss is an important mechanism in the late stage of development of oral leucoplakia into oral squamous cell carcinoma. *Histopathology* **2018**, *72*, 330–338. [CrossRef] [PubMed]
35. Starzyńska, A.; Adamska, P.; Sejda, A.; Sakowicz-Burkiewicz, M.; Adamski, J.; Marvaso, G.; Wychowański, P.; Jereczek-Fossa, B.A. Any Role of PIK3CA and PTEN Biomarkers in the Prognosis in Oral Squamous Cell Carcinoma? *Life* **2020**, *10*, 325. [CrossRef] [PubMed]
36. Iizumi, S.; Uchida, F.; Nagai, H.; Takaoka, S.; Fukuzawa, S.; Kanno, N.I.; Yamagata, K.; Tabuchi, K.; Yanagawa, T.; Bukawa, H. MicroRNA 142-5p promotes tumor growth in oral squamous cell carcinoma via the PI3K/AKT pathway by regulating PTEN. *Heliyon* **2021**, *7*, e08086. [CrossRef]

current issues in molecular biology

Article

Establishment of a Novel Anti-CD44 Variant 10 Monoclonal Antibody C_{44}Mab-18 for Immunohistochemical Analysis against Oral Squamous Cell Carcinomas

Kenichiro Ishikawa [1,†], Hiroyuki Suzuki [1,2,*,†], Mika K. Kaneko [1,2] and Yukinari Kato [1,2,*]

[1] Department of Molecular Pharmacology, Tohoku University Graduate School of Medicine, 2-1 Seiryo-machi, Aoba-ku, Sendai 980-8575, Japan; ken.ishikawa.r3@dc.tohoku.ac.jp (K.I.); k.mika@med.tohoku.ac.jp (M.K.K.)
[2] Department of Antibody Drug Development, Tohoku University Graduate School of Medicine, 2-1 Seiryo-machi, Aoba-ku, Sendai 980-8575, Japan
* Correspondence: hiroyuki.suzuki.b4@tohoku.ac.jp (H.S.); yukinari.kato.e6@tohoku.ac.jp (Y.K.); Tel.: +81-22-717-8207 (H.S. & Y.K.)
† These authors contributed equally to this work.

Abstract: Head and neck squamous cell carcinoma (HNSCC) is the most common type of head and neck cancer, and has been revealed as the second-highest expression of CD44 in cancers. CD44 has been investigated as a cancer stem cell marker of HNSCC and plays a critical role in tumor malignant progression. Especially, splicing variant isoforms of CD44 (CD44v) are overexpressed in cancers and considered a promising target for cancer diagnosis and therapy. We developed monoclonal antibodies (mAbs) against CD44 by immunizing mice with CD44v3–10-overexpressed PANC-1 cells. Among the established clones, C_{44}Mab-18 (IgM, kappa) reacted with CHO/CD44v3–10, but not with CHO/CD44s and parental CHO-K1 using flow cytometry. The epitope mapping using peptides that cover variant exon-encoded regions revealed that C_{44}Mab-18 recognized the border sequence between variant 10 and the constant exon 16-encoded sequence. These results suggest that C_{44}Mab-18 recognizes variant 10-containing CD44v, but not CD44s. Furthermore, C_{44}Mab-18 could recognize the human oral squamous cell carcinoma (OSCC) cell line, HSC-3, in flow cytometry. The apparent dissociation constant (K_D) of C_{44}Mab-18 for CHO/CD44v3–10 and HSC-3 was 1.6×10^{-7} M and 1.7×10^{-7} M, respectively. Furthermore, C_{44}Mab-18 detected CD44v3–10 but not CHO/CD44s in Western blotting, and endogenous CD44v10 in immunohistochemistry using OSCC tissues. These results indicate that C_{44}Mab-18 is useful for detecting CD44v10 in flow cytometry and immunohistochemistry.

Keywords: CD44; CD44v10; monoclonal antibody; oral squamous cell carcinoma; immunohistochemistry

Citation: Ishikawa, K.; Suzuki, H.; Kaneko, M.K.; Kato, Y. Establishment of a Novel Anti-CD44 Variant 10 Monoclonal Antibody C_{44}Mab-18 for Immunohistochemical Analysis against Oral Squamous Cell Carcinomas. *Curr. Issues Mol. Biol.* **2023**, *45*, 5248–5262. https://doi.org/10.3390/cimb45070333

Academic Editors: Violeta Popovici and Emma Adriana Ozon

Received: 25 May 2023
Revised: 14 June 2023
Accepted: 19 June 2023
Published: 21 June 2023

1. Introduction

Head and neck cancer is the seventh most common cancer type globally, and exhibits a profound impact on patients and their quality of life after surgical ablation and therapies [1]. Head and neck squamous cell carcinoma (HNSCC) is the most common type of head and neck cancer. The treatment of HNSCC includes surgery, chemotherapy, radiation therapy, immunotherapy, molecular targeted therapy, or a combination of those modalities [2]. Although survival can be improved through the development of treatments, cancer metastasis and resistance to drugs remain the main causes of death [3]. The rate of 5-year survival remains stagnant at approximately 50% [4].

CD44 is a multifunctional type I transmembrane glycoprotein that mediates metastasis and drug resistance in tumor cells. HNSCC is the second-highest CD44-expressing tumor in the Pan-Cancer Atlas [5]. The alternative splicing of CD44 mRNA produces the various isoforms [6]. The constant exons including the first five (1 to 5) and the last five (16 to 20) are present in all CD44 variants and make up the standard isoform (CD44s). The CD44 variant (CD44v) isoforms are generated by the alternative splicing of variant exons (v1 to v10) with the constant exons of CD44s [7]. Both CD44s and CD44v (pan-CD44) attach to the extracellular matrix, including hyaluronic acid (HA), and facilitate the activation of metastasis-associated intracellular signaling pathways [8].

Tumor metastasis includes multiple processes called the invasion–metastasis cascade. The processes contain (i) dissemination from primary sites, (ii) the acquisition of migration/invasion phenotype, (iii) intra/extravasation, (iv) survival in circulation, and (v) adaptation and colonization in distant organs [9]. Moreover, (vi) cancer-associated fibroblasts and tumor-infiltrating lymphocytes in the tumor microenvironment involve in the promotion of the invasion–metastasis cascade [10]. CD44 mediates the multiple steps of the invasion–metastasis cascade through interaction with HA [11] and CD44v-specific functions [12].

CD44 has been studied as a marker of cancer stem-like cells (CSCs) in tumors [13]. Anti-CD44s or CD44v monoclonal antibodies (mAbs) are used to collect the CD44-high CSCs [13]. The CD44-high population showed the increased self-renewing property, resistance to drugs, and metastatic colonization *in vivo* [13]. CD44 is the first applied CSC marker to isolate CSCs derived from HNSCC [14]. Notably, CD44-high CSCs from HNSCC showed the features of epithelial to mesenchymal transition (EMT). The EMT program activation confers tumor cells the stemness and the ability to migrate, invade, and extravasate [15]. Moreover, CD44-high cells could make colonization in the lungs of immunodeficient mice, compared to CD44-low, which failed to form the metastatic colonization [16].

Furthermore, CD44v8–10 mediates the resistance to treatment. The v8–10-encoded region binds to and stabilizes a cystine–glutamate transporter (xCT), which enhances cystine uptake and glutathione synthesis [17]. The elevation of reduced glutathione (GSH) mediates the defense to reactive oxygen species (ROS) [17], radiation [18], and chemotherapeutic drugs [19]. The expression of CD44v8–10 is associated with the xCT-mediated redox status and the poor prognosis of patients [18]. Therefore, the establishment of each CD44v-specific mAb is essential to reveal the function and develop CD44-targeting cancer therapy. However, the tissue distribution or function of the variant 10-containing CD44 has not been fully understood.

In our previous work, we developed an anti-pan-CD44 mAb, C_{44}Mab-5 (IgG_1, kappa) [20] using the Cell-Based Immunization and Screening (CBIS) method. Additionally, another anti-pan-CD44 mAb, C_{44}Mab-46 [21], was established by immunizing mice with the CD44v3–10 ectodomain. Both C_{44}Mab-5 and C_{44}Mab-46 have the epitopes within the constant exon 2- and 5-encoding sequences [22–24] and could be applied to immunohistochemistry in oral squamous cell carcinomas (OSCC) [20] and esophageal SCC [21], respectively. Furthermore, we produced a class-switched and defucosylated type of recombinant C_{44}Mab-5 (5-mG_{2a}-f) using fucosyltransferase 8 (Fut8)-deficient ExpiCHO-S cells and investigated the antitumor activity in OSCC xenograft-transplanted mice [25]. We have developed various anti-CD44v mAbs, including C_{44}Mab-6 (an anti-CD44v3 mAb) [26], C_{44}Mab-108 (an anti-CD44v4 mAb) [27], C_{44}Mab-3 (an anti-CD44v5 mAb) [28], C_{44}Mab-9 (an anti-CD44v6 mAb) [29], C_{44}Mab-34 (an anti-CD44v7/8 mAb) [30], and C_{44}Mab-1 (an anti-CD44v9 mAb) [31]. The generation of anti-CD44 mAbs, which recognize all variant exons, is important for the comprehensive analysis of human tumors.

In this study, we established a novel anti-CD44v10 mAb, C_{44}Mab-18 (IgM, kappa), using the CBIS method, and evaluated its applications via flow cytometry, Western blotting, and immunohistochemical analyses of OSCC tissues.

2. Materials and Methods

2.1. Cell Lines

A human pancreatic cancer cell line (PANC-1, the Cell Resource Center for Biomedical Research Institute of Development, Aging Sendai, Japan), a mouse multiple myeloma P3x63Ag8U.1 (P3U1) and Chinese hamster ovary (CHO)-K1 cell lines (the American Type Culture Collection, Manassas, VA, USA) were cultured using RPMI-1640 medium

(Nacalai Tesque, Inc., Kyoto, Japan) containing 10% heat-inactivated fetal bovine serum (FBS; Thermo Fisher Scientific, Inc., Waltham, MA, USA) and antibiotics (100 U/mL penicillin, 100 μg/mL streptomycin, and 0.25 μg/mL amphotericin B). A human OSCC cell line (HSC-3, the Japanese Collection of Research Bioresources, Osaka, Japan) was cultured in Dulbecco's Modified Eagle Medium (DMEM) (Nacalai Tesque, Inc., Kyoto, Japan) supplemented as indicated above. All cell lines were grown in a humidified incubator at 37 °C with 5% CO_2.

The cDNAs of CD44v3–10 and CD44s were obtained as described previously [20]. The cDNAs were cloned into pCAG-zeo-ssPA16 and pCAG-neo-ssPA16 vectors with a signal sequence and N-terminal PA16 tag (GLEGGVAMPGAEDDVV). The PA16 tag can be detected by NZ-1 mAb, which was originally developed as an anti-human podoplanin (PDPN) mAb [32]. Stable transfectants including PANC-1/CD44v3–10, CHO/CD44v3–10, and CHO/CD44s were established by introducing corresponding vectors into the cells using a Neon transfection system (Thermo Fisher Scientific, Inc.).

2.2. *Production of Hybridoma Cells*

PANC-1/CD44v3–10 (1×10^8 cells) was intraperitoneally administrated into the 6-week-old female BALB/c mice (CLEA Japan, Tokyo, Japan) with Imject Alum (Thermo Fisher Scientific Inc.). Additional three times immunizations of PANC-1/CD44v3–10 (1×10^8 cells) and a booster injection of PANC-1/CD44v3–10 (1×10^8 cells) two days before the sacrifice was performed. Hybridomas were produced as described previously [28]. The supernatants were selected by flow cytometer (SA3800 Cell Analyzer) and SA3800 software (ver. 2.05, Sony Corp. Tokyo, Japan).

2.3. *Enzyme–Linked Immunosorbent Assay (ELISA)*

Thirty-four peptides, which cover the variant region of CD44v3–10 [22], were obtained from Sigma-Aldrich Corp. (St. Louis, MO, USA), and immobilized on Nunc Maxisorp immunoplates (Thermo Fisher Scientific Inc.) at 20 μg/mL. After the blocking with 1% (*w*/*v*) bovine serum albumin (BSA) in phosphate-buffered saline (PBS) containing 0.05% (*v*/*v*) Tween 20 (PBST; Nacalai Tesque, Inc.), C_{44}Mab-18 (1 μg/mL) was added to each well. The wells were further treated with anti-mouse immunoglobulins peroxidase-conjugate (1:2000 diluted; Agilent Technologies Inc., Santa Clara, CA, USA). The enzymatic reaction was performed using an ELISA POD Substrate TMB Kit (Nacalai Tesque, Inc.) The optical density (655 nm) was measured using an iMark microplate reader (Bio-Rad Laboratories, Inc., Berkeley, CA, USA).

2.4. *Flow Cytometry*

CHO/CD44v3–10, CHO-K1, and HSC-3 cells (1×10^5 cells/sample) were incubated with C_{44}Mab-18, C_{44}Mab-46, or blocking buffer (0.1% BSA in PBS; control) for 30 min at 4 °C. The cells were further treated with anti-mouse IgG conjugated with Alexa Fluor 488 (1:2000; Cell Signaling Technology, Inc. Danvers, MA, USA), and analyzed as indicated above.

2.5. *Determination of Apparent Dissociation Constant (K_D) via Flow Cytometry*

The serially diluted C_{44}Mab-18 at the indicated concentrations was suspended with 2×10^5 of HSC-3 and CHO/CD44v3–10 cells. The cells were further treated with anti-mouse IgG conjugated with Alexa Fluor 488 (1:200). Fluorescence data were analyzed, and the apparent dissociation constant (K_D) was determined by fitting binding isotherms to built-in one-site binding models of GraphPad Prism 8 (GraphPad Software, Inc., La Jolla, CA, USA).

2.6. *Western Blot Analysis*

The SDS-polyacrylamide gel for electrophoresis and transfer onto polyvinylidene difluoride membranes was achieved as described previously [28]. After the blocking in PBST containing 4% skim milk (Nacalai Tesque, Inc.), the membranes were incubated with 10 μg/mL of C_{44}Mab-46, 10 μg/mL of C_{44}Mab-18, or 0.5 μg/mL of an anti-β-actin mAb (AC-15; Sigma-Aldrich Corp.). The membranes were further treated with peroxidase-conjugated anti-mouse immunoglobulins (diluted 1:1000; Agilent Technologies, Inc.). Finally, the chemiluminescence signal was obtained using ImmunoStar LD (FUJIFILM Wako Pure Chemical Corporation, Osaka, Japan) and was detected using a Sayaca-Imager (DRC Co., Ltd., Tokyo, Japan).

2.7. Immunohistochemical Analysis of Formalin-Fixed Paraffin-Embedded (FFPE) Tissues

Antigen retrieval of an OSCC tissue array (OR601c; US Biomax Inc., Rockville, MD, USA) was performed using EnVision FLEX Target Retrieval Solution High pH (Agilent Technologies, Inc.). SuperBlock T20 (Thermo Fisher Scientific, Inc.) was used for blocking. The sections were incubated with 1 μg/mL of C_{44}Mab-18 and 1 μg/mL of C_{44}Mab-46 at room temperature for 1 h. The sections were further treated with the EnVision+ Kit for a mouse (Agilent Technologies Inc.) at room temperature for 30 min. The chromogenic reaction and counterstaining were performed using 3,3′-diaminobenzidine tetrahydrochloride (DAB; Agilent Technologies Inc.) and hematoxylin (FUJIFILM Wako Pure Chemical Corporation), respectively.

3. Results

3.1. Establishment of an Anti-CD44 mAbs via Immunization of PANC-1/CD44v3–10 Cells

In our previous work, we have established anti-CD44 mAbs, including C_{44}Mab-5 (pan-CD44) [20], C_{44}Mab-6 (v3) [26], C_{44}Mab-3 (v5) [28], C_{44}Mab-9 (v6) [29], and C_{44}Mab-1 (v9) [31], using CHO/CD44v3–10 cells as an immunogen. In this study, we established another stable transfectant (PANC-1/CD44v3–10 cells) (Figure 1A). Mice were immunized with PANC-1/CD44v3–10 cells (Figure 1B), and hybridomas were produced via fusion between the splenocyte and P3U1 cells (Figure 1C). The supernatants, which were reactive to CHO/CD44v3–10 cells, but not to CHO-K1, were selected via flow cytometry-based high throughput screening (Figure 1D). After cloning by limiting dilution, anti-CD44 mAb-producing clones were finally established (Figure 1E).

3.2. Flow Cytometric Analysis of C_{44}Mab-18- to CD44-Expressing Cells

In this study, established clones, the epitope of which includes CD44v10, were mainly determined to be IgM, although all mAbs against other CD44 variants are IgG [26–31]. Among those clones, we examined the reactivity of C_{44}Mab-18 (IgM, kappa) against CHO/CD44v3–10 and CHO/CD44s cells via flow cytometry. C_{44}Mab-18 dose-dependently recognized CHO/CD44v3–10 cells (Figure 2A). In contrast, C_{44}Mab-18 recognized neither CHO/CD44s (Figure 2B) nor CHO-K1 (Figure 2C) cells. We confirmed that an anti-pan-CD44 mAb, C_{44}Mab-46 [21], recognized CHO/CD44s cells, but not CHO-K1 cells (Supplementary Figure S1). Furthermore, C_{44}Mab-18 could recognize HSC-3 cells (Figure 2D) in a dose-dependent manner. These results indicated that C_{44}Mab-18 recognizes the variant exon-encoded region between v3 and v10 (Figure 1A).

3.3. Epitope Mapping of C_{44}Mab-18 by ELISA

To determine the epitope of C_{44}Mab-18, we performed the ELISA using synthetic peptides, which cover the variant exon-encoded region between v3 and v10 [22]. As shown in Figure 3, C_{44}Mab-18 recognized the CD44 p551–570 peptide (SNSNVNRSLSGDQDTFHPSG), which corresponds to variant 10 and constant exon 16-encoded sequence (Supplementary Table S1). In contrast, C_{44}Mab-18 never recognized other v3- and v10-encoded peptides. This and the results in Figure 2 indicate that C_{44}Mab-18 specifically recognizes the variant 10-containing CD44.

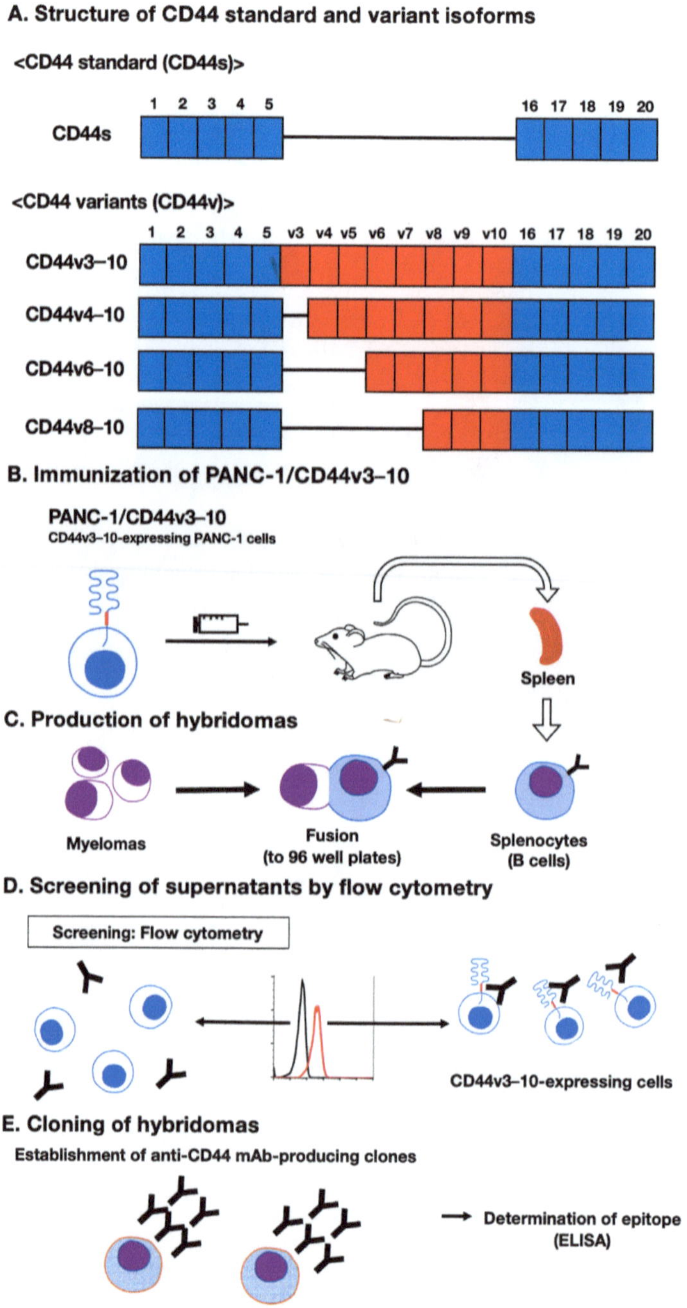

Figure 1. A schematic illustration of the CBIS method to establish anti-human CD44 mAbs. (**A**) Structure of CD44. The CD44s mRNA contains the constant exons (1 to 5) and (16 to 20). The CD44v including CD44v3–10, CD44v4–10, CD44v6–10, and CD44v8–10 are produced via the alternative splicing of variant exons. (**B**) PANC-1/CD44v3–10 cells were injected into BALB/c mice intraperitoneally. (**C**) Hybridomas were produced via the fusion of the splenocytes and P3U1 cells (**D**) The screening was performed via flow cytometry using CHO/CD44v3–10 and parental CHO-K1 cells. (**E**) A clone C_{44}Mab-18 (IgM, kappa) was established after cloning.

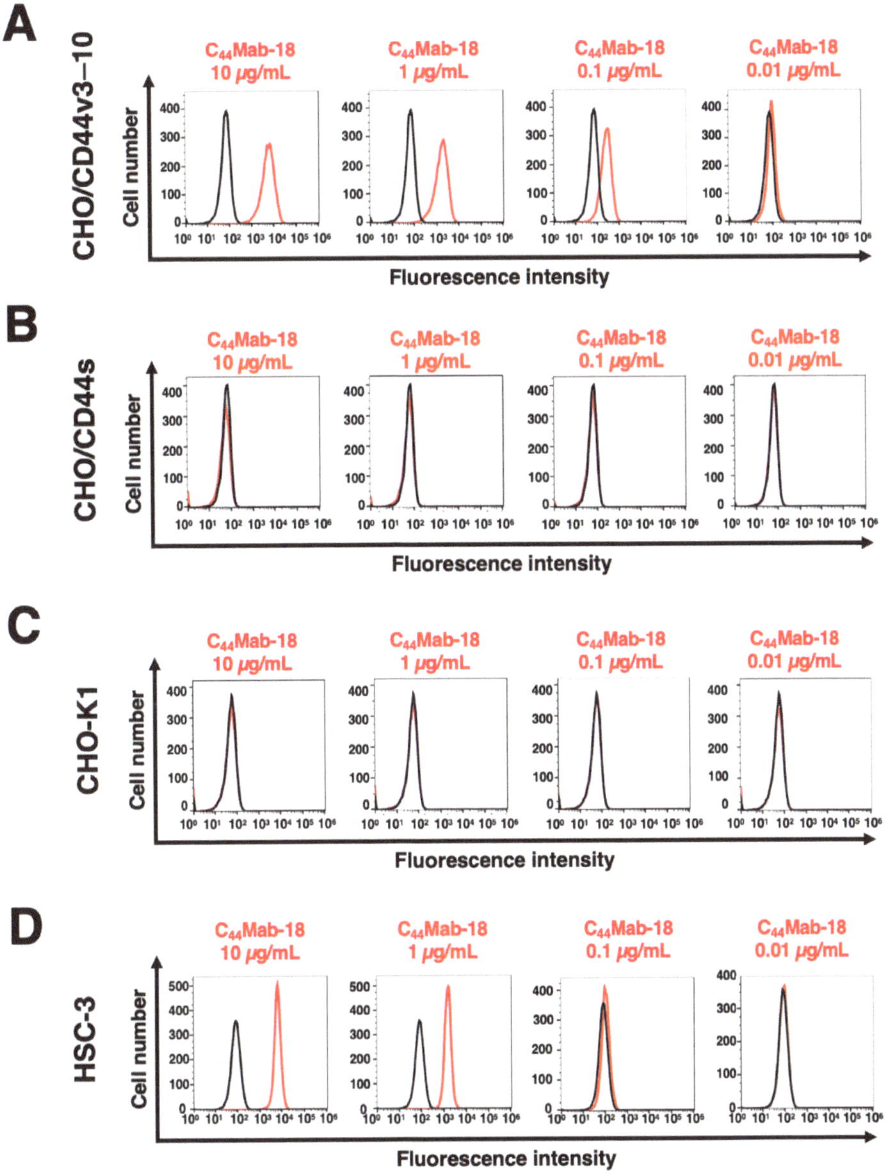

Figure 2. Flow cytometry using C_{44}Mab-18. CHO/CD44v3–10 (**A**), CHO/CD44s (**B**), CHO-K1 (**C**), and HSC-3 (**D**) cells were treated with 0.01–10 µg/mL of C_{44}Mab-18. Then, cells were treated with Alexa Fluor 488-conjugated anti-mouse IgG (Red line). The black line represents the negative control (blocking buffer).

3.4. Determination of the Apparent Binding Affinity of C_{44}Mab-18 via Flow Cytometry

We measured the apparent binding affinity of C_{44}Mab-18 to CHO/CD44v3–10 and HSC-3 cells. The apparent dissociation constant (K_D) of C_{44}Mab-18 for CHO/CD44v3–10 (Figure 4A) and HSC-3 (Figure 4B) was 1.6×10^{-7} M and 1.7×10^{-7} M, respectively. These results indicated that C_{44}Mab-18 possesses a moderate binding affinity for CD44v3–10 or endogenous CD44v10-expressing cells.

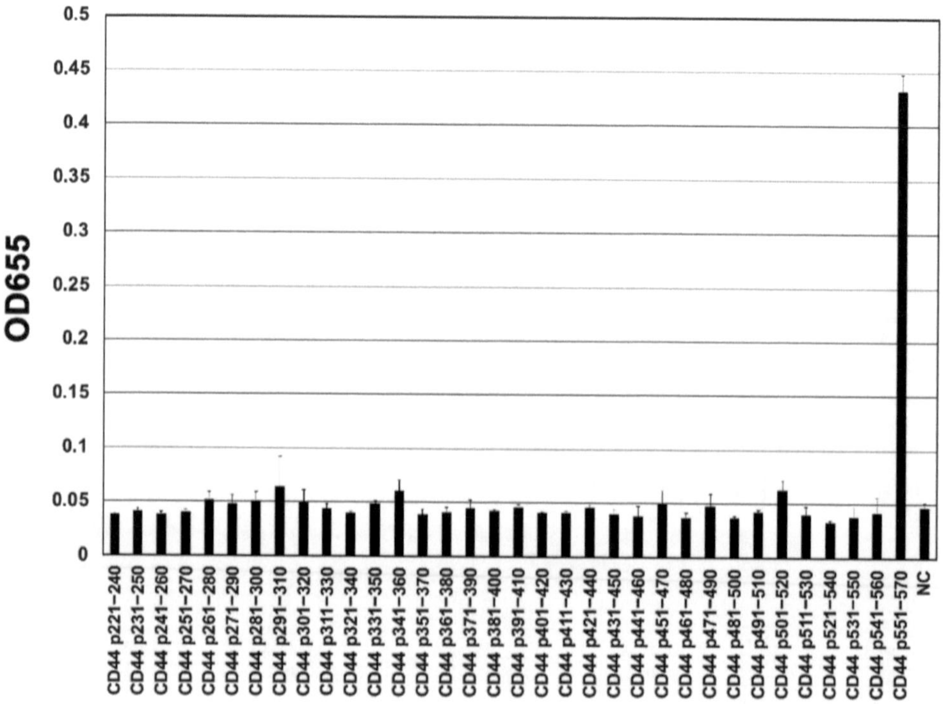

Figure 3. Determination of C_{44}Mab-18 epitope using ELISA. The synthesized peptides, that cover the variant exon-encoded region between v3 and v10, were immobilized on immunoplates. The plates were incubated with C_{44}Mab-18, followed by incubation with anti-mouse immunoglobulins -conjugated with peroxidase. Optical density (655 nm) was measured. The CD44 p551–570 sequence (SNSNVNRSLSGDQDTFHPSG) corresponds to variant 10 and the constant exon 16-encoded sequence. Error bars represent means ± SDs. NC, negative control (0.1% DMSO [solvent] in PBS).

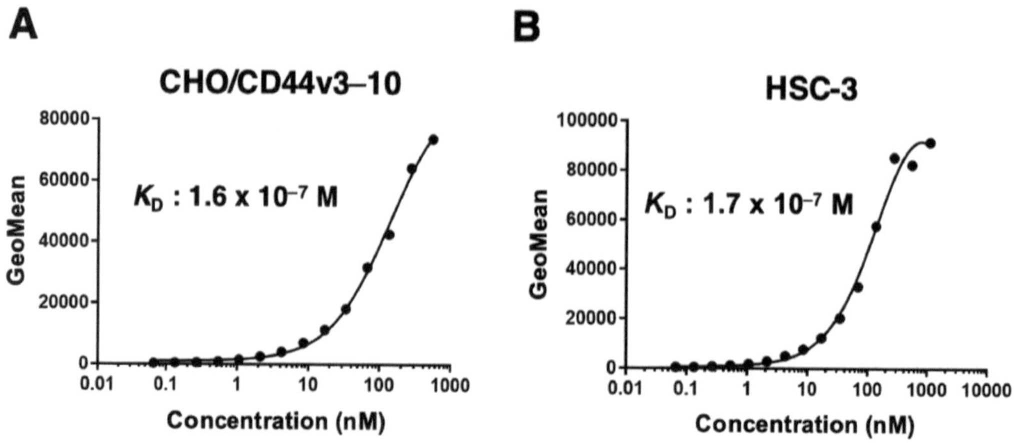

Figure 4. The determination of the apparent dissociation constant (K_D) of C_{44}Mab-18. C_{44}Mab-18 was treated with CHO/CD44v3–10 at indicated concentrations (**A**) and with HSC-3 (**B**). The cells were treated with anti-mouse IgG conjugated with Alexa Fluor 488. Fluorescence data were collected, followed by the calculation of K_D using GraphPad PRISM 8.

3.5. Western Blot Analysis

To assess the sensitivity of C_{44}Mab-18 in Western blot analysis, we analyzed the cell lysates from CHO-K1, CHO/CD44s, and CHO/CD44v3–10. C_{44}Mab-18 mainly detected CD44v3–10 as more than 180-kDa and ~70-kDa bands. However, C_{44}Mab-18 did not detect any bands from lysates of CHO-K1 and CHO/CD44s cells (Figure 5A). An anti-pan-CD44 mAb, C_{44}Mab-46, recognized both CD44v3–10 (>180 kDa) and CD44s (~75 kDa) bands in the lysates of CHO/CD44v3–10 and CHO/CD44s, respectively (Figure 5B). We used β-actin as a loading control (Figure 5C). These results indicate that C_{44}Mab-18 can detect exogenous CD44v3–10.

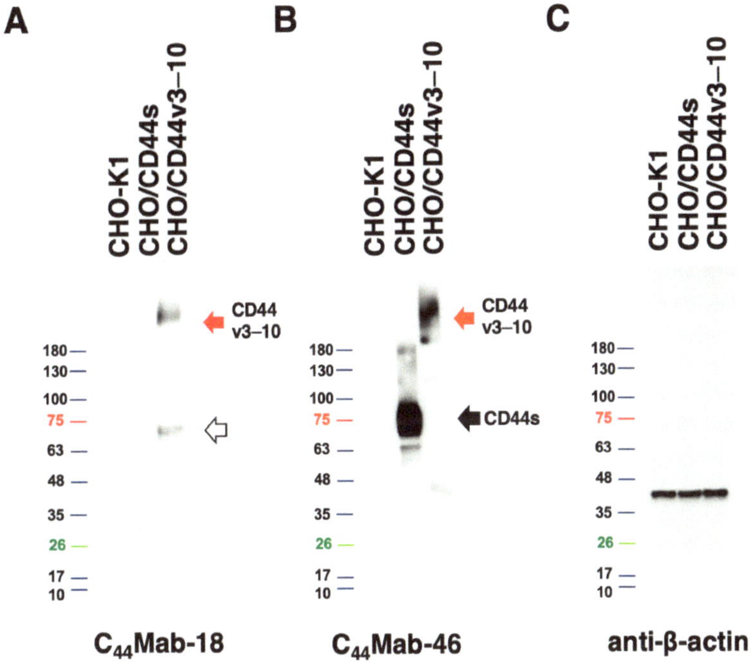

Figure 5. Western blot analysis using C_{44}Mab-18. The total cell lysates (10 µg of protein) were separated and transferred onto polyvinylidene difluoride (PVDF) membranes. The membranes were incubated with 10 µg/mL of C_{44}Mab-18 (**A**), 10 µg/mL of C_{44}Mab-46 (**B**), or 0.5 µg/mL of an anti-β-actin mAb (**C**), followed by incubation with peroxidase-conjugated anti-mouse immunoglobulins. The red arrows indicate the CD44v3–10 (>180 kDa). The black arrow indicates the CD44s (~75 kDa). The white arrow indicates a lower molecular weight band recognized by C_{44}Mab-18 in CHO/CD44v3–10 lysate (~70 kDa).

3.6. Immunohistochemical Analysis Using C_{44}Mab-18 against Tumor Tissues

Since HNSCC is revealed as the second highest CD44-expressing tumor in the Pan-Cancer Atlas [5], we examined the reactivity of C_{44}Mab-18 and C_{44}Mab-46 in immunohistochemical analyses using FFPE sections of OSCC tissue array. As shown in Figure 6, C_{44}Mab-18 was able to distinguish tumor cells from stromal tissues. In contrast, C_{44}Mab-46 stained both. We summarized the data of immunohistochemical analyses in Table 1; C_{44}Mab-18 stained 41 out of 50 cases (82%) in OSCC. These results indicate that C_{44}Mab-18 applies to the immunohistochemical analysis of FFPE tumor sections.

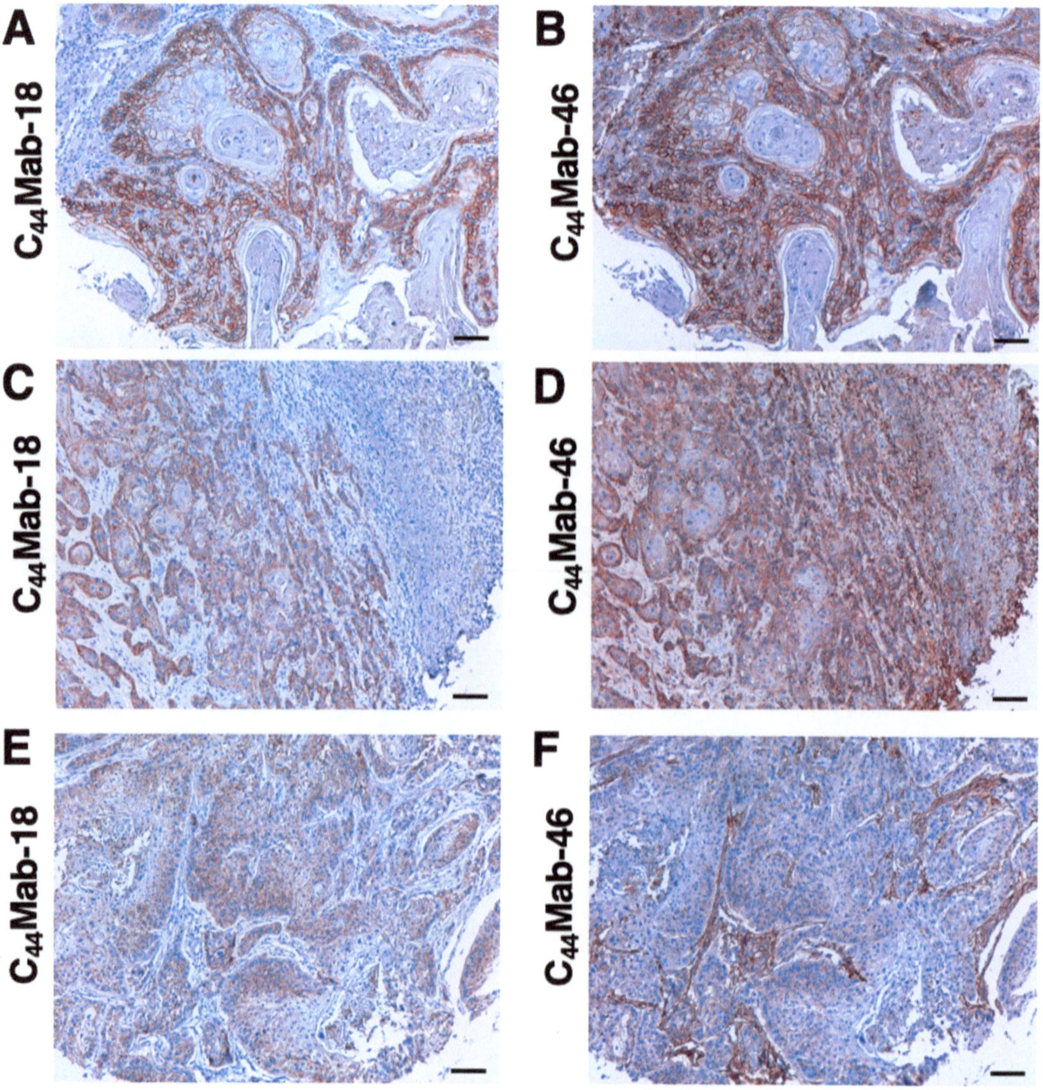

Figure 6. Immunohistochemical analysis using C_{44}Mab-18 and C_{44}Mab-46 against FFPE OSCC tissues. (**A–F**) Serial sections of the OSCC tissue array (OR601c) were incubated with 1 µg/mL of C_{44}Mab-18 or C_{44}Mab-46 followed by treatment with the Envision+ kit. The chromogenic reaction and counterstaining were performed using 3,3′-diaminobenzidine tetrahydrochloride and hematoxylin, respectively. Scale bar = 100 µm.

Table 1. Immunohistochemical analysis using C_{44}Mab-18 against OSCC tissue array.

No.	Age	Sex	Organ /Anatomic Site	Pathology Diagnosis	TNM	C_{44}Mab-18	C_{44}Mab-46
1	78	M	Tongue	SCC of tongue	T2N0M0	+	+
2	40	M	Tongue	SCC of tongue	T2N0M0	+	++
3	75	F	Tongue	SCC of tongue	T2N0M0	-	+

Table 1. *Cont.*

No.	Age	Sex	Organ /Anatomic Site	Pathology Diagnosis	TNM	C_{44}Mab-18	C_{44}Mab-46
4	35	F	Tongue	SCC of tongue	T2N0M0	++	++
5	61	M	Tongue	SCC of tongue	T2N0M0	++	+++
6	41	F	Tongue	SCC of tongue	T2N0M0	+	+
7	64	M	Tongue	SCC of right tongue	T2N2M0	++	++
8	76	M	Tongue	SCC of tongue	T1N0M0	++	++
9	50	F	Tongue	SCC of tongue	T2N0M0	++	++
10	44	M	Tongue	SCC of tongue	T2N1M0	++	+++
11	53	F	Tongue	SCC of tongue	T1N0M0	+	++
12	46	F	Tongue	SCC of tongue	T2N0M0	++	+
13	50	M	Tongue	SCC of root of tongue	T3N1M0	++	+
14	36	F	Tongue	SCC of tongue	T1N0M0	++	+++
15	63	F	Tongue	SCC of tongue	T1N0M0	+	+
16	46	M	Tongue	SCC of tongue	T2N0M0	+	-
17	58	M	Tongue	SCC of tongue	T2N0M0	+	+
18	64	M	Lip	SCC of lower lip	T1N0M0	+	+++
19	57	M	Lip	SCC of lower lip	T2N0M0	+	+++
20	61	M	Lip	SCC of lower lip	T1N0M0	+	++
21	60	M	Gum	SCC of gum	T3N0M0	++	+
22	60	M	Gum	SCC of gum	T1N0M0	+++	+++
23	69	M	Gum	SCC of upper gum	T3N0M0	++	++
24	53	M	Bucca cavioris	SCC of bucca cavioris	T2N0M0	++	+
25	55	M	Bucca cavioris	SCC of bucca cavioris	T1N0M0	+++	+
26	58	M	Tongue	SCC of base of tongue	T1N0M0	++	++
27	63	M	Oral cavity	SCC	T1N0M0	+++	++
28	48	F	Tongue	SCC of tongue	T1N0M0	+	+
29	80	M	Lip	SCC of lower lip	T1N0M0	+++	+++
30	77	M	Tongue	SCC of base of tongue	T2N0M0	++	++
31	59	M	Tongue	SCC of tongue	T2N0M0	+	-
32	77	F	Tongue	SCC of tongue	T1N0M0	+	++
33	56	M	Tongue	SCC of root of tongue	T2N1M0	+	+
34	60	M	Tongue	SCC of tongue	T2N1M0	++	++
35	62	M	Tongue	SCC of tongue	T2N0M0	+	++
36	67	F	Tongue	SCC of tongue	T2N0M0	-	++
37	47	F	Tongue	SCC of tongue	T2N0M0	+++	+++
38	37	M	Tongue	SCC of tongue	T2N1M0	-	-
39	55	F	Tongue	SCC of tongue	T2N0M0	+	+
40	56	F	Bucca cavioris	SCC of bucca cavioris	T2N0M0	+	+
41	49	M	Bucca cavioris	SCC of bucca cavioris	T1N0M0	-	-
42	45	M	Bucca cavioris	SCC of bucca cavioris	T2N0M0	-	-

Table 1. *Cont.*

No.	Age	Sex	Organ /Anatomic Site	Pathology Diagnosis	TNM	C_{44}Mab-18	C_{44}Mab-46
43	42	M	Bucca cavioris	SCC of bucca cavioris	T3N0M0	+++	++
44	44	M	Jaw	SCC of right drop jaw	T1N0M0	+	+++
45	40	F	Tongue	SCC of base of tongue	T2N0M0	-	++
46	49	M	Bucca cavioris	SCC of bucca cavioris	T1N0M0	++	+++
47	56	F	Tongue	SCC of base of tongue	T2N0M0	-	+
48	42	M	Bucca cavioris	SCC a of bucca cavioris	T3N0M0	+++	+++
49	87	F	Face	SCC a of left face	T2N0M0	-	+
50	50	M	Gum	SCC of gum	T2N0M0	-	-

-, No stain; +, Weak intensity; ++, Moderate intensity; +++, Strong intensity.

4. Discussion

In our previous work, we established anti-CD44 mAbs using CHO/CD44v3–10 [20,26,28,29,31] and purified CD44v3–10 ectodomain [21,30] as immunogens. In this study, we used PANC-1/CD44v3–10 as another immunogen. We have compiled a list with this information in "Antibody Bank" (see Supplementary Materials). In this study, we listed a novel anti-CD44v antibody C_{44}Mab-18, which recognizes the border sequence between variant 10 and constant exon 16 (Figure 3). Furthermore, C_{44}Mab-18 could recognize CHO/CD44v3–10, but not CHO/CD44s in flow cytometry (Figure 2) and Western blot analyses (Figure 5). Moreover, C_{44}Mab-18 could stain tumor cells, but not stromal tissues, which could be stained by C_{44}Mab-46, an anti-pan-CD44 mAb (Figure 6). These results indicate that C_{44}Mab-18 is an anti-CD44v10 mAb.

The VFF series anti-human CD44v mAbs were previously established via the immunization of glutathione *S*-transferase fused CD44v3–10 produced by bacteria [33,34]. The clones, VFF-8 (anti-CD44v5), VFF-18 (anti-CD44v6), VFF-9 (anti-CD44v7), VFF-17 (anti-CD44v7/8), and VFF-14 (anti-CD44v10) have been used for various applications [35]. Although VFF-14 was shown to apply to immunohistochemistry [36], the detailed binding epitope of VFF-14 has not been reported. In this study, we determined the epitope of C_{44}Mab-18 as the CD44 p551–570 peptide (<u>SNSNVNRSLSG</u>DQDTFHPSG), which corresponds to the variant 10 (underlined) and constant exon 16-encoded region. In contrast, C_{44}Mab-18 never recognizes the p541–560 peptide (FGVTAVTVGDSNSNVNRSLS) in the variant 10 region. Therefore, C_{44}Mab-18 could have the epitope in the border region, but the inclusion of variant 10 is essential for the recognition.

Since the CD44 protein is modified by a variety of *N*-glycans and *O*-glycans, the molecular weight of CD44v isoforms surpasses 200-kDa [37]. C_{44}Mab-18 recognized both more than 180-kDa and ~70-kDa bands (Figure 5A) in the lysate from CHO/CD44v3–10. The 70 kDa is approximately identical to the predicted molecular weight of CD44v3–10 from the amino acid sequence. Therefore, C_{44}Mab-18 could recognize CD44v3–10 regardless of the glycosylation. The detailed epitope mapping and the influence of glycosylation on C_{44}Mab-18 recognition should be investigated in future studies.

CD44v8–10 was shown to interact with xCT, a glutamate–cystine transporter, and regulate the level of reduced glutathione in tumor cells. The interaction is important for the stabilization of xCT on the cell surface, which promotes the defense against reactive oxygen species [17]. Furthermore, the interaction failed in CD44v8–10 (S301A), an *N*-linked glycosylation consensus motif (Asn-X-Ser/Thr) mutant in the variant 10-encoded region [17]. Therefore, it is worthwhile to investigate whether C_{44}Mab-18 interferes with the interaction between CD44v8–10 and xCT in future studies. Furthermore, several studies have revealed that CD44v9 is used as a predictive marker for recurrence [38] and a biomarker for patient selection and efficacy of xCT inhibitors, sulfasalazine in gastric cancer [39]. Further investigations are also required to clarify the clinical significance of CD44v10 expression using C_{44}Mab-18.

The mAbs against CD44 have been considered a therapeutic option for solid tumors and leukemia [12]. However, anti-pan-CD44 mAbs can affect normal tissues such as the epithelium and hematopoiesis. In a preclinical study using a murine thymoma model, a comparative study between an anti-pan-CD44 mAb (IM-7) and an anti-murine CD44v10 mAb (K926) was conducted in CD44v10-transfected EL4 thymoma (EL4-v10) [40]. The results showed that a blockade of CD44v10 by K926 was superior to that of IM-7 in intra-marrow EL4-v10 growth retardation. Furthermore, K926 hardly disturbed the hematopoietic stem cell (HSC) interaction with the bone marrow stroma. In contrast, IM-7 strongly affected the embedding of HSC in the bone marrow stroma [40]. These results indicated that the therapeutic use of anti-pan-CD44 mAbs should be avoided in favor of CD44v-specific mAbs as far as leukemic cells express CD44v isoforms.

In a humanized mouse model, CD44v8–10 was elevated during chronic myeloid leukemia progression from chronic phase to blast crisis [41]. Furthermore, increased transcription of CD44 mRNA was observed in human acute myeloid leukemia (AML) patients with *FLT3* or *DNMT3A* mutations through the suppression of CpG islands methylation in the promoter [42]. An anti-CD44v6 mAb (BIWA-8) derived from VFF-18 [43] was engineered to develop chimeric antigen receptors (CARs) for AML with *FLT3* or *DNMT3A* mutations. The CD44v6 CAR-T cells exhibited potent anti-leukemic effects [42]. We have established class-switched and defucosylated IgG_{2a} recombinant mAbs and evaluated the antitumor activity in xenograft models [44]. Therefore, the production of class-switched and defucosylated C_{44}Mab-18 is an important strategy to evaluate the antitumor effect in preclinical models.

Since anti-pan-CD44 and anti-CD44v mAbs still have the possibility of causing side effects by affecting normal tissues, the clinical applications are limited. This study used tumor cell-expressed CD44v3–10 as an immunogen. This strategy is critical for the development of cancer-specific mAbs (CasMabs). We developed podocalyxin-targeting CasMabs [45] and PDPN-targeting CasMabs [46], which react with the aberrantly glycosylated targets selectively expressed in cancer [47]. Anti-PDPN-CasMabs have been applied to CAR-T therapy in preclinical studies [48–50]. For CasMab development, we should perform a further selection of our established anti-CD44 mAbs by comparing the reactivity against normal cells and tissues. Anti-CD44 CasMabs could be applicable for designing the modalities, including antibody-drug conjugates and CAR-T.

5. Conclusions

In this study, we established an anti-CD44v10 mAb (C_{44}Mab-18). We also established an anti-CD44v8 mAb (C_{44}Mab-94) (manuscript submitted, see Supplementary Materials). Therefore, we have established an anti-CD44 mAb library that covers almost all CD44 variants. This library could contribute to the diagnosis of not only carcinoma, but also hematopoietic malignancies. Since we have already cloned the V_H and V_L cDNA of anti-CD44 mAbs, the production of recombinant mAbs or CARs could contribute to the development of novel tumor therapies.

Supplementary Materials: The following supporting information can be downloaded at: https://www.mdpi.com/article/10.3390/cimb45070333/s1. Supplementary Figure S1, Recognition of CHO/CD44s and CHO/CD44v3–10 by C_{44}Mab-46 by flow cytometry. Supplementary Table S1, The determination of the binding epitope of C_{44}Mab-18 by ELISA. The information on anti-CD44 mAbs in our laboratory is available in "Antibody Bank" [http://www.med-tohoku-antibody.com/topics/001_paper_antibody_PDIS.htm#CD44 (accessed on 14 June 2023)].

Author Contributions: K.I. and H.S. performed the experiments. M.K.K. and Y.K. designed the experiments. K.I. and H.S. analyzed the data. K.I., H.S. and Y.K. wrote the manuscript. All authors have read and agreed to the published version of the manuscript.

Funding: This research was supported in part by the Japan Agency for Medical Research and Development (AMED) under grant numbers: JP23ama121008 (to Y.K.), JP23am0401013 (to Y.K.), and JP23ck0106730 (to Y.K.); and by the Japan Society for the Promotion of Science (JSPS) Grants-in-Aid for Scientific Research (KAKENHI) grant numbers 22K06995 (to H.S.), 21K07168 (to M.K.K.), and 22K07224 (to Y.K.).

Institutional Review Board Statement: The animal study protocol was approved by the Animal Care and Use Committee of Tohoku University (Permit number: 2019NiA-001) for studies involving animals.

Informed Consent Statement: Not applicable.

Data Availability Statement: All related data and methods are presented in this paper. Additional inquiries should be addressed to the corresponding authors.

Conflicts of Interest: The authors declare no conflict of interest involving this article.

References

1. Mody, M.D.; Rocco, J.W.; Yom, S.S.; Haddad, R.I.; Saba, N.F. Head and neck cancer. *Lancet* **2021**, *398*, 2289–2299. [CrossRef]
2. Xing, D.T.; Khor, R.; Gan, H.; Wada, M.; Ermongkonchai, T.; Ng, S.P. Recent Research on Combination of Radiotherapy with Targeted Therapy or Immunotherapy in Head and Neck Squamous Cell Carcinoma: A Review for Radiation Oncologists. *Cancers* **2021**, *13*, 5716. [CrossRef]
3. Muzaffar, J.; Bari, S.; Kirtane, K.; Chung, C.H. Recent Advances and Future Directions in Clinical Management of Head and Neck Squamous Cell Carcinoma. *Cancers* **2021**, *13*, 338. [CrossRef]
4. Johnson, D.E.; Burtness, B.; Leemans, C.R.; Lui, V.W.Y.; Bauman, J.E.; Grandis, J.R. Head and neck squamous cell carcinoma. *Nat. Rev. Dis. Prim.* **2020**, *6*, 92. [CrossRef]
5. Ludwig, N.; Szczepanski, M.J.; Gluszko, A.; Szafarowski, T.; Azambuja, J.H.; Dolg, L.; Gellrich, N.C.; Kampmann, A.; Whiteside, T.L.; Zimmerer, R.M. CD44(+) tumor cells promote early angiogenesis in head and neck squamous cell carcinoma. *Cancer Lett.* **2019**, *467*, 85–95. [CrossRef] [PubMed]
6. Ponta, H.; Sherman, L.; Herrlich, P.A. CD44: From adhesion molecules to signalling regulators. *Nat. Rev. Mol. Cell Biol.* **2003**, *4*, 33–45. [CrossRef] [PubMed]
7. Chen, C.; Zhao, S.; Karnad, A.; Freeman, J.W. The biology and role of CD44 in cancer progression: Therapeutic implications. *J. Hematol. Oncol.* **2018**, *11*, 64. [CrossRef] [PubMed]
8. Slevin, M.; Krupinski, J.; Gaffney, J.; Matou, S.; West, D.; Delisser, H.; Savani, R.C.; Kumar, S. Hyaluronan-mediated angiogenesis in vascular disease: Uncovering RHAMM and CD44 receptor signaling pathways. *Matrix Biol.* **2007**, *26*, 58–68. [CrossRef]
9. Valastyan, S.; Weinberg, R.A. Tumor metastasis: Molecular insights and evolving paradigms. *Cell* **2011**, *147*, 275–292. [CrossRef] [PubMed]
10. de Visser, K.E.; Joyce, J.A. The evolving tumor microenvironment: From cancer initiation to metastatic outgrowth. *Cancer Cell* **2023**, *41*, 374–403. [CrossRef]
11. Zöller, M. CD44, Hyaluronan, the Hematopoietic Stem Cell, and Leukemia-Initiating Cells. *Front. Immunol.* **2015**, *6*, 235. [CrossRef] [PubMed]
12. Hassn Mesrati, M.; Syafruddin, S.E.; Mohtar, M.A.; Syahir, A. CD44: A Multifunctional Mediator of Cancer Progression. *Biomolecules* **2021**, *11*, 1850. [CrossRef] [PubMed]
13. Zöller, M. CD44: Can a cancer-initiating cell profit from an abundantly expressed molecule? *Nat. Rev. Cancer* **2011**, *11*, 254–267. [CrossRef] [PubMed]
14. Prince, M.E.; Sivanandan, R.; Kaczorowski, A.; Wolf, G.T.; Kaplan, M.J.; Dalerba, P.; Weissman, I.L.; Clarke, M.F.; Ailles, L.E. Identification of a subpopulation of cells with cancer stem cell properties in head and neck squamous cell carcinoma. *Proc. Natl. Acad. Sci. USA* **2007**, *104*, 973–978. [CrossRef]
15. Yang, J.; Antin, P.; Berx, G.; Blanpain, C.; Brabletz, T.; Bronner, M.; Campbell, K.; Cano, A.; Casanova, J.; Christofori, G.; et al. Guidelines and definitions for research on epithelial-mesenchymal transition. *Nat. Rev. Mol. Cell Biol.* **2020**, *21*, 341–352. [CrossRef] [PubMed]
16. Davis, S.J.; Divi, V.; Owen, J.H.; Bradford, C.R.; Carey, T.E.; Papagerakis, S.; Prince, M.E. Metastatic potential of cancer stem cells in head and neck squamous cell carcinoma. *Arch. Otolaryngol. Head Neck Surg.* **2010**, *136*, 1260–1266. [CrossRef] [PubMed]
17. Ishimoto, T.; Nagano, O.; Yae, T.; Tamada, M.; Motohara, T.; Oshima, H.; Oshima, M.; Ikeda, T.; Asaba, R.; Yagi, H.; et al. CD44 variant regulates redox status in cancer cells by stabilizing the xCT subunit of system xc(-) and thereby promotes tumor growth. *Cancer Cell* **2011**, *19*, 387–400. [CrossRef]
18. Kagami, T.; Yamade, M.; Suzuki, T.; Uotani, T.; Tani, S.; Hamaya, Y.; Iwaizumi, M.; Osawa, S.; Sugimoto, K.; Baba, S.; et al. High expression level of CD44v8-10 in cancer stem-like cells is associated with poor prognosis in esophageal squamous cell carcinoma patients treated with chemoradiotherapy. *Oncotarget* **2018**, *9*, 34876–34888. [CrossRef]

19. Hagiwara, M.; Kikuchi, E.; Tanaka, N.; Kosaka, T.; Mikami, S.; Saya, H.; Oya, M. Variant isoforms of CD44 involves acquisition of chemoresistance to cisplatin and has potential as a novel indicator for identifying a cisplatin-resistant population in urothelial cancer. *BMC Cancer* **2018**, *18*, 113. [CrossRef]
20. Yamada, S.; Itai, S.; Nakamura, T.; Yanaka, M.; Kaneko, M.K.; Kato, Y. Detection of high CD44 expression in oral cancers using the novel monoclonal antibody, C(44)Mab-5. *Biochem. Biophys. Rep.* **2018**, *14*, 64–68. [CrossRef]
21. Goto, N.; Suzuki, H.; Tanaka, T.; Asano, T.; Kaneko, M.K.; Kato, Y. Development of a Novel Anti-CD44 Monoclonal Antibody for Multiple Applications against Esophageal Squamous Cell Carcinomas. *Int. J. Mol. Sci.* **2022**, *23*, 5535. [CrossRef] [PubMed]
22. Takei, J.; Asano, T.; Suzuki, H.; Kaneko, M.K.; Kato, Y. Epitope Mapping of the Anti-CD44 Monoclonal Antibody (C44Mab-46) Using Alanine-Scanning Mutagenesis and Surface Plasmon Resonance. *Monoclon. Antib. Immunodiagn. Immunother.* **2021**, *40*, 219–226. [CrossRef] [PubMed]
23. Asano, T.; Kaneko, M.K.; Takei, J.; Tateyama, N.; Kato, Y. Epitope Mapping of the Anti-CD44 Monoclonal Antibody (C44Mab-46) Using the REMAP Method. *Monoclon. Antib. Immunodiagn. Immunother.* **2021**, *40*, 156–161. [CrossRef]
24. Asano, T.; Kaneko, M.K.; Kato, Y. Development of a Novel Epitope Mapping System: RIEDL Insertion for Epitope Mapping Method. *Monoclon. Antib. Immunodiagn. Immunother.* **2021**, *40*, 162–167. [CrossRef] [PubMed]
25. Takei, J.; Kaneko, M.K.; Ohishi, T.; Hosono, H.; Nakamura, T.; Yanaka, M.; Sano, M.; Asano, T.; Sayama, Y.; Kawada, M.; et al. A defucosylated antiCD44 monoclonal antibody 5mG2af exerts antitumor effects in mouse xenograft models of oral squamous cell carcinoma. *Oncol. Rep.* **2020**, *44*, 1949–1960. [CrossRef]
26. Suzuki, H.; Kitamura, K.; Goto, N.; Ishikawa, K.; Ouchida, T.; Tanaka, T.; Kaneko, M.K.; Kato, Y. A Novel Anti-CD44 Variant 3 Monoclonal Antibody C(44)Mab-6 Was Established for Multiple Applications. *Int. J. Mol. Sci.* **2023**, *24*, 8411. [CrossRef]
27. Suzuki, H.; Tanaka, T.; Goto, N.; Kaneko, M.K.; Kato, Y. Development of a Novel Anti-CD44 Variant 4 Monoclonal Antibody C44Mab-108 for Immunohistochemistry. *Curr. Issues Mol. Biol.* **2023**, *45*, 1875–1888. [CrossRef]
28. Kudo, Y.; Suzuki, H.; Tanaka, T.; Kaneko, M.K.; Kato, Y. Development of a Novel Anti-CD44 variant 5 Monoclonal Antibody C44Mab-3 for Multiple Applications against Pancreatic Carcinomas. *Antibodies* **2023**, *12*, 31. [CrossRef]
29. Ejima, R.; Suzuki, H.; Tanaka, T.; Asano, T.; Kaneko, M.K.; Kato, Y. Development of a Novel Anti-CD44 Variant 6 Monoclonal Antibody C(44)Mab-9 for Multiple Applications against Colorectal Carcinomas. *Int. J. Mol. Sci.* **2023**, *24*, 4007. [CrossRef]
30. Suzuki, H.; Ozawa, K.; Tanaka, T.; Kaneko, M.K.; Kato, Y. Development of a Novel Anti-CD44 Variant 7/8 Monoclonal Antibody, C44Mab-34, for Multiple Applications against Oral Carcinomas. *Biomedicines* **2023**, *11*, 1099. [CrossRef]
31. Tawara, M.; Suzuki, H.; Goto, N.; Tanaka, T.; Kaneko, M.K.; Kato, Y. A Novel Anti-CD44 Variant 9 Monoclonal Antibody C44Mab-1 was Developed for Immunohistochemical Analyses Against Colorectal Cancers. *Curr. Issues Mol. Biol.* **2023**, *45*, 3658–3673. [CrossRef] [PubMed]
32. Kato, Y.; Kaneko, M.K.; Kuno, A.; Uchiyama, N.; Amano, K.; Chiba, Y.; Hasegawa, Y.; Hirabayashi, J.; Narimatsu, H.; Mishima, K.; et al. Inhibition of tumor cell-induced platelet aggregation using a novel anti-podoplanin antibody reacting with its platelet-aggregation-stimulating domain. *Biochem. Biophys. Res. Commun.* **2006**, *349*, 1301–1307. [CrossRef] [PubMed]
33. Heider, K.H.; Sproll, M.; Susani, S.; Patzelt, E.; Beaumier, P.; Ostermann, E.; Ahorn, H.; Adolf, G.R. Characterization of a high-affinity monoclonal antibody specific for CD44v6 as candidate for immunotherapy of squamous cell carcinomas. *Cancer Immunol. Immunother.* **1996**, *43*, 245–253. [CrossRef]
34. Heider, K.H.; Mulder, J.W.; Ostermann, E.; Susani, S.; Patzelt, E.; Pals, S.T.; Adolf, G.R. Splice variants of the cell surface glycoprotein CD44 associated with metastatic tumour cells are expressed in normal tissues of humans and cynomolgus monkeys. *Eur. J. Cancer* **1995**, *31a*, 2385–2391. [CrossRef] [PubMed]
35. Gansauge, F.; Gansauge, S.; Zobywalski, A.; Scharnweber, C.; Link, K.H.; Nussler, A.K.; Beger, H.G. Differential expression of CD44 splice variants in human pancreatic adenocarcinoma and in normal pancreas. *Cancer Res.* **1995**, *55*, 5499–5503.
36. Beham-Schmid, C.; Heider, K.H.; Hoefler, G.; Zatloukal, K. Expression of CD44 splice variant v10 in Hodgkin's disease is associated with aggressive behaviour and high risk of relapse. *J. Pathol.* **1998**, *186*, 383–389. [CrossRef]
37. Mishra, M.N.; Chandavarkar, V.; Sharma, R.; Bhargava, D. Structure, function and role of CD44 in neoplasia. *J. Oral Maxillofac. Pathol.* **2019**, *23*, 267–272. [CrossRef]
38. Hirata, K.; Suzuki, H.; Imaeda, H.; Matsuzaki, J.; Tsugawa, H.; Nagano, O.; Asakura, K.; Saya, H.; Hibi, T. CD44 variant 9 expression in primary early gastric cancer as a predictive marker for recurrence. *Br. J. Cancer* **2013**, *109*, 379–386. [CrossRef]
39. Shitara, K.; Doi, T.; Nagano, O.; Imamura, C.K.; Ozeki, T.; Ishii, Y.; Tsuchihashi, K.; Takahashi, S.; Nakajima, T.E.; Hironaka, S.; et al. Dose-escalation study for the targeting of CD44v(+) cancer stem cells by sulfasalazine in patients with advanced gastric cancer (EPOC1205). *Gastric Cancer* **2017**, *20*, 341–349. [CrossRef]
40. Erb, U.; Megaptche, A.P.; Gu, X.; Büchler, M.W.; Zöller, M. CD44 standard and CD44v10 isoform expression on leukemia cells distinctly influences niche embedding of hematopoietic stem cells. *J. Hematol. Oncol.* **2014**, *7*, 29. [CrossRef]
41. Holm, F.; Hellqvist, E.; Mason, C.N.; Ali, S.A.; Delos-Santos, N.; Barrett, C.L.; Chun, H.J.; Minden, M.D.; Moore, R.A.; Marra, M.A.; et al. Reversion to an embryonic alternative splicing program enhances leukemia stem cell self-renewal. *Proc. Natl. Acad. Sci. USA* **2015**, *112*, 15444–15449. [CrossRef] [PubMed]
42. Tang, L.; Huang, H.; Tang, Y.; Li, Q.; Wang, J.; Li, D.; Zhong, Z.; Zou, P.; You, Y.; Cao, Y.; et al. CD44v6 chimeric antigen receptor T cell specificity towards AML with FLT3 or DNMT3A mutations. *Clin. Transl. Med.* **2022**, *12*, e1043. [CrossRef] [PubMed]

43. Verel, I.; Heider, K.H.; Siegmund, M.; Ostermann, E.; Patzelt, E.; Sproll, M.; Snow, G.B.; Adolf, G.R.; van Dongen, G.A. Tumor targeting properties of monoclonal antibodies with different affinity for target antigen CD44V6 in nude mice bearing head-and-neck cancer xenografts. *Int. J. Cancer* **2002**, *99*, 396–402. [CrossRef] [PubMed]
44. Li, G.; Suzuki, H.; Ohishi, T.; Asano, T.; Tanaka, T.; Yanaka, M.; Nakamura, T.; Yoshikawa, T.; Kawada, M.; Kaneko, M.K.; et al. Antitumor activities of a defucosylated anti-EpCAM monoclonal antibody in colorectal carcinoma xenograft models. *Int. J. Mol. Med.* **2023**, *51*, 18. [CrossRef]
45. Kaneko, M.K.; Ohishi, T.; Kawada, M.; Kato, Y. A cancer-specific anti-podocalyxin monoclonal antibody (60-mG(2a)-f) exerts antitumor effects in mouse xenograft models of pancreatic carcinoma. *Biochem. Biophys. Rep.* **2020**, *24*, 100826. [CrossRef]
46. Kato, Y.; Kaneko, M.K. A cancer-specific monoclonal antibody recognizes the aberrantly glycosylated podoplanin. *Sci. Rep.* **2014**, *4*, 5924. [CrossRef] [PubMed]
47. Suzuki, H.; Kaneko, M.K.; Kato, Y. Roles of Podoplanin in Malignant Progression of Tumor. *Cells* **2022**, *11*, 575. [CrossRef]
48. Ishikawa, A.; Waseda, M.; Ishii, T.; Kaneko, M.K.; Kato, Y.; Kaneko, S. Improved anti-solid tumor response by humanized anti-podoplanin chimeric antigen receptor transduced human cytotoxic T cells in an animal model. *Genes Cells* **2022**, *27*, 549–558. [CrossRef]
49. Chalise, L.; Kato, A.; Ohno, M.; Maeda, S.; Yamamichi, A.; Kuramitsu, S.; Shiina, S.; Takahashi, H.; Ozone, S.; Yamaguchi, J.; et al. Efficacy of cancer-specific anti-podoplanin CAR-T cells and oncolytic herpes virus G47Δ combination therapy against glioblastoma. *Mol. Ther.-Oncolytics* **2022**, *26*, 265–274. [CrossRef]
50. Shiina, S.; Ohno, M.; Ohka, F.; Kuramitsu, S.; Yamamichi, A.; Kato, A.; Motomura, K.; Tanahashi, K.; Yamamoto, T.; Watanabe, R.; et al. CAR T Cells Targeting Podoplanin Reduce Orthotopic Glioblastomas in Mouse Brains. *Cancer Immunol. Res.* **2016**, *4*, 259–268. [CrossRef]

current issues in molecular biology

MDPI

Review

Head and Neck Squamous Cell Carcinoma Vaccine: Current Landscape and Perspectives

Piero Giuseppe Meliante [1], Carla Petrella [2], Marco Fiore [2], Antonio Minni [1,3,*] and Christian Barbato [2,*]

1 Department of Sense Organs, Sapienza University of Rome, Viale del Policlinico 155, 00161 Rome, Italy
2 Institute of Biochemistry and Cell Biology (IBBC), National Research Council (CNR), Department of Sense Organs, Sapienza University of Rome, Viale del Policlinico 155, 00161 Rome, Italy
3 Division of Otolaryngology-Head and Neck Surgery, Ospedale San Camillo de Lellis, ASL Rieti-Sapienza University, Viale Kennedy 1, 02100 Rieti, Italy
* Correspondence: antonio.minni@uniroma1.it (A.M.); christian.barbato@cnr.it (C.B.)

Abstract: The treatment of unresectable or metastatic Head and Neck Squamous Cell Carcinoma (HNSCC) has traditionally relied on chemotherapy or radiotherapy, yielding suboptimal outcomes. The introduction of immunotherapy has significantly improved HNSCC treatment, even if the long-term results cannot be defined as satisfactory. Its mechanism of action aims to counteract the blockade of tumor immune escape. This result can also be obtained by stimulating the immune system with vaccines. This review scope is to comprehensively gather existing evidence and summarize ongoing clinical trials focused on therapeutic vaccines for HNSCC treatment. The current landscape reveals numerous promising drugs in the early stages of experimentation, along with a multitude of trials that have been suspended or abandoned for years. Nonetheless, there are encouraging results and ongoing experiments that instill hope for potential paradigm shifts in HNSCC therapy.

Keywords: head and neck squamous cell carcinoma (HNSCC); therapeutic cancer vaccine; Epstein–Barr virus (EBV); human papilloma virus (HPV); mRNA vaccine

Citation: Meliante, P.G.; Petrella, C.; Fiore, M.; Minni, A.; Barbato, C. Head and Neck Squamous Cell Carcinoma Vaccine: Current Landscape and Perspectives. *Curr. Issues Mol. Biol.* **2023**, *45*, 9215–9233. https://doi.org/10.3390/cimb45110577

Academic Editors: Emma Adriana Ozon and Violeta Popovici

Received: 21 September 2023
Revised: 14 November 2023
Accepted: 15 November 2023
Published: 16 November 2023

1. Introduction

Head and neck squamous cell carcinoma (HNSCC) affects 450,000 individuals per year, accounting for an estimated 890,000 new cases, i.e., roughly 4.5% of all cancer diagnoses. The HNSCC incidence includes 380,000 cases of cancer of the oral cavity and the lip, 133,000 of the nasopharynges, 98,000 of the oropharynges, 84,000 of the hypopharynxes, 185,000 of the larynxes, and 54,000 of the salivary glands. The incidence and mortality rates of HNSCC are distributed across geographic regions and demographic traits, with a higher occurrence in men than in women and a male-to-female ratio of 2:1 [1].

In situations where surgical intervention is not viable, chemotherapy and radiation therapy are used. However, results are not optimal, particularly for recurrent or metastatic malignancies. The introduction of immunotherapy has improved those outcomes. Currently approved drugs for the treatment of HNSCC are pembrolizumab (KEYTRUDA, Merck and Co., Rahway, NJ, USA) and nivolumab (OPDIVO, Bristol Myers Squibb, New York, NY, USA), which have improved traditional chemotherapy results. Checkmate-141, Keynote-040, and Keynote-048 trials are milestones in HNSCC treatment and set a new standard of results for non-surgical therapy. However, they still are not satisfactory with long-term efficacy in 20 to 30% of patients only. The discussion of resistance mechanisms to immune checkpoint inhibitors is beyond the scope of this manuscript, but we can state that solutions need to be found to achieve better therapy performance [2–4].

In this scenario, the advancement of novel non-surgical therapies appears to be imperative. HNSCC are categorized as HPV-negative and HPV-positive. Tobacco consumption is the primary risk factor for the development of HPV-negative HNSCC. In addition, betel quid, areca nut, exposure to environmental pollutants, or excessive alcohol consumption is known to synergize with tobacco use to promote carcinogenesis [5,6]. While the tobacco- and alcohol-related neoplasms are decreasing, viral infection is on the rise. HPV is a common risk factor for HNSCC, being associated with oropharyngeal cancers (>70%) [7]. HPV vaccines have been approved for cervical cancer prevention in females, and their impact on HNSCC incidence has been observed [6]. However, therapeutic vaccines for HNSCC are still undergoing investigation. Their objective is to induce immunogenicity against HNSCC cells employing a range of mechanisms, including traditional approaches like cell-mediated cytotoxicity induced by antigens, as well as innovative strategies to counteract tumor immune escape mechanisms or stimulate the immune system's cytotoxic activity against neoplastic cells [5]. Here we show an in-depth review of the limitations of current studies and future perspectives in immunotherapy for the treatment of HNSCC to provide a comprehensive overview.

2. Materials and Methods

A literature search was conducted across PubMed, Embase, and Cochrane Central Register of Controlled Trials databases, without any time restrictions. The search terms specifically focused on head and neck cancer vaccines, and each author independently performed the search and analysis. In addition to the database search, the authors manually screened the reference lists of retrieved articles for further relevant studies. We excluded non-English language papers. We hand-searched the ClinicalTrials.gov database for any relevant trial and checked the actual status of each of those considered. All authors discussed results with conflicts solved by our senior author, A.M. The evidence extracted from these papers was organized into coherent paragraphs. Furthermore, a thorough examination of the current studies' limitations and future perspectives was conducted in order to provide a comprehensive overview (Figure 1). Furthermore, it is important to note that Cetuximab was approved in 2006 as the first monoclonal antibody (mAb) approved by the Food and Drug Administration (FDA), directed against the epidermal growth factor receptor (EGFR) in patients with HNSCC. Therefore, the literature timeline and considered studies utilized to perform this analysis were mainly from the period of 2016 to 2022.

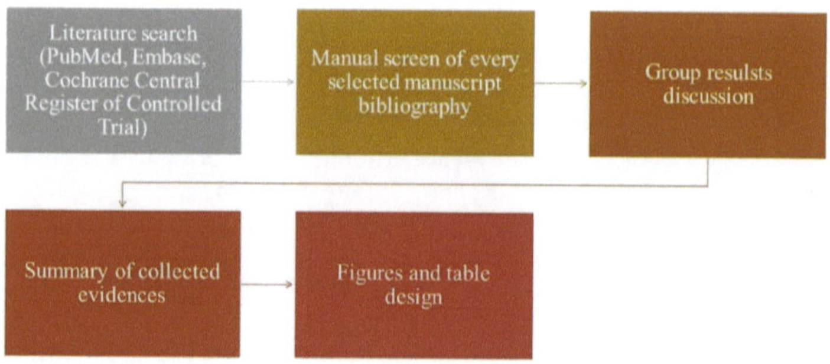

Figure 1. Article selection and discussion process.

2.1. Immunosurveillance and Immune Escape Mechanisms

The immune system plays a crucial role in the uncontrolled growth and spread of neoplastic cells, as seen via immunosurveillance [7]. It can eliminate cancer cells that are constantly produced throughout an individual's life. During neoplastic progression, several mutations in cancer cell DNA permit the acquisition of the ability to evade the immune response through the downregulation of HLA antigens, a decrease in or loss of expression of tumor-associated antigens, and the production of immunosuppressive cytokines [8].

Tumor antigens are taken up by antigen-presenting cells (APCs), such as dendrites, and are exposed to host T lymphocytes, which in turn form effector and memory T lymphocytes. Cytotoxic T lymphocytes and natural killer cells circulate in the peripheral blood and lymphoid and non-lymphoid tissues, as well as in tumoral tissues [9]. T cells encounter tumor cells equipped with these same antigens and become activated and kill them through a cytotoxic mechanism. Cancer cell death is induced via the release of cytokines such as interferon (IFN)-γ, tumor necrosis factor (TNF)-α, granzyme, perforins, and IL-2 [10,11]. To proliferate freely, tumors may develop immune escape mechanisms via cytotoxic T-lymphocyte protein 4 (CTLA4) and programmed cell death protein 1 (PD-1). Immunotherapy drugs have been developed to inhibit these molecules and allow the immune system to act on tumors [2,3]. In addition to the molecular mechanisms of immune escape expressed on the surface of tumor cells, there are several others that involve the tumor microenvironment. These include increases in the percentage of immunosuppressive cells in the tumor matrix (myeloid-derived suppressor cells, tumor-associated macrophages, and T regulatory cells); the secretion of immunosuppressive molecules, such as transforming growth factor-β (TGF-β); and the signal transducer and stimulator of transcription (STAT)-3, as well as the formation of physical barriers and an intricate vascular network that physically prevents the penetration of immune cells into the tumor matrix [5,12].

2.2. Anticancer Vaccine Categories

The anticancer vaccines under study can be divided into two groups: traditional vaccines that induce a T cell-mediated immune response against specific tumor antigens, and less conventional vaccines targeting immune escape mechanisms [5,11].

Autologous vaccines, which utilize patient-specific antigens, offer a highly targeted and specific immune response. However, the complex process, including the extraction and inactivation of tumor cells, as well as the associated development costs and lack of standardization, pose challenges to their widespread use. As a result, recombinant vaccines, which are generated through the laboratory synthesis of tumor antigens, are being subjected to more extensive trials. These recombinant vaccines offer potential advantages in terms of standardization and scalability [5].

2.3. Anticancer Vaccine Antigens

The antigens expressed by tumors are divided into vague categories, and some antigens may belong to more than one. According to the classification of Zarour et al., those categories are oncofetal, oncoviral, overexpressed/accumulated, cancer-testis, linear-restricted, mutated, post-translationally altered, or idiotypic antigens (Table 1) [13–16]. Tumor antigens recognized by T lymphocytes can also be classified as shared antigens, tumor-associated antigens, and tumor-specific antigens, according to Coulie et al. [17].

Table 1. Categories of tumor antigens.

Categories of Tumor Antigens	Description
	Classification by Zarour et al.
Oncofetal	Usually expressed in fetal tissues
Oncoviral	Encoded by virus DNA/RNA
Overexpressed/Accumulated	Expressed in both healthy and neoplastic tissues with higher levels in cancer cells
Cancer-testis	Expressed in adult reproductive tissues physiologically and in neoplastic cells
Linear-restricted	Expressed by specific cancer histotypes
Mutated	Only expressed by cancer
Post translationally altered	Post-transcriptional alteration of molecules
Idiotypic	Highly polymorphic genes expressed in a specific "clonotype" in cancer tissues
	Classification by Coulie et al.
Shared antigens	Expressed both by tumor and healthy cells
Tumor associated antigens	Antigens expressed by tumor and healthy cells that are upregulated in cancers
Tumor specific antigens	Expressed only by tumor cells

2.4. Anticancer Vaccine Platforms

Devaraja et al. [5] conducted a classification of anticancer vaccine platforms, elucidating the advantages and disadvantages associated with each of them (Figure 2).

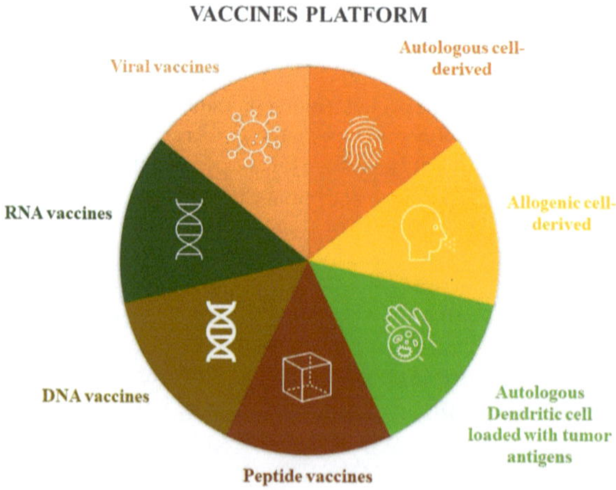

Figure 2. Vaccine platforms.

These vaccines involve extracting tumor cells from the patient, which are then inactivated in the laboratory. Subsequently, these inactivated tumor cells are combined with immunostimulant substances before being administered back to the patient. Another approach mentioned is the utilization of allogeneic tumor lines, where cells from different individuals are inactivated and used for vaccine development. Instead of using the entire tumor cell, some authors have experimented with the use of tumor antigens loaded inside the patient's own dendritic cells, which are then reinfused to stimulate the immune response. Protein components have also been used to produce these drugs, such as peptide vaccines based on epitopes obtained through the combination of MHC class I and tumor antigens. Like proteins, nucleic acids have also been the basis of vaccines, both for DNA and RNA. Finally, viruses with low pathogenicity have also been modified to express neoplastic antigens and induce immunogenicity (Table 2) [5].

Table 2. Advantages and disadvantages of vaccine platforms.

Vaccines Platform	Advantages	Disadvantages
Autologous cell-derived	Exposed to all patient tumor antigens Vaccine designed for specific patient disease	Difficult to manufacture. not standardizable. requires sufficient tissue biopsy
Allogenic cell-derived	More potential antigens available; standardization; lower costs	Less personalization
Autologous dendritic cell loaded with tumor antigens	Dendritic cells are the most powerful antigen-presenting cells	Require leukaphereses; require cell culture processing
Peptide vaccines	Easy to produce; easy to store; no viral component	Easy tolerance; rapid degradation in human body; usually require immunogenic adjuvants
DNA vaccines	Use of multiple genes; can be combined with immunostimulatory agents	Modest efficacy; risk of genetic recombination
RNA vaccines	Low levels of side effects; low levels f autoimmune disease	Rapid degradation
Viral vaccines	Induce immune and cell-mediated responses	

2.5. Virus Infection-Based Cancer Vaccines

2.5.1. Epstein–Barr Virus (EBV)-Related Nasopharyngeal Carcinoma (NPC) Vaccines

Therapeutic vaccination has been extended to NPC in view of its association with EBV. The approaches developed are based on a dendritic cell-based strategy and use virus-based vaccines. Virus-induced malignancies have multiple therapeutic targets due to non-self-origin.

The Epstein–Barr virus nuclear antigen-1 (EBNA1) and the Epstein–Barr virus latent membrane proteins 1/2 (LMP1/2) are target antigens. Four trials (NCT01256853, NCT01800071, NCT01147991, NCT01094405) investigated the efficacy of EBNA1 C-terminal /LMP2 chimeric protein-expressing recombinant modified vaccinia, the Ankara vaccine (MVA) [18].

The study of Trabecutel (Atara Biotherapeutics) and allogenic EBV-T-cell immunotherapy was suspended after phase 1B by the sponsor and phase 2 was never conducted. The trial NCT03769467 was a multicenter, open-label, single-arm phase 1B/2 study to assess the safety and efficacy of Trabecutel in combination with pembrolizumab for the treatment of platinum-pretreated patients with recurrent/metastatic EBV+ NPC (Figure 3).

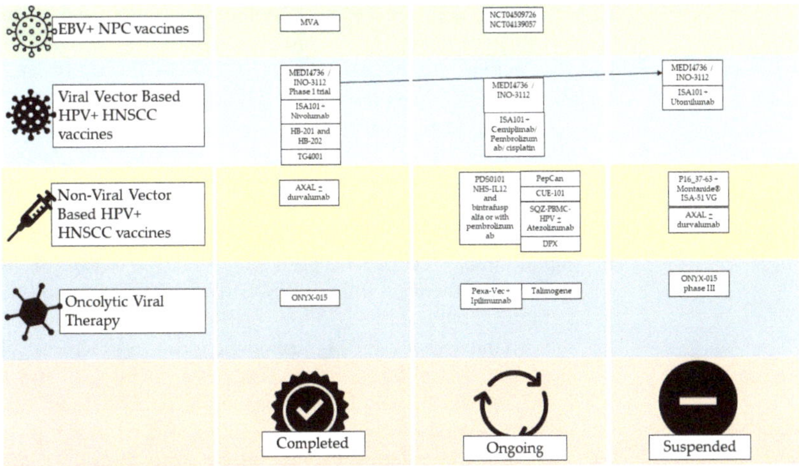

Figure 3. Current status of virus based HNSCC vaccine trials on humans. Unless otherwise stated, the studies considered were phase 1 or 2. Completion does not indicate the success of the therapy, but only the end of the study and the publication of its data. EBV = Epstein–Barr virus; NPC = nasopharyngeal carcinoma; HPV = human papilloma virus; HNSCC = head and neck squamous cell carcinoma; MVA = modified Ankara vaccine; DPX = DepoVax^{TM}.

The NCT04139057 trial is recruiting patients for a phase 1 study on the administration of EBV-specific engineered T cells bearing a TCR (TCR-T) anti-PD-1. The estimated enrollment will be 18 participants affected by EBV+ HNSCC with a single-arm trial design. The estimated study completion date is 1 January 2024 (Figure 3).

TCR-Ts are the subject of a trial that is about to end (August 2023). In this single-arm study, the TCR-Ts are specific for EBV and are equipped with a cytokine-secreting system. The rationale is that cytokines, by activating both innate immunity with NK cells and adaptive immunity, promote the immune response against cancer. The study has an estimated enrollment of 20 patients pre-conditioned with chemotherapy who will be infused with EBV-specific TCR-T cells with cytokine auto-secreting element (NCT04509726) (Figure 3).

2.5.2. HPV+ HNSCC Vaccines

Viral Vector-Based HPV+ HNSCC Vaccines

HPV+ HNSCC vaccines are different from prophylactic HPV vaccines such as Gardasil (Merck and Co., Rahway, NJ, USA) and Cervarix (GlaxoSmithKline Biologicals, Rixensart, Belgium), which target the L1 capsid protein of the virus. Infected cancer cells do not express L1, but they need the oncoproteins E6 and E7, which are induced by the virus. The therapeutic vaccines under development target those proteins from HPV-16 and -18.

For example, MEDI4736 also known as INO-3112 is a DNA-based vaccine with two components, one targeting E6 and E7 antigens from HPV-16 and -18 and another that encodes for a recombinant interleukin IL-12. The vaccine has been studied in a phase Ib/II trial involving 18 HNSCC HPV+ patients and 18 out of 21 showed antigen-specific T cell activity and persistent cellular response after 1 year. The authors concluded that INO-3112 can generate durable peripheral and tumor immune responses and hypothesized that it could be used in association with immune checkpoint inhibitors [19]. INO-3112 was studied in the suspended NCT04001413 studies and two studies in combination with durvalumab in the treatment of recurrent or metastatic HNSCC. In the first, 35 patients were enrolled, but 17 patients died during the study and 13 did not complete follow-up (NCT03162224); the second, preliminary unpublished data, and the study population was composed of any HPV+ cancer (not just head and neck) and to date, the study is indicated as 'active' and 'not recruiting' (NCT03439085) (Figure 3).

ISA101, a synthetic long-peptide HPV-16 vaccine inducing HPV-specific T cells, was studied in combination with Nivolumab in 24 patients, including 22 with oropharyngeal cancer (phase 2). The authors observed an overall response rate of 33%, with a median duration of response of 10.3 months and a median overall survival of 17.5 months, and only two grade 3 or 4 toxicity events were reported [20]. The efficacy of ISA101 in combination with Cemiplimab, Pembrolizumab, cisplatin, or Utomilumab, is ongoing (phase2), but no preliminary results have been published (NCT03669718, NCT04369937, NCT04398524). In addition, the association of ISA101 and Utolimumab (NCT03258008) was discontinued (Figure 3).

Choriomeningitis lymphocytic virus and Pichinde virus were used as two vaccines against the HPV16 E6E7 fusion protein. HB-201 and HB-202 were evaluated in the NCT04180215 trial. It is interesting to observe how intratumoral administration is being evaluated for these vaccines, alone or combined with systemic administration, as well as parenteral administration. Furthermore, the authors also experimented with the alternating administration of the two drugs, observing greater immunogenicity than with the exclusive use of one of the two (Figure 3) [21].

The vaccine TG4001 (Tipapkinogene sovacivec) is formed via an attenuated viral vector expressing the coding sequences for the E6 and E7 proteins of HPV-16 and -18, and IL-2. A phase 1B/2 study was conducted, and among nine patients enrolled, five with head and neck cancer, only three showed T-mediated peripheral immunity against E6/E7, and four showed increased CD8 infiltrate and/or T-reg/CD8 ratio in the neoplastic tissue (Figure 3) [22].

Non-Viral Vector-Based HPV+ HNSCC Vaccines

SQZ-PBMC-HPV (SQZ Biotechnologies, Watertown, MA, USA) is a vaccine produced using a proprietary technology called cell squeeze technology, which acts on circulating mononuclear cells. Phase 1 trial NCT04084951 evaluated its safety and efficacy in monotherapy and association with atezolizumab (Tecentriq) or any other ICI, in patients with locally advanced, recurrent, or metastatic HPV+ cancers, including HNSCC. Preliminary results showed good tolerability and immune response, even though the HNSCC population was only 3 out of the total 12 (Figure 3) [23].

PDS0101 is a liposomal-based vaccine against HPV16 E6 and E7 proteins that also contain R-DOTAP, a lipid under evaluation for anti-HPV+ and HNSCC activity. This vaccine is under study in combination with NHS-IL12 and bintrafusp alfa or with pembrolizumab.

NHS-IL12 is an immunocytokine that results in IL-12, and bintrafusp alfa is a molecule obtained by combining a human IgG1 against PD-L1 and the extracellular domain of the TGF-β receptor type II, and the result is an action against TGFβRII. Their associations with PDS0101 are under evaluation in the NCT04287868 trial, whose recruitment is expected to end on 1 January 2024, and no preliminary results have been published yet. The case of the association with pembrolizumab in the trials NCT04260126 (VERSATILE002) and NCT05232851 is different, as preliminary results have been disclosed. The phase II study VERSATILE002 has a population of patients affected by recurrent or metastatic HNSCC, who are positive for both HPV 16 and PD-L1. The phase 1/2 trial NCT05232851, however, has a population of patients with locally advanced squamous cell carcinoma of the oropharynx. PDS0101 and pembrolizumab were well tolerated with no significant toxicity in the enrolled population [24]. Their combination showed significant anti-tumor activity and the FDA granted the Fast Track designation to this association for use in recurrent or metastatic HPV16+ and HNSCC (Figure 3) [25].

AXDS 11-001, also known as Axalimogene Filolisbac or AXAL (Advaxis Inc., Princeton, NJ, USA), is based on the bacterium Listeria monocytogenes listeriolysin O, modified to secrete the HPV-E7 tumor antigen as a fusion protein called LLO-E7 [26]. The NCT02002182 trial enrolled 15 patients divided into two groups, in one of which the vaccine was administered before transoral robotic surgery in the treatment of squamous cell carcinoma of the larynx, while the other group was directly subjected to surgery to evaluate the immune response induced by the vaccine. Only nine patients completed the study, five in the experimental group and four in the control group. ADXS 11-001 showed increased systemic immune response and CD4+ and CD8+ T cell infiltration. At the same time, the vaccinated subjects had an incidence of adverse events of 55.5% compared to 16.7% in the control group [26]. The suspicion of adverse events associated with this type of drug seems to have been increased by two further trials, NCT02291055 and NCT01598792. The first involved a combination of AXAL and durvalumab and was put on hold due to the death of a patient. The second, concerning HPV16+ oropharyngeal carcinoma, had only two patients enrolled and was suspended because one experienced dose-limiting toxicity (Figure 3) [27].

The DepoVax^{TM} (DPX)-E7 (IMV Inc., Dartmouth, NS, Canada) is an HPV-16 E711-19 nanomer that demonstrated antigen-induced effective anti-cancer immunity in mice models [28]. It is under study in a phase 1b/2 trial for HPV+ head and neck, cervical, or anal cancer (positive for HLA-A*02) in the clinical trial NCT02865135. Eleven patients have been enrolled and are currently under follow-up, the estimated completion time is December 2023. No preliminary data have been published yet (Figure 3).

Another peptide-based vaccine which, to date, has only been tested in animal models is based on the intratumoral injection of the E7 long peptide. This practice effectively controlled buccal TC-1 cancers in mice models and enhanced E7-specific CD8 intratumoral and circulating cells. The immune reaction induced in those animals was not dependent on CD4+ T cells (Figure 3) [29].

A novel technique for vaccine development is Immuno-STAT. These are fusion proteins built to deliver cytokines to achieve specific CD8+ T cell activation. CUE-101 is the first Immuno-STAT-based vaccine in a clinical trial, and it is composed of a human leukocyte antigen (HLA) complex (HLA-A*0201), an HPV16 E7 protein-derived peptide epitope, and four reduced-affinity IL-2. The vaccine was designed to induce HPV-16-specific CD8+ T cell activation. The first phase 1 trial is going to be completed in December 2023 with a population of 85 patients with HPV16+ recurrent/metastatic HNSCC as a monotherapy treatment in the second line or combination therapy with Pembrolizumab in the first line (NCT03978689). Partial data from this trial suggest the selective expansion of HPV-16-specific CD8+ T cells and good tolerability. Data about efficacy are still limited with 1 of 14 patients exhibiting a partial response and 6 of 14 patients exhibiting stable disease for more than 12 weeks in the CUE-101 arm, and 2 of 7 patients exhibiting a partial response and 2 of 7 patients exhibiting stable disease in the combination arm (Figure 3) [30].

The combination of HPV-16 E6 peptides and Candida skin test reagent as a novel adjuvant was used to create the PepCan vaccine [31,32]. For the positive treatment of cervical cancer [33], this vaccine is being studied for HNSCC in the NCT03821272. The investigators are giving PepCan or placebo to patients affected by HNSCC who achieved remission for a period of 2 years.

The University of Southampton, in collaboration with BioNTech SE, is carrying out the HPV Anti-CD40 RNA vaccinE (HARE-40) phase 1/2 vaccine dose escalation study, in which they are analyzing the BNT113 (an anti-CD40 RNA vaccine from BioNTech SE, Mainz, Germany) as a monotherapy. The trial has two arms, the 1A is an intrapatient dose escalation in patients with previously treated for HPV16+ head and neck cancer using two dose cohorts to establish a safe, tolerable, and recommended dose of the HPV vaccine. Arm 1B will perform a dose escalation in patients with advanced HPV16+ cancer (head and neck, anogenital, penile, cervical, and other) using a single cohort to establish a safe, tolerable, and recommended dose of HPV vaccine. The estimate study completion date is 30 April 2025. However, recruitment is still suspended due to the COVID-19 pandemic (NCT03418480).

Classically, HPV E6 and E7 have been used as target antigens in HPV+ HNSCC development. Some studies observed the hyperexpression of $p16^{INK4a}$ and studied it as a target. The NCT02526316 VICORYX-2 trial evaluated the combination of P16_37-63 peptide combined with Montanide® ISA-51 once a week for 4 weeks. The trial included patients with HPV+ cancer with a diffuse expression of $p16^{INK4a}$ (not only head and neck but also anogenital cancers). The investigators enrolled 11 patients, and the study was completed in May 2017. No data have been published yet. A phase 1/2a trial studied the combination of the peptide P16_37-63 and Montanide® ISA-51 VG regarding safety and efficacy. A total of 26 patients with HPV+ SCC (anogenital and head and neck) were enrolled and after an initial safety assessment of 10 of these, the researchers studied the efficacy of the medication. A total of 20 patients received at least four doses of the vaccine and were evaluated for immune response. CD4+ cells were induced in 11 out of 20 patients, CD8+ in 4 out of 20, and antibodies in 14 out of 20. None of the patients healed, but 10 of them had stable disease, of whom 3 were stable for the whole duration of the follow-up (NCT01462838). The trial was prematurely discontinued due to premature death or progressive disease in most of the patients (Figure 3) [34].

2.6. Oncolytic Viral Therapy

The use of viruses as weapons to kill cancer cells was pioneered over 20 years ago. ONYX-015 first entered clinical trials in 1996; it is an adenovirus with a deletion of the E1B gene engineered to selectively lyse p53-deficient neoplastic cells and not attack healthy cells. The drug has been tested using intratumor administration. Post-treatment biopsies showed the presence and/or replication of the virus in 7 of 11 patients in the tumor but not in the immediately adjacent tissues. A total of 21% of patients showed tumor regression with a volume greater than 50% and no alterations of the surrounding tissues [35,36]. Intratumoral ONYX-015 has also been studied in combination with the systemic administration of cisplatin and 5-fluorouracil. The rationale behind Khuri et al. attempting this approach was that while ONYX-015 demonstrated efficacy in HNSCC, the disease rapidly relapsed. The scholars observed a response in all patients treated with the combination, while the group treated only with traditional chemotherapy underwent progression. Again, the intratumor replication of the virus was confirmed by biopsies [37]. In 2001, the use of ONYX-015 was also tested intravenously in patients with metastatic solid tumors. The authors observed an increase in neutralizing antibodies and several inflammatory cytokines. But, in this study, only two of the patients had HNSCC [38]. Given the promising results of the phase 2 studies regarding ONYX-015, a phase 3 study has been reported to have taken place more than 20 years ago, but no data are available [39–41].

Pexa-Vec is an oncolytic virus vaccine derived from the Wyeth-strain that has been genetically modified to express the huma GM-CSF. The mechanism of action includes the activation of dendritic cells and the enhancement of the tumor immune infiltrating cells. The association of Ipilimumab with the treatment of locally advanced, recurrent, or metastatic solid cancers, including HNSCC, is under evaluation in the NCT02977156 trial.

Talimogene is derived from the herpes virus carrying GM-CSF, and its association with pembrolizumab demonstrated a tolerable safety profile, but the efficacy was similar to that of pembrolizumab monotherapy in historical HNSCC trials (Figure 3) [42].

H101 is another oncolytic adenovirus-based vaccine like ONYX-015. Its intratumorally administration associated with systemic chemotherapy has been compared with chemotherapy alone. The combination arm showed a higher response rate (79% vs. 39.6%, $p < 0.001$). In 2005, the Chinese government approved the H101 vaccine in combination with cisplatin-based chemotherapy for the treatment of nasopharyngeal carcinoma (Figure 3).

2.7. Cancer Testis Antingen-Based Vaccines

The most frequently over-expressed cancer testis antigens in HNSCC are from the MAGE group [43]. A pilot study using Trojan vaccines demonstrated acceptable toxicity and systemic immune responses against HLA-II-restricted epitopes in five MAGE-A3/HPV 16+ patients with recurrent or metastatic (R/M) HNSCC. Montanide ISA 51 and GM-CSF were used as adjuvants to facilitate dendritic cell migration to the vaccination site and enhance antigen presentation [44]. A phase 1 trial (NCT00257738), involving additional cases of progressive recurrent or metastatic HNSCC (HLA A2+), confirmed the feasibility and safety of these vaccines. Unfortunately, the trial, originally intended to enroll 90 cases, prematurely closed due to poor accrual after enrolling only 17 patients (Figure 4). Any immunized patients in both studies demonstrated partial or complete clinical response. The efficacy of a dual-antigenic peptide vaccine comprising MAGED4B and four-jointed box 1 (FJX1) was studied, evidencing strong immunogenic responses with the peptide combination compared to individual use. These have only been studied in vitro or in mouse models (Figure 4) [45,46].

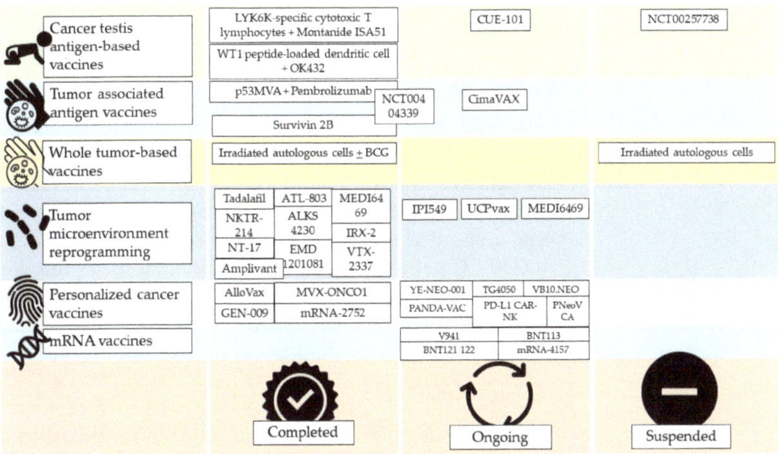

Figure 4. Current status of non-virus-based HNSCC vaccine trials on humans. Unless otherwise stated, the studies considered were phase 1 or 2. Completion does not indicate the success of the therapy but only the end of the study and the publication of its data. MVA = modified Ankara vaccine.

Another peptide that has been studied in vaccine development is LY6K. It is over-expressed in HNSCC and undetectably low in normal cells. A vaccine based on LYK6K-specific cytotoxic T lymphocytes has been studied in 37 patients affected by recurrent or metastatic HNSCC along with Montanide ISA51 as an adjuvant. This therapy was demonstrated to be more effective than the best supportive care. The authors observed an antigen-specific immune response and found that it was related to overall survival (Figure 4) [47].

In a T phase 1/2 trial conducted in HNSCC patients, a WT1 peptide-loaded dendritic cell-based vaccine in combination with the OK-432 adjuvant and chemotherapy, was administered. It demonstrated feasibility, safety, and promising clinical efficacy in patients with recurrent or metastatic HNSCC (Figure 4) [48]. CUE-102, an Immuno-STAT, shares remarkable similarities with vaccine-CUE-101. This vaccine is currently being evaluated in clinical trials for various solid malignancies (Figure 4) [5].

2.8. Tumor-Associated Antigen Vaccines

In HPV-HNSCC, a vaccine against a mutated epitope of p53 requires custom development, whereas, for the wild-type p53 gene, it could be produced on a large scale. In the phase 1 clinical trial (NCT00404339), the intranodal injection of autologous dendritic cells loaded with wild-type p53 as a tumor peptide-specific p53 vaccine was found to be safe and effective. The two-year disease-free survival rate in a cohort including patients with advanced HNSCC was 88%, and the three-year survival rate was 80%, which outperformed the disease-free survival rate of 70% observed in a similar cohort treated with chemoradiation alone. Although the trial aimed to enroll 50 patients, only 17 were recruited [49]. A phase 1 study (NCT02432963) involving patients with high p53 expression, including one HNSCC, demonstrated the efficacy of p53-expressing modified vaccinia virus Ankara (MVA) (p53MVA) vaccination in combination with pembrolizumab, leading to clinical benefits in select patients. Furthermore, the loss of p53 function can also be targeted for oncolytic therapy using ONYX-15, as discussed earlier [50].

EGFR overexpression is typical of HPV- HNSCC. A vaccine based on dendritic cells containing EGFR fused to a glutathione-S-transferase induced a significant immunity response in mice. A phase 1/2 trial using a recombinant human EGF-rP64K/Montanide ISA 51 vaccine (CIMAvax) and nivolumab for patients with metastatic non-small cell lung cancer or HNSCC is ongoing.

A phase 1 trial (UMIN000000976) showed the safety and advantageous therapeutic potential of survivin-2B peptide vaccination in HLA-A*2402 patients with unresectable, locally advanced, or recurrent oropharyngeal squamous cell carcinoma [51].

2.9. Whole Tumor-Based Vaccines

Irradiated NDV-modified autologous tumor cells have been injected intradermally to induce anti-cancer immunity in 20 heterogeneous HNSCC patients 3 months after surgery. The authors reported a 5-year overall survival of 61% and confirmed peripheral immunity after 5 years of disease-free patients [52]. The injection of irradiated autologous tumor cells associated with BCG and vaccine-primed lymph node cells demonstrated efficacy in HNSCC patients [53]. Serial immunological studies demonstrated significant immune responses in vaccinated HNSCC patients with autologous tumors, but it was withdrawn due to not enrolling enough patients [52,53].

Another trial used apoptotic autologous tumor cells fused with dendric cells and administered them to patients with locally advanced HNSCC who had been successfully treated with first-line therapy but were at risk of recurrence or developing a second primary tumor. Serial immunological studies demonstrated measurable immune responses in vaccinated HNSCC patients, specifically targeting the autologous tumor. However, the study has been withdrawn due to impossibility of enrolling enough patients [54].

2.10. Tumor Microenvironment Reprogramming

One of the immune escape mechanism of HNSCC is the immunity suppression in the tumor microenvironment. Cancer cells use several immune escape mechanisms, such as PD-1, CTLA-4, IL-6, IL-10, TGF-beta, and STAT-3. The result is the suppression of the CD8+ T Cells and the increase in the Treg and myeloid-derived suppressor cell populations [2,3].

Since the inhibition of myeloid-derived suppressors using a phosphodiesterase-5 (PDE5) inhibitor restores the CD8* cells' activity, they became a potential target for vaccines [55]. The role of PDE5 inhibitor in potentiating nonspecific and tumor-specific immune responses in HNSCC confirmed by two randomized, double-blinded, placebo-controlled clinical trials (NCT00894413 and NCT00843635), which investigated the use of tadalafil as a PDE5 inhibitor. Another phase 2 trial (NCT01697800) evaluated the combination of tadalafil with conventional therapy in 40 patients with HNSCC between September 2012 and July 2014; however, the results of this trial have not been published. According to ClinicalTrials.gov, among the 25 patients in the tadalafil group, one patient experienced mortality compared to none in the placebo group. A randomized phase 1/2 clinical trial (NCT02544880) was started by the same research group in April 2016, aimed to evaluate the efficacy of tadalafil treatment and Anti-MUC1 in patients with recurrent or second primary HNSCC. Preliminary data reported a safety profile of PDE5 inhibition in HNSCC. This study was motivated by the lack of significant efficacy, as observed in tadalafil monotherapy in previous studies despite the positive enhancement of anti-tumor immune responses [56].

UCPVax is a vaccine against some novel major histocompatibility complexes class II derived from the human telomerase reverse transcriptase that is usually overexpressed in HVP+ HNSCC. This mediation is under study in the VolATIL phase 2 trial (NCT03946358). A similar mechanism of action is used by the vaccine under evaluation in the FOCUS phase 2 trial (NCT05075122) [57].

OX40 is expressed by T cells and enhances their survival and activity, and it can be considered an antagonist of the tumor-suppressive microenvironment. OX40 agonists are under evaluation in HNSCC treatment. Neoadjuvant Anti-OX40 (MEDI6469) demonstrated promising results [57,58]. Another OX40 agonist is under study in a phase 1 trial (NCT04198766 and NCT03739931).

Macrophages are part of the tumor microenvironment, and there are two categories of tumor-associated macrophages, M1 and M2. M1 can kill cancer cells and destroy the extracellular matrix, and M2 has a tumor-promoting action. The transition from M1 to M2 is induced by IL-4, and the opposite switch is induced by IFN-γ [59]. The gamma isoform of phosphoinositide 3-kinase (PI3Kγ) inhibition has been effective in inducing the M1 macrophage expression in animal models [60]. The IPI-549 is a PI3Kγ inhibitor under study as monotherapy or in association with immune checkpoint inhibitor nivolumab for patients with HNSCC in a phase 2 trial (NCT03795610).

Several interleukins, such as IL-15, -2, -7, -12, etc., have been used as targets for HNSCC vaccines development. N-803 (ANKTIVA, ImmunityBio Inc., El Segundo, CA, USA), also known as ATL-803 or Nogapendekin alfa, is an IL-15 agonist bound with its receptor. IT is under evaluation in combination with immune checkpoint inhibitor in a phase 2b study (NCT03228667) with promising preliminary data. Other trials are using N-803 in association with the chimeric antigen receptor T (NCT04847466) or the anti-PD-L1/TGF-beta 'Trap' with Bintrafusp alfa (M7824) plus the TriAd vaccine (ETBX-011, ETBX-051, and ETBX-061) (NCT04247282). NKTR-214 (Nektar Therapeutics, San Francisco, CA, USA and Bristol Myers Squibb, New York, NY, USA), also known as Bempegaldesleukin, is an IL-2 pathway stimulator under study in the NCT04936841 phase 2 trial for HNSC. Similarly, ALKS 4230 (Alkermes, Inc., Dublin, Ireland), also known as Nemvaleukin alfa, showed good tolerability in the (NCT04144517) trial [61]. NT-17 is a recombinant ILO-7 called Efineptakin alfa (NeoImmuneTech, NeoImmuneTech, Rockville, MD, USA) under study (NCT04588038). Edodekin alfa, a recombinant Il-12, showed great immunity response in combination with cetuximab in a phase 1/2 trial [62]. The combination of several cytokines in the IRX-2 showed safety and efficacy as a neoadjuvant therapy [63].

TLR stimulation induces natural killer cells activation and antibody cytotoxicity against cancer. Moltolimod, also known as VTX-2337, (APExBio, Houston, TX, USA), is a TRLT8 agonist that increases cetuximab efficacy [64]. Active8 was a multicenter, randomized, double-blinded, placebo-controlled clinical trial comparing the ETREME regimen with placebo or Moltolimod. They observed that adding the vaccine did not significantly improved overall survival and disease-free survival, but a significant benefit was observed in the HPV+ sub-population [65]. EMD 1201081 (Aceragen Inc, Cambridge, MA, USA), also known as HYB-2055, IMO-2055, or IMOxine, is a TLR9 agonist that has been studied in association with cetuximab with no improvement seen in oncological outcomes [66]. Amplivant (AV) (ISA Pharmaceuticals, Leiden, Oegstgeest, The Netherlands) is a TLR-2 agonist that has been conjugated with the HPV E6 to create the HESPeCTA (HPV E Six Peptide Conjugated To Amplivant) vaccine. Its intradermal administration showed safety and efficacy in eliciting immune responses, and further studies are needed the define clinical efficacy [67].

2.11. Personalized Cancer Vaccines

Thanks to genome sequencing, it is now possible to analyze the genomic profile of a patient's cancer and develop a vaccination based on it. YE-NEO-001 (NantBioScience, Inc., Los Angeles, CA, USA) is a recombinant yeast-based vaccine that expresses antigens derived from the patient's tumor and is under study in a phase 1 trial (NCT03552718). TG4050, an MVA-based therapeutic vaccine based on the myvacTM platform, is under evaluation for locally advanced HNSCC in a phase 1 trial (NCT04183166). AlloVax is a chaperone-rich cell lysate prepared from a patient's cancer cells associated with AlloStim^{TM} as an adjuvant. This association shows promising results and good tolerability (NCT01998542). MVX-ONCO-1 is made from irradiated autologous tumor cells with a genetically modified cell line called MVX-1 that releases GM-CSF, which exhibited safety and efficacy in HNSCC patients previously treated with nivolumab- or cisplatin-based chemotherapy (NCT02193503). PANDA-VAC is defined as a personalized and adjusted neoantigen peptide vaccine and its association with pembrolizumab is the center of the NCT04266730 phase I clinical trial. VB10.NEO (Nykode Therapeutics ASA, Norway) and NKTR-214, immunotherapy is under evaluation in the NCT03548467 phase 1/2a trial. ATLAS^{TM} is a technology platform for neoantigen selection from tumors. It has been used to make GEN-009, a neoantigen mix made with this technology. It has been administered in association with immune checkpoint inhibitors in a phase 1 trial. It demonstrated good tolerability and promising efficacy [68]. PNeoVCA is a personalized neoantigen peptide-based vaccine under evaluation in association with pembrolizumab (NCT05269381). mRNA-2752 is an mRNA-based vaccine encoding OX40L, IL-23, and IL36γ. It is under evaluation in monotherapy and association with durvalumab (NCT03739931). A different approach uses the in vitro expansion of anti-tumor T-cells extracted from the patient. The phase 2 trial NCT04847466 concerns the association between PD-L1 CAR-NK cells, pembrolizumab, and N-803.

2.12. mRNA Vaccines

mRNA can be used to induce the expression of neoantigen peptides and break the immune tolerance to cancer. V941 is a vaccine developed by Moderna and Merck using mRNA-5671, which targets G12D, G12V, G13D, and G12C (the most common KRAS mutations in solid tumors). The NCT03948763 phase 1 trial aims to assess its safety and tolerability either as monotherapy or in combination with pembrolizumab. BNT113 is a HPV16 E7 mRNA and it is currently under study in the phase 1/2 NCT03418480, in combination with HARE-40 (an anti-CD40) and in association with pembrolizumab in the phase 2 NCT04534205 trial. The mRNA used in a vaccine can also encode antibodies. One example of this use with a potential application in head and neck cancer is BNT142, which encodes molecules targeting CD3xCLDN6 (NCT05262530). This vaccine is developed using bispecific T cell engagers called BiTEs, which are bispecific antibodies without the FC region.

The use of mRNA in personalized medicine requires the analysis of tumor antigen expression and the MHC profiling of the patient. Some machine learning algorithms have been used to predict it [9]. Several platforms, such as iNeST (BioNTech SE), have been used to develop BNT121 and BNT122, which are under analysis in solid tumors (NCT02035956,

NCT03289962). Moderna has developed the mRNA-4157, which is undergoing testing (NCT03313778).

3. Results and Discussion

HPV+ HNSCC usually has a better response to therapy than the HPV- forms [5,69]. The different behavior is due to the carcinogenesis mechanisms induced by the virus or by cigarette smoke and which is easily observable by analyzing gene expression profiles. HPV+ tumors display PIK3CA amplification and CDKN2a or p16 overexpression, they do not overexpress epidermal growth factor receptor (EGFR), and they have the wild type p53. Conversely, non-HPV-related tumors have p53 loss-of-function mutations and EGFR overexpression [69]. A higher percentage of immune cells was also observed in the tumor microenvironment of HPV+ neoplasms [70,71]. The observation that the inflammatory infiltrate is greater in tumors with the best prognosis, i.e., HPV+, correlates with the observations of Zhang et al. in 2021. They observed that HNSCC can be divided into three groups. The first has a greater inflammatory infiltration, less stimulated oncogenic signaling, a greater response to therapy (both chemotherapy and immunotherapy), and, consequently, a better prognosis. The third group consists of those tumors with opposite characteristics, and thus less inflammatory infiltration, more oncogenic mutations, less response to therapy, and worse prognosis [72]. Comparing the characteristics of HPV+ tumors with those of HPV-, regarding the lack of mutation of p53 and EGFR, as well as the greater inflammatory infiltration into HPV+ tumors, it is easy to explain why HPV+ tumors have a better prognosis with medical therapy [69]. Whereby, HPV-positive head and neck cancers have better outcomes compared to HPV-negative diseases. The significant improvement in the application of chemotherapy and radiotherapy protocols has led to good levels of patient treatment and obtaining satisfactory results. [73]. The massive development of vaccines targeting E6 and E7 molecules should consider and compete with those optimal results. Surgery, both traditional and robotic, helps to heal patients in many cases [74], with suggesting the comparison of the results obtained from vaccines with those of the new therapeutic frontiers.

Regarding NPC EBV+ vaccines, clinical efficacy data are limited given the early stage of the trials, yet those medications seem to be well tolerated and able to elicit a selective immune response against the targeted antigens [75]. Obviously, this does not mean that we can declare them to be an effective therapy against HNSCC.

Tumor-associated antigens are not specific to cancer cells but can also be expressed at lower levels in normal tissues. Therefore, vaccines targeting those molecules have low specificity [9].

In addition to evaluating drugs individually, a basic technique for enhancing their efficacy is to combine multiple vaccines. As has already been suggested by Huang et al., we think that future studies should focus on combination therapies with the association of several vaccines or vaccines and other medications, such as the traditional chemotherapy or the new immunotherapy [75]. For example, of the four trials (NCT01256853, NCT01800071, NCT01147991, NCT01094405) that investigated the efficacy of the EBNA1 C-terminal/LMP2 MVA vaccine, only one was concerned with the clinical efficacy of the drug against cancer (NCT01094405). The recruitment was completed, but the results have not been published. Similarly in clinical trial NCT04180215, an improved immune response was observed using the combination of HB-201 and HB-202 vaccines [21].

In our opinion, this observation pushes the future frontiers of research towards the study of the combinations of drugs that prove to be individually effective when creating a therapy that is strong in generating immunogenicity and complete in the antigenic pool. The combination approach also makes sense with the increased action of immunotherapy against immune checkpoints. The actions of drugs such as pembrolizumab are limited by the poor immune responses they generate, despite their actions against a tumor's immune escape mechanisms. The enhancement of cytotoxic lymphocyte activity through the combination of immune checkpoint inhibitors and therapeutic vaccines is promising, as suggested by the combination of PDS0101 and Pembrolizumab, which has demonstrated such efficacy as to have received Fast Track designation from the FDA in the treatment of HPV16+ in HNSCC [24,25]. Furthermore, we should consider that having an effective drug in monotherapy does not mean that it is more effective than those already on the market. For example, T-VEC was tested in combination with pembrolizumab and the data itself was positive, but the authors who studied the effectiveness of the drug had the foresight to compare the association with the historical data of pembrolizumab alone and showed that there were no significant differences [42]. Consequently, there appears to be no apparent benefit to adding T-VEC to pembrolizumab therapy, and the therapeutic efficacy measured in the trial could be that of pembrolizumab alone. From these data, we can reach two important conclusions: (1) The demonstration of the in vivo efficacy of an immune response against a tumor does not equate to a response in therapeutic terms, and it does not mean that the drug is effective in curing the disease. (2) It is essential to compare these drugs with the current therapeutic gold standard and understand if they are comparable to therapies already in use.

The main goal of vaccine therapy is to induce an immune response against cancer cells. Researchers are still focusing their research on finding vaccines able to activate immunity against tumor-specific antigens. The activation of immune response does not mean therapeutic efficacy. Intuitively, the immune escape mechanisms adopted by cancer cells before vaccine administration could be responsible for tumor cells' survival after immunity activation against them. For this reason, several researchers are testing anticancer therapeutic vaccines in combination with immune checkpoint escape inhibitors such as Pembrolizumab [24,25,42].

A vaccine to be effective needs to induce an immune response. Adjuvant molecules are extensively used to increase immune activation. Several examples can be used, such as the Candida skin test reagent in the PepCan Vaccine [31,32], Montanide ISA 51 and GM-CSF in LY6K [44,47,58], and AlloStimTM in AlloVax. Their impact on T cell activation may be helpful in increase vaccine efficacy.

Because these drugs act selectively on the immune system and specific antigens, reduced efficacy may occur overall, and it would be important to identify the subpopulation of patients most suitable for receiving the therapeutic vaccine. This principle is the basis of the personalized medicine towards which we are moving. Indeed, the result is that Moltolimod is not effective in the HNSCC population, but the benefit it provides in association with the extreme protocol is significant in HPV+ HNSCC patients [65]. Properly defined personalized vaccines are currently all in experimental stages too early to be able to give a real definition of their clinical efficacy (AlloVax, ATLAS™, PNeoVCA, MVX-ONCO-1, YE-NEO-001, TG4050, VB10.NEO, mRNA-2752, and PANDA-VAC), but certainly, even once their effectiveness has been demonstrated, the widespread diffusion of these therapies is not easy since it is not a matter of distributing a pre-manufactured drug. Furthermore, the question of the diffusion of the technologies necessary for its realization should be mentioned.

Taking, for example, the case of vaccination carried out after surgery with autologous cells: the population taken into consideration is extremely heterogeneous in terms of tumor origin and staging, so it is not easy to evaluate the real clinical efficacy of this practice [52]. Obviously, the list is not limited to this, but even the survivin vaccine, although it showed an immune reaction, did not show therapeutic efficacy. Only one out of ten vaccinated patients showed a partial clinical response, and six out of eight evaluated patients exhibited a noticeable increase in peptide-specific CTLs. However, the investigators noted that the induced CTL response by the vaccine was insufficient to achieve tumor regression [51].

It is premature to make comparisons between standard chemotherapy and vaccine therapy since no phase 3 trials seem to have been completed at the time of writing. For the same reason, vaccine therapy for HNSCC cannot be considered a first-line treatment today. Promising results came from the combination of therapies; for example, the intratumoral administration of ONYX-015 and 5-fluorouracile and cisplatin demonstrated better results than traditional chemotherapy [37]. Those promising results were not confirmed in the phase 3 study that was started in 2001, as no results have been published, even after more than 20 years. The COVID-19 pandemic increased difficulties in conducting clinical trials, making follow-up and experimental protocol application more challenging, sometimes resulting in trial suspensions (NCT03418480).

Future studies must examine not only the conventional efficacy of parenteral administrations but also new administration routes. The NCT04180215 trial is studying, for example, the efficacy of intratumorally administration associated with systemic administration.

4. Conclusions

HPV+ HNSCC has better outcomes than HPV. We need to compare the HPV+ HNSCC vaccines with actual results. Data on the clinical efficacy of EBV+ HNSCC are limited given the early stage of the studies. On the other hand, a correct approach to personalized medicine for a population susceptible to the vaccine could produce greater therapeutic advantages.

The actual effectiveness of each new vaccine will have to be compared with the therapeutic successes and health costs of current therapies. Finally, much of data are fragmentary, and numerous studies concerning these vaccines have been aborted, which is a relevant problem in the evaluation of therapeutic vaccines.

Author Contributions: Conceptualization, P.G.M.; methodology, P.G.M.; validation, M.F. and C.P.; formal analysis, M.F. and C.P.; resources, P.G.M.; data curation, P.G.M.; writing—original draft preparation, P.G.M.; writing—review and editing, P.G.M., A.M. and C.B.; project administration, C.B. and A.M. All authors have read and agreed to the published version of the manuscript.

Funding: This research was funded by the grant 'Progetto Medio Ateneo', Univ. Sapienza, no. RM11916B88DF74E7, awarded to A.M.

Acknowledgments: The authors thank the Division of Otolaryngology-Head and Neck Surgery, Ospedale San Camillo de Lellis, ASL Rieti-Sapienza University, for its support.

Conflicts of Interest: The authors declare no conflict of interest.

References

1. Sung, H.; Ferlay, J.; Siegel, R.L.; Laversanne, M.; Soerjomataram, I.; Jemal, A.; Bray, F. Global Cancer Statistics 2020: GLOBOCAN Estimates of Incidence and Mortality Worldwide for 36 Cancers in 185 Countries. *CA Cancer J. Clin.* **2021**, *71*, 209–249. [CrossRef] [PubMed]
2. Meliante, P.G.; Barbato, C.; Zoccali, F.; Ralli, M.; Greco, A.; de Vincentiis, M.; Colizza, A.; Petrella, C.; Ferraguti, G.; Minni, A.; et al. Programmed Cell Death-Ligand 1 in Head and Neck Squamous Cell Carcinoma: Molecular Insights, Preclinical and Clinical Data, and Therapies. *Int. J. Mol. Sci.* **2022**, *23*, 15384. [CrossRef] [PubMed]
3. Meliante, P.G.; Zoccali, F.; de Vincentiis, M.; Ralli, M.; Petrella, C.; Fiore, M.; Minni, A.; Barbato, C. Diagnostic Predictors of Immunotherapy Response in Head and Neck Squamous Cell Carcinoma. *Diagnostics* **2023**, *13*, 862. [CrossRef] [PubMed]
4. Botticelli, A.; Cirillo, A.; Strigari, L.; Valentini, F.; Cerbelli, B.; Scagnoli, S.; Cerbelli, E.; Zizzari, I.G.; Della Rocca, C.; D'Amati, G.; et al. Anti–PD-1 and Anti–PD-L1 in Head and Neck Cancer: A Network Meta-Analysis. *Front. Immunol.* **2021**, *12*, 705096. [CrossRef]
5. Devaraja, K.; Aggarwal, S.; Singh, M. Therapeutic Vaccination in Head and Neck Squamous Cell Carcinoma—A Review. *Vaccines* **2023**, *11*, 634. [CrossRef] [PubMed]
6. Harder, T.; Wichmann, O.; Klug, S.J.; van der Sande, M.A.B.; Wiese-Posselt, M. Efficacy, effectiveness and safety of vaccination against human papillomavirus in males: A systematic review. *BMC Med.* **2018**, *16*, 110. [CrossRef] [PubMed]
7. Rettig, E.M.; D'Souza, G. Epidemiology of Head and Neck Cancer. *Surg. Oncol. Clin. N. Am.* **2015**, *24*, 379–396. [CrossRef] [PubMed]
8. Burnet, M. Cancer-A Biological Approach I. The Processes of Control. *Br. Med. J.* **1957**, *1*, 779–786. [CrossRef]
9. Sun, H.; Zhang, Y.; Wang, G.; Yang, W.; Xu, Y. mRNA-Based Therapeutics in Cancer Treatment. *Pharmaceutics* **2023**, *15*, 622. [CrossRef]
10. Shibata, H.; Xu, N.; Saito, S.; Zhou, L.; Ozgenc, I.; Webb, J.; Fu, C.; Zolkind, P.; Egloff, A.M.; Uppaluri, R. Integrating CD4+ T cell help for therapeutic cancer vaccination in a preclinical head and neck cancer model. *Oncoimmunology* **2021**, *10*, 1958589. [CrossRef]
11. Song, Q.; Zhang, C.D.; Wu, X.H. Therapeutic cancer vaccines: From initial findings to prospects. *Immunol. Lett.* **2018**, *196*, 11–21. [CrossRef]

12. Tan, Y.S.; Sansanaphongpricha, K.; Xie, Y.; Donnelly, C.R.; Luo, X.; Heath, B.R.; Zhao, X.; Bellile, E.; Hu, H.; Chen, H.; et al. Mitigating SOX2-potentiated Immune Escape of Head and Neck Squamous Cell Carcinoma with a STING-inducing Nanosatellite Vaccine. *Clin. Cancer Res.* **2018**, *24*, 4242–4255. [CrossRef] [PubMed]
13. Zarour, H.M.; DeLeo, A.; Finn, O.J. Categories of Tumor Antigens. In *Holland-Frei Cancer Medicine*, 6th ed.; BC Decker: Hamilton, ON, Canada, 2003.
14. Wang, R.F.; Rosenberg, S.A. Human tumor antigens for cancer vaccine development. *Immunol. Rev.* **1999**, *170*, 85–100. [CrossRef] [PubMed]
15. Renkvist, N.; Castelli, C.; Robbins, P.F.; Parmiani, G. A listing of human tumor antigens recognized by T cells. *Cancer Immunol. Immunother.* **2001**, *50*, 3–15. [CrossRef] [PubMed]
16. De Smet, C.; Lurquin, C.; De Plaen, E.; Brasseur, F.; Zarour, H.; De Backer, O.; Coulie, P.G.; Boon, T. Genes coding for melanoma antigens recognised by cytolytic T lymphocytes. *Eye* **1997**, *11*, 243–248. [CrossRef] [PubMed]
17. Coulie, P.G.; Van Den Eynde, B.J.; Van Der Bruggen, P.; Boon, T. Tumour antigens recognized by T lymphocytes: At the core of cancer immunotherapy. *Nat. Rev. Cancer* **2014**, *14*, 135–146. [CrossRef]
18. Taylor, G.S.; Jia, H.; Harrington, K.; Lee, L.W.; Turner, J.; Ladell, K.; Price, D.A.; Tanday, M.; Matthews, J.; Roberts, C.; et al. A recombinant modified vaccinia ankara vaccine encoding Epstein-Barr Virus (EBV) target antigens: A phase I trial in UK patients with EBV-positive cancer. *Clin. Cancer Res.* **2014**, *20*, 5009–5022. [CrossRef]
19. Aggarwal, C.; Cohen, R.B.; Morrow, M.P.; Kraynyak, K.A.; Sylvester, A.J.; Knoblock, D.M.; Bauml, J.M.; Weinstein, G.S.; Lin, A.; Boyer, J.; et al. Immunotherapy targeting HPV16/18 generates potent immune responses in HPV-associated head and neck cancer. *Clin. Cancer Res.* **2019**, *25*, 110–124. [CrossRef]
20. Massarelli, E.; William, W.; Johnson, F.; Kies, M.; Ferrarotto, R.; Guo, M.; Feng, L.; Lee, J.J.; Tran, H.; Kim, Y.U.; et al. Combining Immune Checkpoint Blockade and Tumor-Specific Vaccine for Patients with Incurable Human Papillomavirus 16-Related Cancer: A Phase 2 Clinical Trial. *JAMA Oncol.* **2019**, *5*, 67–73. [CrossRef]
21. Edwards, D.; Schwendinger, M.; Katchar, K.; Schlienger, K.; Orlinger, K.; Matushansky, I.; Lauterbach, H. Abstract 3284: HB-201 and HB-202, an arenavirus-based immunotherapy, induces tumor T cell infiltration in patients with HNSCC and other HPV16+ tumors. *Cancer Res.* **2022**, *82*, 3284. [CrossRef]
22. Le Tourneau, C.; Delord, J.-P.; Cassier, P.; Loirat, D.; Tavernaro, A.; Bastien, B.; Bendjama, K. Phase Ib/II trial of TG4001 (Tipapkinogene sovacivec), a therapeutic HPV-vaccine, and Avelumab in patients with recurrent/metastatic (R/M) HPV-16+ cancers. *Ann. Oncol.* **2019**, *30*, v494–v495. [CrossRef]
23. Jimeno, A.; Baranda, J.C.; Mita, M.M.; Gordon, M.S.; Taylor, M.H.; Iams, W.T.; Janku, F.; Matulonis, U.A.; Bernstein, H.; Loughhead, S.; et al. Initial results of a first-in-human, dose escalation study of a cell-based vaccine in HLA A*02+ patients (pts) with recurrent, locally advanced or metastatic HPV16+ solid tumors: SQZ-PBMC-HPV-101. *J. Clin. Oncol.* **2021**, *39*, 2536. [CrossRef]
24. Wood, L.; Chintakuntlawar, A.V.; Price, K.; Kaczmar, J.; Conn, G.; Bedu-Addo, F.K.; Weiss, J. Preliminary Safety of PDS0101 (Versamune + HPVmix) and Pembrolizumab Combination Therapy in Subjects with Recurrent/Metastatic Human Papillomavirus-16 Positive Oropharyngeal Squamous Cell Carcinoma (OPSCC). *Int. J. Radiat. Oncol.* **2022**, *112*, e37–e38. [CrossRef]
25. PDS Biotechnology Granted FDA Fast Track Designation for Lead Candidate PDS0101. Available online: https://www.globenewswire.com/news-release/2022/06/02/2455153/37149/en/PDS-Biotechnology-Granted-FDA-Fast-Track-Designation-for-Lead-Candidate-PDS0101.html (accessed on 5 September 2023).
26. Krupar, R.; Imai, N.; Miles, B.; Genden, E.; Misiukiewicz, K.; Saenger, Y.; Demicco, E.G.; Patel, J.; Herrera, P.C.; Parikh, F.; et al. Abstract LB-095: HPV E7 antigen-expressing Listeria-based immunotherapy (ADXS11-001) prior to robotic surgery for HPV-positive oropharyngeal cancer enhances HPV-specific T cell immunity. *Cancer Res.* **2016**, *76*, LB-095. [CrossRef]
27. Schneider, K.; Grønhøj, C.; Hahn, C.H.; von Buchwald, C. Therapeutic human papillomavirus vaccines in head and neck cancer: A systematic review of current clinical trials. *Vaccine* **2018**, *36*, 6594–6605. [CrossRef]
28. Karkada, M.; Quinton, T.; Blackman, R.; Mansour, M. Tumor Inhibition by DepoVax-Based Cancer Vaccine Is Accompanied by Reduced Regulatory/Suppressor Cell Proliferation and Tumor Infiltration. *ISRN Oncol.* **2013**, *2013*, 753427. [CrossRef]
29. Yang, M.C.; Yang, A.; Qiu, J.; Yang, B.; He, L.; Tsai, Y.C.; Jeang, J.; Wu, T.C.; Hung, C.F. Buccal injection of synthetic HPV long peptide vaccine induces local and systemic antigen-specific CD8+ T-cell immune responses and antitumor effects without adjuvant. *Cell Biosci.* **2016**, *6*, 17. [CrossRef] [PubMed]
30. Christine, S.I.P.; Chung, H.; Colevas, A.D.; Adkins, D.; Michael Gibson, K.; Rodriguez, C.P.; Sukari, A.; Bauman, J.E.; Wirth, L.J.; Johnson, F.M.; et al. A phase 1 dose-escalation and expansion study of CUE-101, a novel HPV16 E7-pHLA-IL2-Fc fusion protein, given alone and in combination with pembrolizumab in patients with recurrent/metastatic HPV16+ head and neck cancer. *J. Clin. Oncol.* **2022**, *40*, 6045.
31. Coleman, H.N.; Greenfield, W.W.; Stratton, S.L.; Vaughn, R.; Kieber, A.; Moerman-Herzog, A.M.; Spencer, H.J.; Hitt, W.C.; Quick, C.M.; Hutchins, L.F.; et al. Human papillomavirus type 16 viral load is decreased following a therapeutic vaccination. *Cancer Immunol. Immunother.* **2016**, *65*, 563–573. [CrossRef]
32. Wang, X.; Coleman, H.N.; Nagarajan, U.; Spencer, H.J.; Nakagawa, M. Candida skin test reagent as a novel adjuvant for a human papillomavirus peptide-based therapeutic vaccine. *Vaccine* **2013**, *31*, 5806–5813. [CrossRef]
33. Greenfield, W.W.; Stratton, S.L.; Myrick, R.S.; Vaughn, R.; Donnalley, L.M.; Coleman, H.N.; Mercado, M.; Moerman-Herzog, A.M.; Spencer, H.J.; Andrews-Collins, N.R.; et al. A phase I dose-escalation clinical trial of a peptide-based human papillomavirus

therapeutic vaccine with Candida skin test reagent as a novel vaccine adjuvant for treating women with biopsy-proven cervical intraepithelial neoplasia 2/3. *Oncoimmunology* **2015**, *4*, e1031439. [CrossRef] [PubMed]
34. Reuschenbach, M.; Rafiyan, M.; Pauligk, C.; Karbach, J.; Kloor, M.; Prigge, E.-S.; Sauer, M.; Jäger, E.; Al-Batran, S.-E.; von Knebel Doeberitz, M. Phase I/IIa trial targeting p16 INK4a by peptide vaccination in patients with human papillomavirus-associated cancer. *J. Clin. Oncol.* **2015**, *33*, e14030. [CrossRef]
35. Nemunaitis, J.; Ganly, I.; Khuri, F.; Arseneau, J.; Kuhn, J.; McCarty, T.; Landers, S.; Maples, P.; Romel, L.; Randlev, B.; et al. Selective replication and oncolysis in p53 mutant tumors with ONYX-015, an E1B-55kD gene-deleted adenovirus, in patients with advanced head and neck cancer: A phase II trial. *Cancer Res.* **2000**, *60*, 6359–6366. [PubMed]
36. Kirn, D.H. A Phase II Trial of Intratumoral Injection with a Selectively Replicating Adenovirus (ONYX-015) in Patients with Recurrent, Refractory Squamous Cell Carcinoma of the Head and Neck. In *Gene Therapy of Cancer*; Humana: Totowa, NJ, USA, 2003; pp. 559–574. [CrossRef]
37. Khuri, F.R.; Nemunaitis, J.; Ganly, I.; Arseneau, J.; Tannock, I.F.; Romel, L.; Gore, M.; Ironside, J.; MacDougall, R.H.; Heise, C.; et al. A controlled trial of intratumoral ONYX-015, a selectively-replicating adenovirus, in combination with cisplatin and 5-fluorouracil in patients with recurrent head and neck cancer. *Nat. Med.* **2000**, *6*, 879–885. [CrossRef] [PubMed]
38. Nemunaitis, J.; Cunningham, C.; Buchanan, A.; Blackburn, A.; Edelman, G.; Maples, P.; Netto, G.; Tong, A.; Randlev, B.; Olson, S.; et al. Intravenous infusion of a replication-selective adenovirus (ONYX-015) in cancer patients: Safety, feasibility and biological activity. *Gene Ther.* **2001**, *8*, 746–759. [CrossRef] [PubMed]
39. Kirn, D.H. The end of the beginning: Oncolytic virotherapy achieves clinical proof-of-concept. *Mol. Ther.* **2006**, *13*, 237–238. [CrossRef] [PubMed]
40. Onyx plans phase III trial of ONYX-015 for head & neck cancer. *Oncologist* **1999**, *4*, 432.
41. Edwards, S.J.; Dix, B.R.; Myers, C.J.; Dobson-Le, D.; Huschtscha, L.; Hibma, M.; Royds, J.; Braithwaite, A.W. Evidence that Replication of the Antitumor Adenovirus ONYX-015 Is Not Controlled by the p53 and p14 ARF Tumor Suppressor Genes. *J. Virol.* **2002**, *76*, 12483–12490. [CrossRef]
42. Harrington, K.J.; Kong, A.; Mach, N.; Chesney, J.A.; Fernandez, B.C.; Rischin, D.; Cohen, E.E.W.; Radcliffe, H.S.; Gumuscu, B.; Cheng, J.; et al. Talimogene Laherparepvec and Pembrolizumab in Recurrent or Metastatic Squamous Cell Carcinoma of the Head and Neck (MASTERKEY-232): A Multicenter, Phase 1b Study. *Clin. Cancer Res.* **2020**, *26*, 5153–5161. [CrossRef] [PubMed]
43. Cuffel, C.; Rivals, J.P.; Zaugg, Y.; Salvi, S.; Seelentag, W.; Speiser, D.E.; Liénard, D.; Monnier, P.; Romero, P.; Bron, L.; et al. Pattern and clinical significance of cancer-testis gene expression in head and neck squamous cell carcinoma. *Int. J. Cancer* **2011**, *128*, 2625–2634. [CrossRef]
44. Voskens, C.J.; Sewell, D.; Hertzano, R.; Desanto, J.; Rollins, S.; Lee, M.; Taylor, R.; Wolf, J.; Suntharalingam, M.; Gastman, B.; et al. InducTION of mage-A3 and HPV-16 immunity by Trojan vaccines in patients with head and neck carcinoma. *Head Neck* **2012**, *34*, 1734–1746. [CrossRef] [PubMed]
45. Chai, S.J.; Fong, S.C.Y.; Gan, C.P.; Pua, K.C.; Lim, P.V.H.; Lau, S.H.; Zain, R.B.; Abraham, T.; Ismail, S.M.; Abdul Rahman, Z.A.; et al. In vitro evaluation of dual-antigenic PV1 peptide vaccine in head and neck cancer patients. *Hum. Vaccines Immunother.* **2019**, *15*, 167–178. [CrossRef] [PubMed]
46. Wang, C.; Zainal, N.S.; Chai, S.J.; Dickie, J.; Gan, C.P.; Zulaziz, N.; Lye, B.K.W.; Sutavani, R.V.; Ottensmeier, C.H.; King, E.V.; et al. DNA Vaccines Targeting Novel Cancer-Associated Antigens Frequently Expressed in Head and Neck Cancer Enhance the Efficacy of Checkpoint Inhibitor. *Front. Immunol.* **2021**, *12*, 4275. [CrossRef] [PubMed]
47. Yoshitake, Y.; Fukuma, D.; Yuno, A.; Hirayama, M.; Nakayama, H.; Tanaka, T.; Nagata, M.; Takamune, Y.; Kawahara, K.; Nakagawa, Y.; et al. Phase II clinical trial of multiple peptide vaccination for advanced head and neck cancer patients revealed induction of immune responses and improved OS. *Clin. Cancer Res.* **2015**, *21*, 312–321. [CrossRef] [PubMed]
48. Ogasawara, M.; Miyashita, M.; Yamagishi, Y.; Ota, S. Phase I/II Pilot Study of Wilms' Tumor 1 Peptide-Pulsed Dendritic Cell Vaccination Combined With Conventional Chemotherapy in Patients With Head and Neck Cancer. *Ther. Apher. Dial.* **2019**, *23*, 279–288. [CrossRef]
49. Schuler, P.J.; Harasymczuk, M.; Visus, C.; De Leo, A.; Trivedi, S.; Lei, Y.; Argiris, A.; Gooding, W.; Butterfield, L.H.; Whiteside, T.L.; et al. Phase i dendritic cell p53 peptide vaccine for head and neck cancer. *Clin. Cancer Res.* **2014**, *20*, 2433–2444. [CrossRef]
50. Chung, V.; Kos, F.J.; Hardwick, N.; Yuan, Y.; Chao, J.; Li, D.; Waisman, J.; Li, M.; Zurcher, K.; Frankel, P.; et al. Evaluation of safety and efficacy of p53MVA vaccine combined with pembrolizumab in patients with advanced solid cancers. *Clin. Transl. Oncol.* **2019**, *21*, 363–372. [CrossRef]
51. Miyazaki, A.; Kobayashi, J.; Torigoe, T.; Hirohashi, Y.; Yamamoto, T.; Yamaguchi, A.; Asanuma, H.; Takahashi, A.; Michifuri, Y.; Nakamori, K.; et al. Phase I clinical trial of survivin-derived peptide vaccine therapy for patients with advanced or recurrent oral cancer. *Cancer Sci.* **2011**, *102*, 324–329. [CrossRef]
52. Karcher, J.; Dyckhoff, G.; Beckhove, P.; Reisser, C.; Brysch, M.; Ziouta, Y.; Helmke, B.H.; Weidauer, H.; Schirrmacher, V.; Herold-Mende, C. Antitumor vaccination in patients with head and neck squamous cell carcinomas with autologous virus-modified tumor cells. *Cancer Res.* **2004**, *64*, 8057–8061. [CrossRef]
53. Chang, A.E.; Li, Q.; Jiang, G.; Teknos, T.N.; Chepeha, D.B.; Bradford, C.R. Generation of vaccine-primed lymphocytes for the treatment of head and neck cancer. *Head Neck* **2003**, *25*, 198–209. [CrossRef]

54. Whiteside, T. Anti-Tumor Vaccines in Head and Neck Cancer: Targeting Immune Responses to the Tumor. *Curr. Cancer Drug Targets* **2007**, *7*, 633–642. [CrossRef] [PubMed]
55. Serafini, P.; Meckel, K.; Kelso, M.; Noonan, K.; Califano, J.; Koch, W.; Dolcetti, L.; Bronte, V.; Borrello, I. Phosphodiesterase-5 inhibition augments endogenous antitumor immunity by reducing myeloid-derived suppressor cell function. *J. Exp. Med.* **2006**, *203*, 2691–2702. [CrossRef] [PubMed]
56. Weed, D.T.; Zilio, S.; Reis, I.M.; Sargi, Z.; Abouyared, M.; Gomez-Fernandez, C.R.; Civantos, F.J.; Rodriguez, C.P.; Serafini, P. The reversal of immune exclusion mediated by tadalafil and an anti-tumor vaccine also induces PDL1 upregulation in recurrent head and neck squamous cell carcinoma: Interim analysis of a phase I clinical trial. *Front. Immunol.* **2019**, *10*, 1206. [CrossRef] [PubMed]
57. Duhen, R.; Ballesteros-Merino, C.; Frye, A.K.; Tran, E.; Rajamanickam, V.; Chang, S.C.; Koguchi, Y.; Bifulco, C.B.; Bernard, B.; Leidner, R.S.; et al. Neoadjuvant anti-OX40 (MEDI6469) therapy in patients with head and neck squamous cell carcinoma activates and expands antigen-specific tumor-infiltrating T cells. *Nat. Commun.* **2021**, *12*, 1047. [CrossRef]
58. Glisson, B.S.; Leidner, R.S.; Ferris, R.L.; Powderly, J.; Rizvi, N.A.; Keam, B.; Schneider, R.; Goel, S.; Ohr, J.P.; Burton, J.; et al. Safety and clinical activity of MEDI0562, a humanized OX40 agonist monoclonal antibody, in adult patients with advanced solid tumors. *Clin. Cancer Res.* **2020**, *26*, 5358–5367. [CrossRef] [PubMed]
59. Chen, J.; Li, M.; Yang, C.; Yin, X.; Duan, K.; Wang, J.; Feng, B. Macrophage phenotype switch by sequential action of immunomodulatory cytokines from hydrogel layers on titania nanotubes. *Colloids Surf. B Biointerfaces* **2018**, *163*, 336–345. [CrossRef]
60. De Henau, O.; Rausch, M.; Winkler, D.; Campesato, L.F.; Liu, C.; Cymerman, D.H.; Budhu, S.; Ghosh, A.; Pink, M.; Tchaicha, J.; et al. Overcoming resistance to checkpoint blockade therapy by targeting PI3Kγ in myeloid cells. *Nature* **2016**, *539*, 443–447. [CrossRef]
61. Gastman, B.; Cheever, M.; Fling, S.; Perez, C.; Patel, M.; Geiger, J.; Li, Z.; Posner, M.; Steuer, C.; D'Amico, L.; et al. 432 Nemvaleukin alfa, a novel engineered IL-2 cytokine, in combination with the anti-PD-1 antibody pembrolizumab in patients with recurrent/metastatic head and neck squamous cell carcinoma (ION-01 study). *J. Immunother. Cancer* **2021**, *9*, A462. [CrossRef]
62. McMichael, E.L.; Benner, B.; Atwal, L.S.; Courtney, N.B.; Mo, X.; Davis, M.E.; Campbell, A.R.; Duggan, M.C.; Williams, K.; Martin, K.; et al. A phase I/II trial of cetuximab in combination with interleukin-12 administered to patients with unresectable primary or recurrent head and neck squamous cell carcinoma. *Clin. Cancer Res.* **2019**, *25*, 4955–4965. [CrossRef]
63. Wolf, G.T.; Fee, W.E.; Dolan, R.W.; Moyer, J.S.; Kaplan, M.J.; Spring, P.M.; Suen, J.; Kenady, D.E.; Newman, J.G.; Carroll, W.R.; et al. Novel neoadjuvant immunotherapy regimen safety and survival in head and neck squamous cell cancer. *Head Neck* **2011**, *33*, 1666–1674. [CrossRef]
64. Chow, L.Q.M.; Morishima, C.; Eaton, K.D.; Baik, C.S.; Goulart, B.H.; Anderson, L.N.; Manjarrez, K.L.; Dietsch, G.N.; Bryan, J.K.; Hershberg, R.M.; et al. Phase Ib trial of the toll-like receptor 8 agonist, motolimod (VTX-2337), combined with cetuximab in patients with recurrent or metastatic SCCHN. *Clin. Cancer Res.* **2017**, *23*, 2442–2450. [CrossRef] [PubMed]
65. Ferris, R.L.; Saba, N.F.; Gitlitz, B.J.; Haddad, R.; Sukari, A.; Neupane, P.; Morris, J.C.; Misiukiewicz, K.; Bauman, J.E.; Fenton, M.; et al. Effect of adding motolimod to standard combination chemotherapy and cetuximab treatment of patients with squamous cell carcinoma of the head and neck the ACTIVE8 randomized clinical trial. *JAMA Oncol.* **2018**, *4*, 1583–1588. [CrossRef] [PubMed]
66. Ruzsa, A.; Sen, M.; Evans, M.; Lee, L.W.; Hideghety, K.; Rottey, S.; Klimak, P.; Holeckova, P.; Fayette, J.; Csoszi, T.; et al. Phase 2, open-label, 1:1 randomized controlled trial exploring the efficacy of EMD 1201081 in combination with cetuximab in second-line cetuximab-naïve patients with recurrent or metastatic squamous cell carcinoma of the head and neck (R/M SCCHN). *Investig. New Drugs* **2014**, *32*, 1278–1284. [CrossRef] [PubMed]
67. Speetjens, F.M.; Welters, M.J.P.; Slingerland, M.; Van Poelgeest, M.I.E.; De Vos Van Steenwijk, P.J.; Roozen, I.; Boekestijn, S.; Loof, N.M.; Zom, G.G.; Valentijn, A.R.P.M.; et al. Intradermal vaccination of HPV-16 E6 synthetic peptides conjugated to an optimized Toll-like receptor 2 ligand shows safety and potent T cell immunogenicity in patients with HPV-16 positive (pre-)malignant lesions. *J. Immunother. Cancer* **2022**, *10*, e005016. [CrossRef]
68. Gillison, M.L.; Awad, M.M.; Twardowski, P.; Sukari, A.; Johnson, M.L.; Stein, M.N.; Hernandez, R.; Price, J.; Mancini, K.J.; Shainheit, M.; et al. Long term results from a phase 1 trial of GEN-009, a personalized neoantigen vaccine, combined with PD-1 inhibition in advanced solid tumors. *J. Clin. Oncol.* **2021**, *39*, 2613. [CrossRef]
69. Devaraja, K.; Aggarwal, S.; Verma, S.S.; Gupta, S.C. Clinico-pathological peculiarities of human papilloma virus driven head and neck squamous cell carcinoma: A comprehensive update. *Life Sci.* **2020**, *245*, 117383. [CrossRef] [PubMed]
70. Mito, I.; Takahashi, H.; Kawabata-Iwakawa, R.; Ida, S.; Tada, H.; Chikamatsu, K. Comprehensive analysis of immune cell enrichment in the tumor microenvironment of head and neck squamous cell carcinoma. *Sci. Rep.* **2021**, *11*, 16134. [CrossRef]
71. Canning, M.; Guo, G.; Yu, M.; Myint, C.; Groves, M.W.; Byrd, J.K.; Cui, Y. Heterogeneity of the head and neck squamous cell carcinoma immune landscape and its impact on immunotherapy. *Front. Cell Dev. Biol.* **2019**, *7*, 52. [CrossRef]
72. Zhang, P.; Li, S.; Zhang, T.; Cui, F.; Shi, J.H.; Zhao, F.; Sheng, X. Characterization of Molecular Subtypes in Head and Neck Squamous Cell Carcinoma With Distinct Prognosis and Treatment Responsiveness. *Front. Cell Dev. Biol.* **2021**, *9*, 711348. [CrossRef]
73. Mehanna, H.; Robinson, M.; Hartley, A.; Kong, A.; Foran, B.; Fulton-Lieuw, T.; Dalby, M.; Mistry, P.; Sen, M.; O'Toole, L.; et al. Radiotherapy plus cisplatin or cetuximab in low-risk human papillomavirus-positive oropharyngeal cancer (De-ESCALaTE HPV): An open-label randomised controlled phase 3 trial. *Lancet* **2019**, *393*, 51–60. [CrossRef]

74. De Virgilio, A.; Costantino, A.; Mercante, G.; Petruzzi, G.; Sebastiani, D.; Franzese, C.; Scorsetti, M.; Pellini, R.; Malvezzi, L.; Spriano, G. Present and Future of De-intensification Strategies in the Treatment of Oropharyngeal Carcinoma. *Curr. Oncol. Rep.* **2020**, *22*, 91. [CrossRef] [PubMed]
75. Huang, J.; Harris, E.; Lorch, J. Vaccination as a therapeutic strategy for Nasopharyngeal carcinoma. *Oral Oncol.* **2022**, *135*, 106083. [CrossRef] [PubMed]

current issues in molecular biology

MDPI

Article

Calcitriol Treatment Decreases Cell Migration, Viability and β-Catenin Signaling in Oral Dysplasia

Daniel Peña-Oyarzún [1,*], Constanza Guzmán [2], Catalina Kretschmar [3], Vicente A. Torres [3,4,5], Andrea Maturana-Ramirez [2], Juan Aitken [2] and Montserrat Reyes [2,*]

1 Faculty of Odontology and Rehabilitation Sciences, Universidad San Sebastián, Los Leones Campus, Santiago 7510157, Chile
2 Department of Pathology and Oral Medicine, Faculty of Odontology, Universidad de Chile, Santiago 8380544, Chile
3 Institute for Research in Dental Sciences, Faculty of Odontology, Universidad de Chile, Santiago 8380544, Chile
4 Millennium Institute on Immunology and Immunotherapy, Universidad de Chile, Santiago 8380453, Chile
5 Advanced Center for Chronic Diseases (ACCDiS), Universidad de Chile, Santiago 8380494, Chile
* Correspondence: daniel.pena@uss.cl (D.P.-O.); mreyes@odontologia.uchile.cl (M.R.)

Abstract: Nearly 90% of oral cancers are characterized as oral squamous cell carcinoma (OSCC), representing the sixth most common type of cancer. OSCC usually evolves from oral potentially malignant disorders that, in some cases, are histologically consistent with a oral dysplasia. The levels of 1α,25 dihydroxyvitamin D3 (1,25-(OH)2D3; calcitriol), the active form of vitamin D3, have been shown to be decreased in patients with oral dysplasia and OSCC. Moreover, treatment with 1,25-(OH)2D3 has been proven beneficial in OSCC by inhibiting the Wnt/β-catenin pathway, a signaling route that promotes cell migration, proliferation, and viability. However, whether this inhibition mechanism occurs in oral dysplasia is unknown. To approach this question, we used dysplastic oral keratinocyte cultures and oral explants (ex vivo model of oral dysplasia) treated with 1,25-(OH)2D3 for 48 h. Following treatment with 1,25-(OH)2D3, both in vitro and ex vivo models of oral dysplasia showed decreased levels of nuclear β-catenin by immunofluorescence (IF) and immunohistochemistry (IHC). Consistently, reduced protein and mRNA levels of the Wnt/β-catenin target gene survivin were observed after treatment with 1,25-(OH)2D3. Moreover, 1,25-(OH)2D3 promoted membranous localization of E-cadherin and nuclear localization of vitamin D receptor (VDR). Functionally, DOK cells treated with 1,25-(OH)2D3 displayed diminished cell migration and viability in vitro.

Keywords: calcitriol; oral cancer; oral dysplasia; β-catenin

Citation: Peña-Oyarzún, D.; Guzmán, C.; Kretschmar, C.; Torres, V.A.; Maturana-Ramirez, A.; Aitken, J.; Reyes, M. Calcitriol Treatment Decreases Cell Migration, Viability and -Catenin Signaling in Oral Dysplasia. *Curr. Issues Mol. Biol.* **2024**, *46*, 3050–3062. https://doi.org/10.3390/cimb46040191

Academic Editors: Emma Adriana Ozon and Violeta Popovici

Received: 9 February 2024
Revised: 26 March 2024
Accepted: 26 March 2024
Published: 2 April 2024

1. Introduction

Oral cancer is a malignant neoplasm that arises as a lesion of primary origin in the oral tissues that line the oral cavity [1,2]. The incidence of oral cancer has increased dramatically worldwide in recent years, and only 40–50% of patients survive at 5 years, leading to the fact that oral cancer is a major problem of global public health [3]. About 90% of oral cancers originate in the stratified, non-keratinized epithelium of the oral mucosa, which is the reason for its denomination as oral squamous cell carcinoma (OSCC), and the main risk factors accounting for it include consumption of tobacco and alcohol [3,4]. OSCC is usually preceded by oral potentially malignant disorders (OPMDs), a group of mucosal abnormalities with increased risk of developing oral cancer, such as leukoplakia, erythroplakia, proliferative verrucous leukoplakia, and oral lichen planus, among others [5]. Oral epithelial dysplasia is a histological grade of some OPMDs characterized by loss of cell polarity, sharp lateral margins, and abnormally large hyperchromatic nuclei, as well as other apoptotic and architectural features [6]. Hence, early detection of oral dysplasia can minimize OSCC morbidity, while avoiding the detrimental side effects associated with OSCC treatment [4,7].

We and others have previously shown that the Wnt/β-catenin pathway, also known as the Wnt canonical pathway, is upregulated in oral dysplasia biopsies and dysplastic oral keratinocytes [8–11]. In normal oral epithelial cells, β-catenin is continuously targeted for proteasomal degradation by a protein destruction complex formed by casein kinase 1α (CK1α), glycogen synthase kinase 3β (GSK-3β), tumor suppressor protein adenomatous polyposis coli (APC) and Axin, resulting in a switched-off Wnt/β-catenin pathway [12,13] β-catenin is targeted for degradation by phosphorylation at Ser33/Ser37/Thr41 [12,13]. Remarkably, in oral dysplastic cells, the Wnt/β-catenin pathway is upregulated, a phenomenon associated with increased Wnt ligand secretion, which ultimately blocks β-catenin phosphorylation and its subsequent degradation [9,14]. Cytoplasmic stabilization of this non-phosphorylated form of β-Catenin (also known as transcriptionally active) is followed by its translocation to the nucleus, allowing the transcription of TCF/LEF target genes, which are associated with cell proliferation and cell viability, such as survivin and cyclin D1 [9]. Indeed, we have demonstrated in both in vitro and ex vivo models of oral dysplasia that massive nuclear localization of β-catenin occurs in a Wnt3a-dependent manner [9,14]. These observations support the notion that aberrant activation of the canonical Wnt signaling pathway is a recurring event in oral tumorigenesis.

Recent studies indicate that vitamin D3 and its most active metabolite, 1α,25 dihydroxyvitamin D3 (also known as calcitriol or "1,25-(OH)2D3"), have multiple therapeutic anti-inflammatory, antioxidant, and anti-carcinogenic effects [15–17]. These actions are mediated by the vitamin D receptor (VDR), a transcriptional factor activated by intracellular binding of 1,25-(OH)2D3 [18–20]. The use of 1,25-(OH)2D3 as a novel adjuvant for cancer treatment has been demonstrated in several neoplasms, including colorectal and breast cancer, among others [15,16,21,22]. Epidemiologic studies have suggested that low vitamin D status (<30 ng/mL) is associated with an increased risk of cancer and poorer prognosis [17,23]. Interestingly, lower serum levels of 1,25-(OH)2D3 have been associated with a higher incidence of oral dysplasia and OSCC [24,25].

The mutual regulation between 1,25-(OH)2D3 and β-catenin signaling has been proposed. 1,25(OH)2D3 inhibits Wnt/β-catenin signaling by several mechanisms at different points along the pathway. Previous work has demonstrated mainly that 1,25(OH)2D3 inhibits β-catenin transcriptional activity by promoting VDR binding to β-catenin, promoting the translocation of β-catenin from the nucleus to the plasma membrane, and inducing of E-cadherin expression in several cancers [15,22,26]. Intriguingly, whether 1,25-(OH)2D3 disrupts the Wnt/β-catenin pathway in oral dysplasia, is unknown. Therefore, this study sought to investigate the effects of 1,25-(OH)2D3 on cell migration, cell viability, and β-catenin-dependent signaling in oral dysplastic models in vitro and ex vivo.

2. Materials and Methods

2.1. Materials

Mouse monoclonal anti-β-catenin (M3539) was from DAKO Agilent (Santa Clara, CA, USA), and rabbit monoclonal anti-non-phosphorylated β-catenin (lacking phosphorylation on residues Ser33/37/Thr41; 88145) was from Cell Signaling Technology (Denver, MA, USA). E-cadherin and VDR mouse monoclonal antibodies were from Santa Cruz Biotechnology (Dallas, TX, USA). Ki-67 (SP6) rabbit monoclonal was from Cell Marque (Rocklin, CA, USA). Alexa Fluor 488 and Alexa Fluor 568 conjugated secondary antibodies were from Invitrogen (Carlsbad, CA, USA). 1α,25-Dihydroxyvitamin D3 and Lithium Chloride (SC203110A) were from Santa Cruz Biotechnology (Dallas, TX, USA). Goat anti-rabbit and goat anti-mouse antibodies coupled to horseradish peroxidase (HRP) were from Bio-Rad Laboratories (Hercules, CA, USA). Tissue culture medium, antibiotics, and fetal bovine serum (FBS) were from Corning Mediatech (Manassas, VA, USA). The EZ-ECL chemiluminescent substrate was from Pierce Chemical (Rockford, IL, USA). Tissue culture medium, antibiotics, and fetal bovine serum (FBS) were from Corning Mediatech (Manassas, VA, USA).

2.2. Cell Culture

The dysplastic oral keratinocytes (DOK) cell line was obtained from Sigma-Aldrich (ECACC #94122104; St. Louis, MO, USA). DOK cells derive from a moderate dysplasia at the tongue of a 57-year-old man with oral squamous cell carcinoma. Cells were cultured in DMEM-high glucose, supplemented with penicillin (10,000 U/mL) and streptomycin (10 μg/mL). For treatment, DOK cells were stimulated with 0.1 μM 1,25-(OH)2D3 or vehicle (isopropanol) for 48 h in culture media.

2.3. Immunofluorescence

Cells were grown on glass coverslips and fixed with 4% formaldehyde in PBS for 15 min. Permeabilization was done with 0.1% Triton X-100 in PBS for 15 min. After washing with PBS, samples were blocked with 5% bovine serum albumin (BSA) in PBS for 1 h and incubated with primary antibodies (diluted in 5% BSA) overnight at 4 °C. Following a PBS wash, samples were incubated with secondary antibodies for 1 h at room temperature, and stained with Hoechst 33342 (diluted 1:10,000 in PBS) for 10 min. Finally, samples were mounted using the DAKO mounting medium and visualized by fluorescence microscopy, using a Nikon C2 Plus confocal microscope.

2.4. Oral Explant

Tissue biopsies were obtained from three donor patients diagnosed with oral dysplasia who were attending the Dental Clinic, Faculty of Dentistry, University of Chile, Santiago, Chile. Ethical human agreement was mandatory, as requested by the Comité Ético Científico, Faculty of Dentistry, Universidad de Chile. All subjects were evaluated for the presence of clinical oral lesions, such as leukoplakia, erythroplakia, and erythroleukoplakia, which were histologically diagnosed as dysplasia.

Oral tissues were divided into two pieces and cultured in DMEM medium with either 0.1 μM 1,25-(OH)2D3 or vehicle for 48 h. Then, samples were prepared for histology and immunohistochemistry.

2.5. Measurement of Serum 25(OH)D

Serum 25-hydroxyvitamin D3 (the precursor of 1,25-(OH)2D3, which is also known as calcidiol or "25(OH)D") was measured in donor patients by electrochemiluminescence immunoassay (ECLIA, Roche, Basel, Switzerland) for total vitamin D (25-OHD, D2, and D3). Vitamin D deficiency was defined as <30 ng/mL 25(OH)D.

2.6. Immunohistochemistry

Oral explants were fixed in 10% formalin and included in paraffin blocks. Sections of the included samples (3 μm) were deparaffinized with xylene and rehydrated with ethanol (decreasing concentrations) and distilled water. Antigen retrieval was done with sodium citrate buffer (pH 6.0), while endogenous peroxidase activity was inactivated with 3% hydrogen peroxide in methanol for 10 min. Sections were blocked with horse serum for 30 min and incubated with primary antibodies against β-catenin, E-cadherin, VDR, and Ki-67 (overnight at 4 °C). Sections were incubated with biotinylated secondary antibodies for 30 min at 37 °C and then with peroxidase-conjugated streptavidin (Universal Detection System Vectastain Elite Kit wide spectrum ABC-HRP, RTU, Vector-USA, EE. UU) for 30 min at 37 °C. The reaction was finally visualized with diaminobenzidine (DAB) and stained with Harris hematoxylin (HE).

2.7. SDS-PAGE and Western Blot

Cells were washed with ice-cold PBS and homogenized in Tissue Protein Extraction Reagent (T-PER; Thermo Fisher Scientific, Waltham, MA, USA) lysis buffer supplemented with protease and phosphatase inhibitors. Total protein extracts were resolved by denaturing polyacrylamide gel electrophoresis (SDS-PAGE) and transferred to nitrocellulose membranes for Western blotting using a Trans-Blot Turbo System (Bio Rad, Hercules, CA,

USA). Membranes were blocked with 5% non-fat milk in 0.1% Tween-TBS for 1 h and then incubated with primary antibodies overnight. Primary antibodies were detected with HRP-conjugated antibodies using a chemiluminescence EZ-ECL system as the HRP enzyme substrate.

2.8. Viability Analysis

DOK cells were seeded in 96-well plates at a density of 10,000 cells per well. After 24 h, a 20:1 mixture of MTS: PMS (MTS® kit; Promega, Madison, WI, USA) was added to each well and incubated for 1 h at 37 °C. The reaction was stopped with 10% SDS. Finally, the reduction of the MTS compound in formazan was determined by measuring the absorbance at 490 nm.

2.9. Migration Assays

Transwell migration assays were performed in Boyden Chambers (Transwell Costar, 6.5 mm diameter and 8 μm pore diameter; Sigma-Aldrich, St. Louis, MO, USA). The outer layers of the inserts were pre-coated with 2 μg/mL fibronectin. 50,000 cells per condition were re-suspended in serum-free DMEM medium and plated into the top chamber. Complete medium with 10% FBS was added to the bottom chamber. After 5 h, inserts were washed with distilled water and simultaneously stained and fixed with 0.1% Crystal Violet in methanol.

2.10. RNA Isolation and RT-qPCR

RNA was extracted with an RNeasy Kit ™ (Qiagen, Hilden, Germany). Reverse transcription of 1 μg RNA per sample was performed using the High Capacity cDNA Reverse Transcription Kit (Applied Biosystems, Waltham, MA, USA). Quantitative amplification of ciclin-D1 (primers 5′-CCACCTGTCCCACTCCTACGAT-3′; 5′-GCAGGGCCGTTGGGTAG AAA-3′), survivin (primers 5′-GCTTCGCTGGAAACCTCTGGA-3′; 5′-TCTGGGCAGATGG CTGTTGG-3′), GAPDH (primers 5′-ACCCACTCCTCCACCTTTGA-3′;5′-CTGTTGCTGTA GCCAAATTCGT-3′) cDNAs was done with Fast SYBR® Green Master Mix (Applied Biosystems, Waltham, MA, USA) and analyzed with a StepOne Real-Time PCR system (Applied Biosystems, Waltham, MA, USA).

2.11. Statistical Analysis

Given that exploratory analysis showed a normal distribution, *t*-test, and ANOVA were used. Values are graphically depicted as the mean ± SEM (standard deviation of the mean; from at least three independent experiments). A $p < 0.05$ was considered statistically significant. GraphPad Prism software version 6 was used for statistical analysis.

3. Results

3.1. Effects of 1,25-(OH)2D3 on β-Catenin Localization and the Expression of Target Genes in Dysplastic Oral Keratinocytes

Following 48 h treatment with 0.1 μM 1,25-(OH)2D3, using isopropanol as a vehicle control, DOK cells were analyzed by immunofluorescence for β-catenin localization. In doing so, we found that 58% of cells showed nuclear β-catenin staining, as compared with the 96% of control vehicle-treated cells (Figure 1A,B). Accordingly, 61% of DOK cells treated with 1,25-(OH)2D3 presented nuclear localization of non-phosphorylated (Ser33/Ser37/Thr41, transcriptionally active) β-catenin, while 94% of cells treated with vehicle showed nuclear localization of non-phosphorylated β-catenin (Figure 1A,C). This suggests that 1,25-(OH)2D3 prevents nuclear translocation of both total and transcriptionally active β-catenin in an in vitro model of oral dysplasia. To confirm these observations, total protein levels of both β-catenin and its target gene, survivin, were assessed by Western blotting of vehicle- and 1,25-(OH)2D3-treated DOK cells. Treatment with 1,25-(OH)2D3 decreased protein levels of both non-phosphorylated β-catenin (relative to total β-catenin; Figure 1D,E) and survivin (relative to GAPDH; Figure 1D,F). Consistently, DOK cells treated with 1,25-(OH)2D3 displayed lower mRNA expression of survivin ($p < 0.001$; Figure 1G) and cyclin D1 ($p = 0.06$; Figure 1H). Taken together, these results suggest that treatment of dysplastic oral cells with 1,25-(OH)2D3 leads to decreased β-catenin activity.

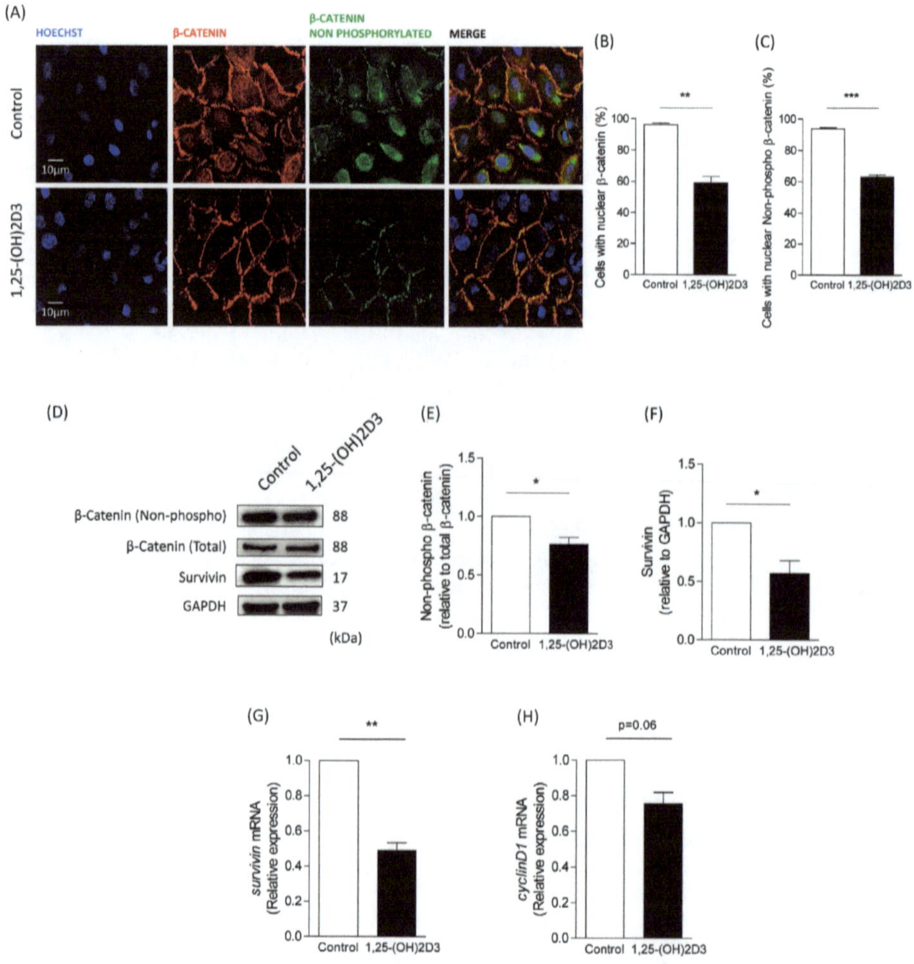

Figure 1. 1,25-(OH)2D3 treatment decreases nuclear localization of β-catenin and expression of target genes in dysplastic oral cells. DOK cells were incubated with 1,25-(OH)2D3 or isopropanol (as vehicle control) for 48 h and used for subsequent analysis. (**A–C**) Subcellular localization of total β-catenin and non-phosphorylated (Ser33/Ser37/Thr41, transcriptionally active) β-catenin in DOK cells exposed to 0.1 μM of 1,25-(OH)2D3. Representative confocal microscope images from three independent experiments are shown in (**A**), using Hoechst for nuclear staining, while the percentage of cells with nuclear localization of total β-catenin and non-phosphorylated (Ser33/Ser37/Thr41, transcriptionally active) β-catenin is graphically depicted in (**B**,**C**), respectively (mean ± SEM; *t*-test; *** $p \leq 0.001$; ** $p \leq 0.01$; $n = 3$). (**D–F**) Western blot analysis of DOK cells treated with 1,25-(OH)2D3. Survivin, total and non-phosphorylated β-catenin, and GAPDH were blotted with specific antibodies. Representative blot images from 3 independent experiments are shown in (**D**), while relative levels of non-phosphorylated β-catenin (relative to total β-catenin) and survivin (relative to GAPDH) were quantified by scanning densitometry and graphically presented in (**E**,**F**), respectively (mean ± SEM; *t*-test; * $p \leq 0.05$; $n = 3$). (**G**,**H**) Relative survivin and cyclin D1 mRNA levels were quantified by RT-qPCR (mean ± SEM; *t*-test; ** $p \leq 0.01$; $n = 3$).

3.2. *Effects of 1,25-(OH)2D3 on E-Cadherin and VDR Localization in Dysplastic Oral Keratinocytes*

We examined the membranous localization of E-cadherin in 1,25-(OH)2D3-treated and vehicle-treated DOK cells by immunofluorescence. We found that following treatment with 1,25-(OH)2D3, 95% of cells displayed membrane-localization of E-cadherin, which differed from the 24% observed in vehicle-treated cells (Figure 2A,B). This indicates that 1,25-(OH)2D3 increases E-cadherin expression at the cell membrane in oral dysplastic cells.

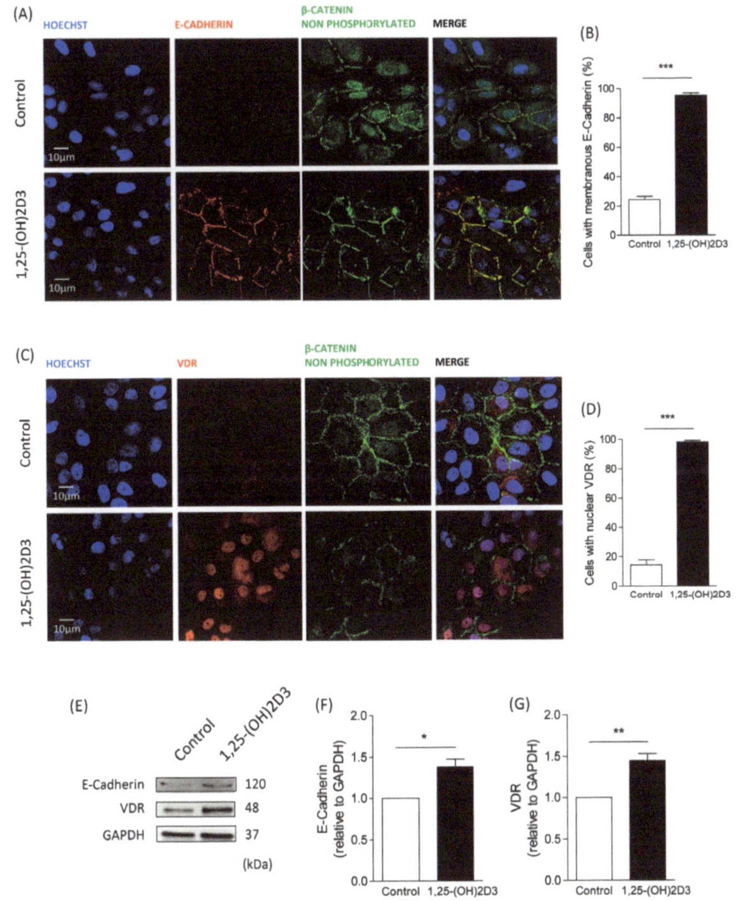

Figure 2. Increased expression of E-cadherin and VDR in oral dysplastic keratinocytes treated with 1,25-(OH)2D3. DOK cells were incubated with 1,25-(OH)2D3 (isopropanol as vehicle control) for 48 h. (**A–D**) Subcellular localization of E-cadherin, VDR, and non-phosphorylated (Ser33/Ser37/Thr41, transcriptionally active) β-catenin in DOK cells exposed to 0.1 μM of 1,25-(OH)2D3. Representative confocal microscope images, using Hoechst for nuclear staining, are shown in (**A**,**C**), while the percentage of cells with membranous localization of E-cadherin and nuclear VDR are graphically depicted in (**B**,**D**), respectively. Quantifications were obtained as described in the materials and methods, using the ImageJ software version 2.15.1, and data are shown as the average from three independent experiments (mean ± SEM; *t*-test; *** $p \leq 0.001$; $n = 3$). (**E–G**) Western blot analysis of DOK cells treated with 1,25-(OH)2D3. E-cadherin, VDR, and GAPDH were blotted with specific antibodies. Representative images from 3 independent experiments are shown in (**E**), while densitometry analysis of E-cadherin and VDR levels (relative to GAPDH) is graphically depicted in (**F**,**G**), respectively (mean ± SEM; *t*-test; * $p \leq 0.05$; ** $p \leq 0.01$; $n = 3$).

We then explored whether 1,25-(OH)2D3 promotes nuclear translocation of the intracellular receptor VDR. Compared to vehicle-treated DOK cells, which displayed 14% of cells with nuclear VDR, treatment of DOK cells with 1,25-(OH)2D3 showed 98% of cells with nuclear VDR (Figure 2C,D). This suggests that 1,25-(OH)2D3 increases nuclear translocation of VDR in oral dysplastic cells. Importantly, increased membranous localization of E-cadherin and nuclear translocation of VDR were found to be correlated with decreased nuclear β-catenin (Figure 2A,C).

Finally, we determined whether 1,25-(OH)2D3 also modulates E-cadherin and VDR total protein levels. To achieve this, DOK cells were stimulated with either 1,25-(OH)2D3 or control vehicle, and samples were analyzed by Western blotting. We observed that 1,25-(OH)2D3 led to increased protein levels of both E-cadherin and VDR (relative to GAPDH; Figure 2E–G). This indicates that 1,25-(OH)2D3 not only regulates E-cadherin and VDR localization in oral dysplastic cells but also enhances their protein levels.

3.3. *Effects of 1,25-(OH)2D3 on Migratory Capacity and Viability in Dysplastic Oral Keratinocytes*

Since 1,25-(OH)2D3 increased membranous localization of E-cadherin in DOK cells (Figure 2A,B), suggesting a more stable cell-to-cell adhesion phenotype, we set out to determine whether cell migration might be affected by 1,25-(OH)2D3. Indeed, treatment of DOK cells with 1,25-(OH)2D3 showed a 50% decrease in cell migration in a Transwell assay, compared with DOK cells treated with vehicle (Figure 3A,B).

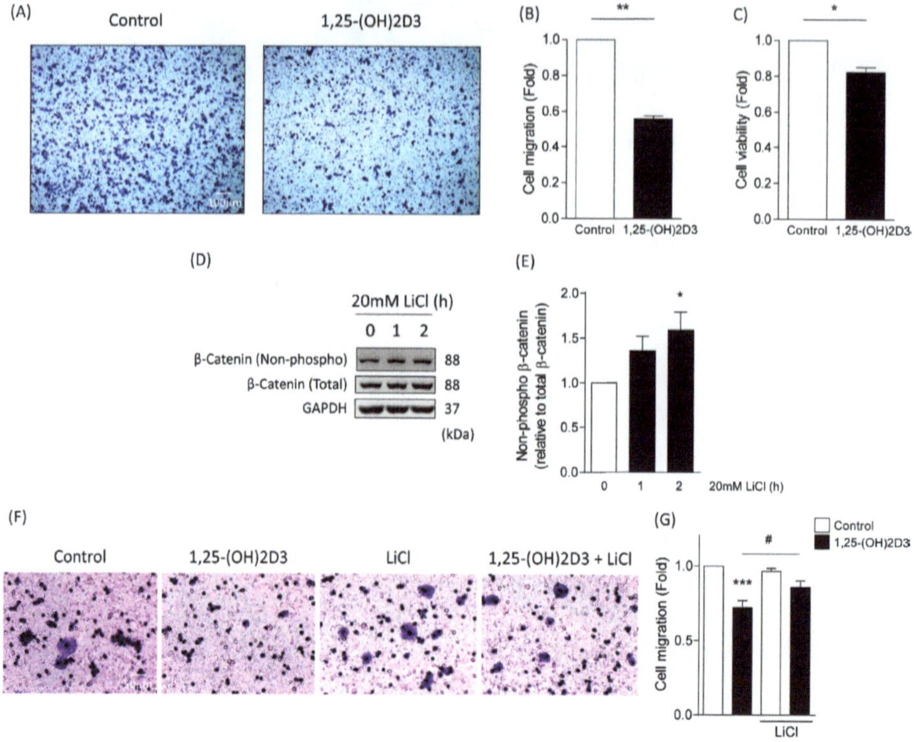

Figure 3. 1,25-(OH)2D3 decreases cell migration and viability in oral dysplastic keratinocytes. (**A**,**B**) DOK cells were allowed to migrate for 300 min in Transwell chambers in the presence of either 0.1 μM 1,25-(OH)2D3 or vehicle. Cells that migrated were stained with crystal violet. Representative

images are shown in (**A**), and graph that represents the averages of three independent experiments is depicted in (**B**) (mean ± SEM; *t*-test; ** $p \leq 0.01$; $n = 3$). (**C**) Cell viability of DOK treated with either 0.1 μM 1,25-(OH)2D3 or vehicle was evaluated with the MTS® kit, and formazan formation was measured at 490 nm (mean ± SEM; *t*-test; * $p < 0.05$; $n = 3$). (**D**,**E**) Western blot analysis of DOK cells treated with 20 mM lithium chloride (LiCl) for 1 or 2 h. Non-phosphorylated β-catenin, total β-catenin, and GAPDH were blotted with specific antibodies. Representative images from 3 independent experiments are shown in (**D**), and relative levels of non-phosphorylated β-catenin (relative to total β-catenin) are depicted in (**E**). (mean ± SEM; ANOVA; * $p \leq 0.05$; $n = 3$). (**F**,**G**) Transwell cell migration assay was performed in DOK cells in the presence of either 0.1 μM 1,25-(OH)2D3 or vehicle control (48 h), and treated with 20 mM LiCl for 2 h. Representative images of cells stained with crystal violet are shown in (**F**), and quantification of migration fold from three independent experiments is graphically presented in (**G**) (mean ± SEM; ANOVA; *** $p \leq 0.001$ vs. control, # $p \leq 0.05$ vs. 1,25-(OH)2D3; $n = 3$).

On the other hand, given that 1,25-(OH)2D3 reduced protein and mRNA levels of survivin (Figure 1E–G), which is a critical protein involved in cell viability and apoptosis, we analyzed whether 1,25-(OH)2D3 impacted cell viability. DOK cells treated with 1,25-(OH)2D3 presented a 20% reduction in cell viability in an MTS assay, compared with vehicle-treated cells (Figure 3C). To assess whether these effects were due to reduced β-catenin activity induced by 1,25-(OH)2D3, we used lithium chloride (LiCl), which is known to inhibit β-catenin degradation. Activation of β-catenin in DOK cells was observed 2 h after treatment with 20 mM LiCl (Figure 3D,E). The use of LiCl prevented the decrease in migration observed following 1,25-(OH)2D3 treatment, suggesting that activation of β-catenin is sufficient to restore migration of cells challenged with 1,25-(OH)2D3 (Figure 3F,G). Collectively, these results suggest that 1,25-(OH)2D3 reduces cell migration and viability of oral dysplastic cells, and that β-catenin activation prevents the effects of 1,25-(OH)2D3 on cell migration.

3.4. Effects of 1,25-(OH)2D3 on Oral Explant from Tissues of Donor Patients with Oral Dysplasia

To corroborate our results, we generated oral explant cultures using tissues from OPMD patients with histological diagnosis of oral dysplasia. Tissue samples obtained from patients were kept in DMEM medium until their arrival at the laboratory. Each sample was cut into two halves: one half was treated with 1,25-(OH)2D3 (0.1 μM) for 48 h, while the other half was incubated with vehicle alone. Samples were then fixed in 10%-buffered formalin for 24 h, before being embedded in paraffin (Figure 4A). 3 μm sections were made, and the subcellular localization of β-catenin, Ki-67 (a cell proliferation marker), E-cadherin, and VDR was determined by immunohistochemistry (Figure 4B).

Compared to explants treated with vehicle, explants treated with 1,25-(OH)2D3 showed reduced nuclear localization of β-catenin (Figure 4C) and expression of Ki-67 (Figure 4D), as well as increased membranous expression of E-cadherin (Figure 4E) and nuclear VDR (Figure 4F). These ex vivo results are consistent with those observed in vitro and collectively suggest that 1,25-(OH)2D3 reduces β-catenin activation, decreases cell proliferation, increases VDR nuclear translocation, and increases cell-to-cell adhesion. The clinical-demographic characteristics of the patients who voluntarily participated in this study are presented in Figure 4G. Interestingly, all three patients had low serum 25(OH)D levels (<30 ng/mL), suggesting that vitamin D deficiency may be associated to oral dysplasia in these patients.

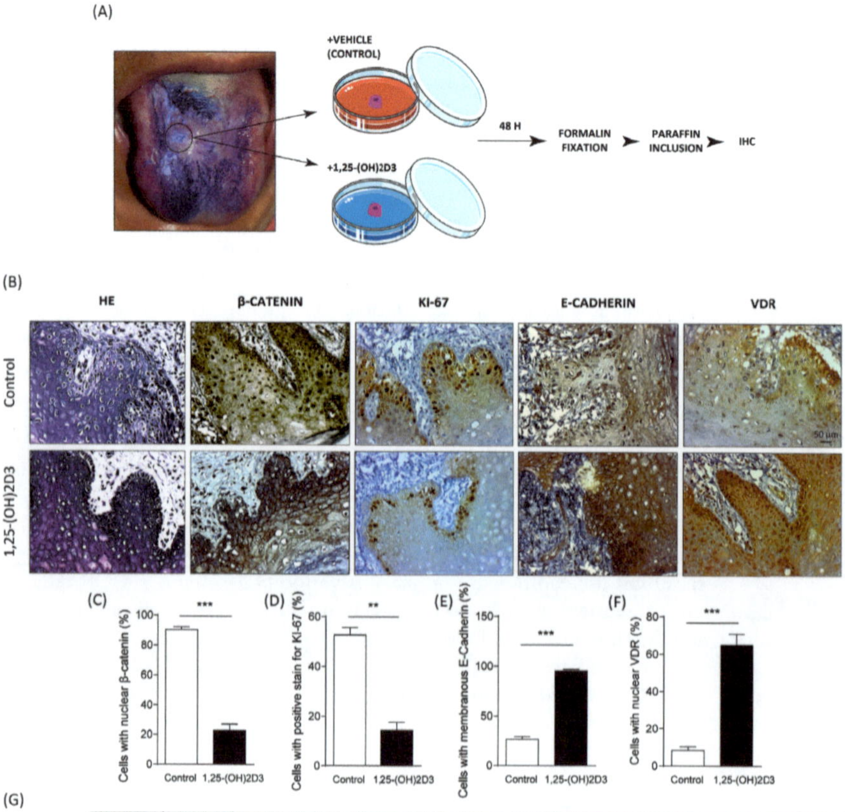

(G)

	Gender	Age	Smoking	Location	Time of evolution	Clinical diagnosis	Histopathological diagnosis	Vitamin D levels
Patient 1	Male	75	No Ex-smoker for 30 years 1 cigarette per day	Tongue	4 months	Proliferative verrucous leukoplakia	Moderate epithelial dysplasia	12.2 ng/ml
Patient 2	Male	74	Yes 10 cigarettes per day	Tongue/ alveolar ridge	1 year	Proliferative leukoplakia	Mild epithelial dysplasia	23.4 ng/ml
Patient 3	Female	58	Yes 15 cigarettes per day	Tongue	8 months	Proliferative verrucous leukoplakia	Moderate epithelial dysplasia	19.1 ng/ml

Figure 4. 1,25-(OH)2D3 decreases nuclear β-catenin and Ki-67 and increases E-cadherin and VDR expression in oral dysplasia tissues. (**A**) Oral tissues from three patients with dysplasia were divided into two pieces and cultured in DMEM medium with either 0.1 μM 1,25-(OH)2D3 or isopropanol (vehicle) for 48 h. Oral explants were fixed in 10% formalin and included in paraffin blocks. (**B–F**) HE and immunohistochemistry of β-catenin, Ki-67, E-cadherin, and VDR in oral dysplasia. Representative images were photographed at 40× using an Olympus CX41 microscope and visualized with the Micrometrics SE Premium software version 4.5.1 (**C**), while quantification of the percentage of cells with nuclear β-catenin (**D**), positive KI-67 staining (**E**), membranous E-Cadherin (**F**), and nuclear VDR (**G**) were analyzed with the ImageJ software version 2.15.1 (** $p \leq 0.01$ and *** $p \leq 0.001$ vs. control; $n = 3$). (**G**) Clinical-demographic characteristics of the patients who voluntarily participated in this study.

4. Discussion

We and others previously showed that aberrant activation of the Wnt/β-catenin pathway is a recurrent phenomenon in oral dysplasia [8,11,27]. However, no information is available about factors that might negatively target this phenomenon, with the possibility of therapeutic applications. In this regard, it becomes relevant to consider the feasibility of 1,25-(OH)2D3 (calcitriol), which has been widely reported to target this pathway in different cancer models [17–20,22,28], although its effects have not been explored in OPMDs with histological diagnosis of oral dysplasia. Unlike other vitamins, such as vitamin A and C, vitamin D has shown remarkable antineoplastic effects in in vitro and in vivo oral cancer models [29]. Indeed, a systematic review that analyzed 80 studies published between 1959 and 2022 [29] showed improvement in the number and severity of oral lichen planus lesions in patients prescribed with vitamin D supplements [30], reduced tumor volume of oral cancer xenografts in mice co-treated with erlotinib (an epidermal growth factor receptor inhibitor) and 1,25-(OH)2D3 [16], and synergic effects of 5-fluorouracil and 1,25(OH)2D3 on human oral squamous cell carcinoma lines (C152) viability [31]. Moreover, other preclinical studies have shown that 1,25-(OH)2D3 can inhibit OSCC growth in vivo [16,21,25]. The purpose of this study was to evaluate the effects of 1,25-(OH)2D3 on cell migration, cell viability, and β-catenin activity in oral dysplasia. We revealed that 1,25(OH)2D3 reduced nuclear localization of both total and transcriptionally active β-catenin, which was in line with reduced expression of β-catenin target genes, including survivin and cyclin D1, at both mRNA and protein levels.

Several reports have demonstrated the inhibitory action of 1,25-(OH)2D3 on the Wnt/β-catenin pathway in different types of tumors, evidenced by the decreased formation of transcriptional TCF-β-catenin complexes and increased expression of Wnt antagonist [26,32]. In this work, we demonstrated that 1,25-(OH)2D3 increased the expression of the tumor suppressor protein E-cadherin, while decreasing the nuclear expression of non-phosphorylated β-catenin, in in vitro and ex vivo models of oral dysplasia. In breast and colon cancer, it has been shown that 1,25-(OH)2D3 increases the expression of E-cadherin, inducing translocation of β-catenin from the nucleus to the membrane, thus decreasing its transcriptional activity [22,26,33]. On the other hand, we found a decrease in the proliferation marker Ki-67 in explants treated with 1,25-(OH)2D3 compared with vehicle. Similar results have been found in skin keratinocytes, where treatment with 1,25(OH)2D3 inhibits β-catenin activity, thereby decreasing cell proliferation [19]. We also demonstrated that 1,25-(OH)2D3 induces expression of the nuclear receptor of vitamin D (VDR) in dysplastic oral cells. VDR is widely expressed in most cell types, and its expression is progressively reduced during tumor progression in many cancer types [18–20,28,34]. Studies in melanoma cells have shown that 1,25-(OH)2D3 inhibits the Wnt/β-catenin pathway and cell growth in vitro and in vivo, and that VDR expression was significantly upregulated post-treatment [35]. By comparing VDR expression levels in normal, benign, and malignant tissues of skin, breast, ovarian and prostate, it was described a negative correlation between VDR expression and tumor malignancy [17,18]. Similarly, in colorectal cancer, low VDR expression is predominantly observed in patients with advanced cancer stages (III and IV), and interestingly, overexpression of VDR reduced β-catenin and Cyclin D1 levels, suggesting that the Wnt/β-catenin pathway is active during colorectal cancer because of reduced VDR [36]. In fact, tumor growth of the colon adenocarcinoma cell line SW480 in mice is impaired when cells overexpress VDR, indicating that VDR acts as a tumor suppressor in colorectal cancer [36]. Nuclear VDR expression has also been associated with better overall survival in lung and urothelial bladder cancer patients [20,23]. In our study, both oral dysplasia cells and oral explants from OPMD patients with histological diagnosis of dysplasia showed decreased nuclear expression of VDR. Interestingly, following treatment with 1,25-(OH)2D3, nuclear expression of VDR increased significantly. These observations suggest that VDR expression may be a valuable biomarker for the early diagnosis of OSCC. To complement these results, the effects of 1,25-(OH)2D3 on cell migration and viability were assessed. Here, we show for the first time that treatment of dysplastic oral cells with 1,25(OH)2D3 significantly decreased both migration and viability. In kidney cancer cells, it was shown that 1,25-(OH)2D3 inhibits migration and invasion by suppressing β-catenin [15]. Consistently, in our model, we showed that lithium-dependent activation of β-catenin prevents the reduced migration observed after treatment with 1,25(OH)2D3. Interestingly, the patients in our study displayed low serum vitamin D levels, which might be associated with the progression of their proliferative verrucous leukoplakias. Although the classical role of vitamin D is to regulate the metabolism of calcium and phosphate, extensive research has suggested that low sunlight exposure and vitamin D deficiency are also associated with an increased risk of cancer [37]. An inverse correlation between serum 25(OH)D levels and the high risk of developing colon, breast, prostate, and gastric cancer, among others, has been proposed. Epidemiologic studies suggest that low vitamin D levels are associated with an increased risk of OSCC [24,38,39]. However, the mechanism by which 1,25-(OH)2D3 inhibits migration is unknown.

The strengths of this study include the use of human in vitro and ex vivo models of oral dysplasia from similar anatomical locations (tongue), as well as the use of 1,25-(OH)2D3 0.1 μM, equivalent to 41.7 ng/mL, within the physiological range for calcitriol in serum [40]. A limitation of this study, however, is that it remains unknown whether the use of 1,25-(OH)2D3 could reduce other β-catenin-dependent malignant traits, such as cell invasion. Future studies will address these possibilities, as well as the role of 1,25-(OH)2D3 during oral carcinogenesis in vivo and in novel in vitro preclinical models. Indeed, DOK spheroids treated with 4-nitroquinoline 1-oxide (4NQO)—a carcinogen that mimics the effects of cigarette smoking—represent an in vitro 3D model of oral carcinogenesis [41]. This model is suitable to evaluate the area and length of the spheroid invasion after challenge with a drug, to assess the chemotherapeutic response in vitro [41]. Thus, it would be interesting to observe whether reduced spheroids invasion occurs after treatment with 1,25-(OH)2D3. Finally, given that 1,25-(OH)2D3 has synergistic effects on cytotoxicity when combined with other cancer treatments, i.e., chemotherapy, the results of this study open novel therapeutic approaches to prevent the progression from oral dysplasia to oral cancer.

5. Conclusions

In summary, this study shows that 1,25-(OH)2D3 treatment decreases migration and viability in oral dysplasia cells. This was accompanied by decreased Wnt/β-catenin activation. The results of our study suggest the potential of using 1,25-(OH)2D3 as a new therapeutic approach to prevent the progression of OPMDs with histological diagnosis of oral dysplasia towards OSCC by decreasing cell proliferation, cell migration, and cell viability, while inhibiting the Wnt/β-catenin pathway.

Author Contributions: Conceptualization, D.P.-O. and M.R.; methodology, D.P.-O. and M.R.; software, D.P.-O. and M.R.; validation, D.P.-O. and M.R.; formal analysis, C.G., C.K., D.P.-O. and M.R.; investigation, C.G., C.K., D.P.-O. and M.R.; resources, V.A.T., D.P.-O. and M.R.; data curation, D.P.-O. and M.R.; writing—original draft preparation, D.P.-O. and M.R.; writing—review and editing, J.A., A.M.-R., D.P.-O. and M.R.; visualization, D.P.-O. and M.R.; supervision, D.P.-O. and M.R.; project administration, D.P.-O. and M.R.; funding acquisition, D.P.-O. and M.R. All authors have read and agreed to the published version of the manuscript.

Funding: This work was supported by the National Fund for Scientific and Technological Development FONDECYT 11240983 (to D.P.-O.), 11221000 (to M.R.) and 1220517 (to V.A.T.); Vice-rectory for Research and Development (VID) at Universidad de Chile UI-024/19 (to M.R.) and PRI-ODO 19-01 (to A.M.); ICN09_016/ICN 2021_045 (to V.A.T.).

Institutional Review Board Statement: The study was conducted in accordance with the Declaration of Helsinki and approved on 3 July2018 by the Ethics Committee of the Faculty of Odontology, Universidad de Chile, under the code 2018/04.

Informed Consent Statement: Informed consent was obtained from all subjects involved in the study.

Data Availability Statement: Data are contained within the article.

Conflicts of Interest: The authors declare no conflicts of interest.

References

1. van der Waal, I. Oral potentially malignant disorders: Is malignant transformation predictable and preventable? *Med. Oral Patol. Oral Cir. Bucal.* **2014**, *19*, 386–390. [CrossRef] [PubMed]
2. Johnson, D.E.; Burtness, B.; Leemans, C.R.; Lui, V.W.Y.; Bauman, J.E.; Grandis, J.R. Head and neck squamous cell carcinoma. *Nat. Rev. Dis. Primers* **2020**, *6*, 92. [CrossRef] [PubMed]
3. Warnakulasuriya, S.; Kerr, A.R. Oral Cancer Screening: Past, Present, and Future. *J. Dent. Res.* **2021**, *100*, 1313–1320. [CrossRef] [PubMed]
4. Nag, R.; Kumar Das, R. Analysis of images for detection of oral epithelial dysplasia: A review. *Oral Oncol.* **2018**, *78*, 8–15. [CrossRef] [PubMed]
5. Warnakulasuriya, S.; Kujan, O.; Aguirre-Urizar, J.M.; Bagan, J.V.; González-Moles, M.; Kerr, A.R.; Lodi, G.; Mello, F.W.; Monteiro, L.; Ogden, G.R.; et al. Oral potentially malignant disorders: A consensus report from an international seminar on nomenclature and classification, convened by the WHO Collaborating Centre for Oral Cancer. *Oral Dis.* **2021**, *27*, 1862–1880. [CrossRef] [PubMed]
6. Odell, E.; Kujan, O.; Warnakulasuriya, S.; Sloan, P. Oral epithelial dysplasia: Recognition, grading and clinical significance. *Oral Dis.* **2021**, *27*, 1947–1976. [CrossRef]
7. Awadallah, M.; Idle, M.; Patel, K.; Kademani, D. Management update of potentially premalignant oral epithelial lesions. *Oral Surg. Oral Med. Oral Pathol. Oral Radiol.* **2018**, *125*, 628–636. [CrossRef]
8. Reyes, M.; Rojas-Alcayaga, G.; Maturana, A.; Aitken, J.P.; Rojas, C.; Ortega, A.V. Increased nuclear beta-catenin expression in oral potentially malignant lesions: A marker of epithelial dysplasia. *Med. Oral Patol. Oral Cir. Bucal.* **2015**, *20*, 540–546. [CrossRef]
9. Reyes, M.; Pena-Oyarzun, D.; Maturana, A.; Torres, V.A. Nuclear localization of beta-catenin and expression of target genes are associated with increased Wnt secretion in oral dysplasia. *Oral Oncol.* **2019**, *94*, 58–67. [CrossRef]
10. Ishida, K.; Ito, S.; Wada, N.; Deguchi, H.; Hata, T.; Hosoda, M.; Nohno, T. Nuclear localization of beta-catenin involved in precancerous change in oral leukoplakia. *Mol. Cancer* **2007**, *6*, 62. [CrossRef]
11. Sato, K.; Okazaki, Y.; Tonogi, M.; Tanaka, Y.; Yamane, G.Y. Expression of beta-catenin in rat oral epithelial dysplasia induced by 4-nitroquinoline 1-oxide. *Oral Oncol.* **2002**, *38*, 772–778. [CrossRef] [PubMed]
12. Clevers, H.; Nusse, R. Wnt/beta-catenin signaling and disease. *Cell* **2012**, *149*, 1192–1205. [CrossRef] [PubMed]
13. Nusse, R. Wnt signaling. *Cold Spring Harb. Perspect. Biol.* **2012**, *4*, 5. [CrossRef] [PubMed]
14. Peña-Oyarzún, D.; Flores, T.; Torres, V.A.; Quest, A.F.G.; Lobos-Gonzalez, L.; Kretschmar, C.; Contreras, P.; Maturana-Ramírez, A.; Criollo, A.; Reyes, M. Inhibition of PORCN blocks Wnt signaling to attenuate progression of oral carcinogenesis. *Clin. Cancer Res.* **2024**, *5*, 209–223. [CrossRef] [PubMed]
15. Xu, S.; Zhang, Z.H.; Fu, L.; Song, J.; Xie, D.D.; Yu, D.X.; Xu, D.X.; Sun, G.P. Calcitriol inhibits migration and invasion of renal cell carcinoma cells by suppressing Smad2/3-, STAT3- and β-catenin-mediated epithelial-mesenchymal transition. *Cancer Sci.* **2020**, *111*, 59–71. [CrossRef]
16. Bothwell, K.D.; Shaurova, T.; Merzianu, M.; Suresh, A.; Kuriakose, M.A.; Johnson, C.S.; Hershberger, P.A.; Seshadri, M. Impact of Short-term 1,25-Dihydroxyvitamin D3 on the Chemopreventive Efficacy of Erlotinib against Oral Cancer. *Cancer Prev. Res.* **2015**, *8*, 765–776. [CrossRef] [PubMed]
17. Al-Azhri, J.; Zhang, Y.; Bshara, W.; Zirpoli, G.; McCann, S.E.; Khoury, T.; Morrison, C.D.; Edge, S.B.; Ambrosone, C.B.; Yao, S. Tumor Expression of Vitamin D Receptor and Breast Cancer Histopathological Characteristics and Prognosis. *Clin. Cancer Res.* **2017**, *23*, 97–103. [CrossRef]
18. Hendrickson, W.K.; Flavin, R.; Kasperzyk, J.L.; Fiorentino, M.; Fang, F.; Lis, R.; Fiore, C.; Penney, K.L.; Ma, J.; Kantoff, P.W.; et al. Vitamin D receptor protein expression in tumor tissue and prostate cancer progression. *J. Clin. Oncol.* **2011**, *29*, 2378–2385. [CrossRef]
19. Hu, L.; Bikle, D.D.; Oda, Y. Reciprocal role of vitamin D receptor on β-catenin regulated keratinocyte proliferation and differentiation. *J. Steroid Biochem. Mol. Biol.* **2014**, *144*, 237–241. [CrossRef]
20. Jóźwicki, W.; Brożyna, A.A.; Siekiera, J.; Slominski, A.T. Expression of Vitamin D Receptor (VDR) Positively Correlates with Survival of Urothelial Bladder Cancer Patients. *Int. J. Mol. Sci.* **2015**, *16*, 24369–24386. [CrossRef]
21. Meier, J.D.; Enepekides, D.J.; Poirier, B.; Bradley, C.A.; Albala, J.S.; Farwell, D.G. Treatment with 1-alpha,25-dihydroxyvitamin D3 (vitamin D3) to inhibit carcinogenesis in the hamster buccal pouch model. *Arch. Otolaryngol. Head Neck Surg.* **2007**, *133*, 1149–1152. [CrossRef] [PubMed]
22. Pálmer, H.G.; González-Sancho, J.M.; Espada, J.; Berciano, M.T.; Puig, I.; Baulida, J.; Quintanilla, M.; Cano, A.; de Herreros, A.G.; Lafarga, M.; et al. Vitamin D(3) promotes the differentiation of colon carcinoma cells by the induction of E-cadherin and the inhibition of beta-catenin signaling. *J. Cell Biol.* **2001**, *154*, 369–387.
23. Srinivasan, M.; Parwani, A.V.; Hershberger, P.A.; Lenzner, D.E.; Weissfeld, J.L. Nuclear vitamin D receptor expression is associated with improved survival in non-small cell lung cancer. *J. Steroid Biochem. Mol. Biol.* **2011**, *123*, 30–36. [CrossRef] [PubMed]
24. Grimm, M.; Cetindis, M.; Biegner, T.; Lehman, M.; Munz, A.; Teriete, P.; Reinert, S. Serum vitamin D levels of patients with oral squamous cell carcinoma (OSCC) and expression of vitamin D receptor in oral precancerous lesions and OSCC. *Med. Oral Patol. Oral Cir. Bucal.* **2015**, *20*, 188–195. [CrossRef] [PubMed]
25. Verma, A.; Vincent-Chong, V.K.; DeJong, H.; Hershberger, P.A.; Seshadri, M. Impact of dietary vitamin D on initiation and progression of oral cancer. *J. Steroid Biochem. Mol. Biol.* **2020**, *199*, 105603. [CrossRef] [PubMed]

26. Pendás-Franco, N.; García, J.M.; Peña, C.; Valle, N.; Pálmer, H.G.; Heinäniemi, M.; Carlberg, C.; Jimenez, B.; Bonilla, F.; Munoz, A.; et al. DICKKOPF-4 is induced by TCF/beta-catenin and upregulated in human colon cancer, promotes tumour cell invasion and angiogenesis and is repressed by 1alpha,25-dihydroxyvitamin D3. *Oncogene* **2008**, *27*, 4467–4477. [CrossRef] [PubMed]
27. Reyes, M.; Pena-Oyarzun, D.; Silva, P.; Venegas, S.; Criollo, A.; Torres, V.A. Nuclear accumulation of beta-catenin is associated with endosomal sequestration of the destruction complex and increased activation of Rab5 in oral dysplasia. *FASEB J.* **2020**, *34*, 4009–4025. [CrossRef] [PubMed]
28. Thill, M.; Fischer, D.; Kelling, K.; Hoellen, F.; Dittmer, C.; Hornemann, A.; Salehin, D.; Diedrich, K.; Friedrich, M.; Becker, S. Expression of vitamin D receptor (VDR), cyclooxygenase-2 (COX-2) and 15-hydroxyprostaglandin dehydrogenase (15-PGDH) in benign and malignant ovarian tissue and 25-hydroxycholecalciferol (25(OH2)D3) and prostaglandin E2 (PGE2) serum level in ovarian cancer patients. *J. Steroid Biochem. Mol. Biol.* **2010**, *121*, 387–390. [PubMed]
29. See, J.K.L.; Liu, X.; Canfora, F.; Moore, C.; McCullough, M.; Yap, T.; Paolini, R.; Celentano, A. The Role of Vitamins in Oral Potentially Malignant Disorders and Oral Cancer: A Systematic Review. *J. Pers. Med.* **2023**, *23*, 1520. [CrossRef]
30. Nazeer, J.; Singh, S.; Jayam, C.; Singh, R.; Iqubal, M.A.; Singh, R. Assessment of the Role of Vitamin D in the Treatment of Oral Lichen Planus. *J. Contemp. Dent. Pract.* **2020**, *1*, 390–395.
31. Dalirsani, Z.; Farajnia, S.; Javadzadeh, Y.; Mehdipour, M.; Koozegari, S. The effects of 5-fluorouracil alone and in combination with 13-cis retinoic acid and vitamin D3 on human oral squamous cell carcinoma lines. *J. Contemp. Dent. Pract.* **2012**, *13*, 345–350. [CrossRef] [PubMed]
32. Larriba, M.J.; González-Sancho, J.M.; Barbáchano, A.; Niell, N.; Ferrer-Mayorga, G.; Muñoz, A. Vitamin D Is a Multilevel Repressor of Wnt/b-Catenin Signaling in Cancer Cells. *Cancers* **2013**, *5*, 1242–1260. [CrossRef] [PubMed]
33. Johnson, A.L.; Zinser, G.M.; Waltz, S.E. Vitamin D3-dependent VDR signaling delays ron-mediated breast tumorigenesis through suppression of β-catenin activity. *Oncotarget* **2015**, *6*, 16304–16320. [CrossRef] [PubMed]
34. Fathi, N.; Ahmadian, E.; Shahi, S.; Roshangar, L.; Khan, H.; Kouhsoltani, M.; Dizaj, S.M.; Sharifi, S. Role of vitamin D and vitamin D receptor (VDR) in oral cancer. *Biomed. Pharmacother.* **2019**, *109*, 391–401. [CrossRef] [PubMed]
35. Muralidhar, S.; Filia, A.; Nsengimana, J.; Poźniak, J.; O'Shea, S.J.; Diaz, J.M.; Harland, M.; Randerson-Moor, J.A.; Reichrath, J.; Laye, J.P.; et al. Vitamin D-VDR Signaling Inhibits Wnt/β-Catenin-Mediated Melanoma Progression and Promotes Antitumor Immunity. *Cancer Res.* **2019**, *79*, 5986–5998. [CrossRef] [PubMed]
36. Yu, J.; Sun, Q.; Hui, Y.; Xu, J.; Shi, P.; Chen, Y.; Chen, Y. Vitamin D receptor prevents tumour development by regulating the Wnt/β-catenin signalling pathway in human colorectal cancer. *BMC Cancer* **2023**, *12*, 336. [CrossRef] [PubMed]
37. Jeon, S.M.; Shin, E.A. Exploring vitamin D metabolism and function in cancer. *Exp. Mol. Med.* **2018**, *50*, 1–14. [CrossRef] [PubMed]
38. Fanidi, A.; Muller, D.C.; Midttun, Ø.; Ueland, P.M.; Vollset, S.E.; Relton, C.; Vineis, P.; Weiderpass, E.; Skeie, G.; Brustad, M.; et al. Circulating vitamin D in relation to cancer incidence and survival of the head and neck and oesophagus in the EPIC cohort. *Sci. Rep.* **2016**, *6*, 36017. [CrossRef]
39. Orell-Kotikangas, H.; Schwab, U.; Österlund, P.; Saarilahti, K.; Mäkitie, O.; Mäkitie, A.A. High prevalence of vitamin D insufficiency in patients with head and neck cancer at diagnosis. *Head Neck* **2012**, *34*, 1450–1455. [CrossRef]
40. Beer, T.M.; Myrthue, A. Calcitriol in cancer treatment: From the lab to the clinic. *Mol. Cancer Ther.* **2004**, *3*, 373–381. [CrossRef]
41. Chitturi Suryaprakash, R.T.; Shearston, K.; Farah, C.S.; Fox, S.A.; Iqbal, M.M.; Kadolsky, U.; Zhong, X.; Saxena, A.; Kujan, O. A Novel Preclinical In Vitro 3D Model of Oral Carcinogenesis for Biomarker Discovery and Drug Testing. *Int. J. Mol. Sci.* **2023**, *24*, 4096. [CrossRef] [PubMed]

current issues in molecular biology

MDPI

Article

p62 Is a Potential Biomarker for Risk of Malignant Transformation of Oral Potentially Malignant Disorders (OPMDs)

Ryo Takasaki [1,2], Fumihiko Uchida [2,*], Shohei Takaoka [2], Ryota Ishii [3], Satoshi Fukuzawa [2], Eiji Warabi [4], Naomi Ishibashi-Kanno [2], Kenji Yamagata [2], Hiroki Bukawa [2] and Toru Yanagawa [2,5]

1 Oral and Maxillofacial Surgery, Clinical Sciences, Graduate School of Comprehensive Human Sciences, University of Tsukuba, 1-1-1 Tennodai, Tsukuba 305-8875, Japan

2 Department of Oral and Maxillofacial Surgery, Institute of Clinical Medicine, Faculty of Medicine, University of Tsukuba, 1-1-1 Tennodai, Tsukuba 305-8575, Japan

3 Department of Biostatistics, Faculty of Medicine, University of Tsukuba, 1-1-1 Tennodai, Tsukuba 305-8575, Japan

4 Department of Anatomy and Rmbryology, Institute of Medicine, University of Tsukuba, 1-1-1 Tennodai, Tsukuba 305-8575, Japan

5 Department of Oral and Maxillofacial Surgery, Ibaraki Prefectural Central Hospital, 6528 Koibuchi, Kasama 309-1793, Japan

* Correspondence: f-uchida@md.tsukuba.ac.jp; Tel./Fax: +81-29-853-3192

Abstract: To determine the intracellular behavior of p62, a marker of selective autophagy, in oral potentially malignant disorders (OPMDs). This retrospective study includes 70 patients who underwent biopsy or surgical resection and were definitively diagnosed with OPMDs. Immunohistochemical staining for p62, XPO1, p53, and ki67 was performed on all samples and positive cell occupancy was calculated. We statistically investigated the correlation between protein expression in OPMDs and the association between malignant transformation, clinicopathological characteristics, and occupancy. ki67 expression was negatively correlated with p62 expression in the nucleus ($p < 0.01$) and positively correlated with p62 expression in the cytoplasm ($p < 0.01$). For malignant transformation, the expression of p62 in the nucleus ($p = 0.03$) was significantly lower in malignant transformation cases, whereas the expression of p62 in the cytoplasm ($p = 0.03$) and the aggregation expression ($p < 0.01$) were significantly higher. Our results suggest that the function of p62 is altered by its subcellular localization. In addition, defects in selective autophagy occur in cases of malignant transformation, suggesting that p62 is a potential biomarker of the risk of malignant transformation of OPMDs.

Keywords: p62; OPMDs; XPO1; ki67; p53; cancer development

Citation: Takasaki, R.; Uchida, F.; Takaoka, S.; Ishii, R.; Fukuzawa, S.; Warabi, E.; Ishibashi-Kanno, N.; Yamagata, K.; Bukawa, H.; Yanagawa, T. p62 Is a Potential Biomarker for Risk of Malignant Transformation of Oral Potentially Malignant Disorders (OPMDs). *Curr. Issues Mol. Biol.* **2023**, *45*, 7630–7641. https://doi.org/10.3390/cimb45090480

Academic Editors: Emma Adriana Ozon and Violeta Popovici

Received: 4 September 2023
Revised: 18 September 2023
Accepted: 18 September 2023
Published: 19 September 2023

1. Introduction

Oral potentially malignant disorders (OPMDs) are a newly defined concept by the World Health Organization (WHO), which defined them in 2017 as "clinical conditions that carry risk of cancer development in the oral cavity, regardless of whether they are clinically definite precursor lesion or normal mucosa" [1]. OPMDs include 12 diseases, including oral leukoplakia, oral lichen planus (OLP), and oral submucosal fibrosis. The malignant conversion rate of OPMDs is reported to be 7.9% [2]. One of the problems associated with OPMDs is their malignant nature; however, the actual causes of their pathogenesis and malignant transformation are not fully understood.

Anil et al. reported the presence and extent of epithelial dysplasia as predictors of the development of oral leukoplakia in cancer [3]. Although WHO has established a classification of epithelial dysplasia, its diagnosis is based on the subjectivity of the observer, thereby making uniform diagnosis difficult [4]. Therefore, to ensure objectivity in diagnosis, biomarkers must be developed to assess the risk of cancer development in OPMDs.

Treatment options for OPMDs include follow-up, removal of disease triggers such as inappropriate restorations and sharp edges of teeth, medication, and surgical resection [5]; however, there are no uniform treatment guidelines based on the subjective risk judgments of malignant transformation by medical professionals [6]. In addition, recurrence or malignant transformation may occur after surgical resection [7]. Identifying the biomarkers of developing cancer from OPMDs may lead to the development of treatments for OPMDs and may also lead to the prevention of oral cancer.

Selective autophagy is a mechanism that selectively degrades bacteria, specific organelles such as mitochondria, and aggregates of polyubiquitinated proteins [8]. Degradation by autophagy is usually nonselective; however, selectivity is achieved by specifically recognizing the degradation products [9]. Sequestosome-1 (p62) is a receptor protein that selectively incorporates ubiquitinated cargo into selective autophagy [10]. p62 contains a ubiquitin-associated domain (UBA) at its C-terminus, through which it can recognize and bind ubiquitinated proteins [11]; Additionally, p62 contains an LC3-interacting region (LIR). LC3 is an autophagy-associated protein that binds to isolated and autophagosome membranes [12]. This allows for ubiquitinated proteins to be selectively directed by p62 to the isolation membrane where they form autolysosomes and are degraded together with p62.

Recently, autophagy has been linked to various diseases, including Alzheimer's disease, Parkinson's disease, and cardiovascular disease [13]. It has also been reported that autophagy is impaired and in various malignant tumors, such as oral squamous cell carcinoma (OSCC), gastric carcinoma, and breast carcinoma, which results in the accumulation of p62 [14–16]. We previously reported that the expression of autophagy-related markers, including p62, at the resection margins of OSCC was associated with tumor recurrence [17]. We also found that p62 expression in the nucleus and p62 aggregation in oral leukoplakia were related to the presence or absence of epithelial dysplasia [18]. This suggests that p62, an adaptor of selective autophagy, may be useful as a marker for assessing the risk of recurrence at the resection margin of OSCC as well as the risk of malignant transformation in precancerous lesions.

Yoshida et al. reported the expression of p62 in oral leukoplakia; however, to the best of our knowledge, there are no studies on the expression of p62 in OPMDs [18]. Lin et al. reported that p62 showed a change in subcellular localization between oral mucosa and OSCC, suggesting that the subcellular localization of p62 changes in association with malignant transformation [19]. The purpose of this study was to reveal the intracellular behavior of p62 in OPMDs, with a focus on selective autophagy. We evaluated the intracellular expression of p62 in OPMDs and compared it with various clinical features, such as the malignant transformation of OPMDs, and biomarkers, such as Exportin-1 (XPO1), p53, and ki67, which are cancer-related factors.

2. Materials and Methods

2.1. Samples

Of the 104 patients who visited the Department of Oral and Maxillofacial Surgery at the University of Tsukuba Hospital between 2014 and 2021, underwent biopsy or surgical resection, and received a definitive diagnosis of OPMDs, 70 patients provided their informed consent to participate in the study. The specimens retrieved from the patients were formalin-fixed, and paraffin-embedded blocks were prepared for pathological examination. For patients who underwent both biopsy and surgical resection, the specimen obtained during surgical resection was used based on its condition. Clinical and clinicopathological data were obtained from medical records. All the specimens were diagnosed by at least two pathologists. Regarding malignant transformation of OPMDs, since OPMDs have the propensity to recur after surgical resection and is similar to OSCC, the following definition of local recurrence of OSCC was used as a reference: a local recurrence is defined as less than 2 cm away from or occurs within 3 years of the primary tumor and a second primary tumor as more than 2 cm from or occurs more than 3 years after the primary tumor in OSCC [20].

Clinicopathological characteristics including age, sex, drinking habits, smoking habits, location, disorder, presence of epithelial dysplasia, and development of cancer in the 70 cases included in the study are shown in Table 1. The median age of the patients included in the study was 64 years (range, 23–90 years) and the average age was 61.7 years at the time of treatment. Of the 70 patients, 50 had oral leukoplakia, and 20 had oral lichen planus. In addition, six cases of cancer developed after biopsy or surgical resection (8.6%).

Table 1. Clinical characteristics of 70 cases.

Characteristics		**Cases (%)**
Sex	Male	47 (67.1)
	Female	23 (32.9)
Age	65>	37 (52.9)
	65≤	33 (47.1)
Drinking	Yes	51 (72.9)
	No	19 (27.1)
Smoking	Yes	43 (61.4)
	No	27 (38.6)
Location	Tongue	24 (34.3)
	Others	46 (65.7)
Disorder	Leukoplakia	50 (71.4)
	OLP	20 (28.6)
Epithelial dysplasia	Positive	53 (75.7)
	Negative	17 (24.3)
Developed Cancer	Yes	6 (8.6)
	No	64 (91.4)

This study was conducted in accordance with the Declaration of Helsinki and approved by the Institutional Review Board of the University of Tsukuba Hospital (approval no. R040-04).

2.2. Immunohistochemistry

All tissue sections had been fixed in formalin, paraffin-embedded, and sectioned to a thickness of 4 μm. Human liver cancer tissue sections were used as positive controls for p62, p53, and ki67, and human endometrioid adenocarcinoma tissue sections were used for XPO1 as specified in the protocol. Staining without primary antibodies was used as the negative control.

After deparaffinization and rinsing, samples were pretreated with a high pH system antigen activation solution (pH 9.0, K8004, Dako, Tokyo, Japan) in a microwave oven at 95 °C for 20 min. Endogenous peroxidase activity was removed by treatment with a mixture of 0.3% hydrogen peroxide in 100% methanol for 10 min. After thorough rinsing, the samples were incubated with primary antibodies against p62 (1:2000; PM045; MBL, Tokyo, Japan), XPO1 (1:250; D6V7N, #46249; Cell Signaling Technology, Tokyo, Japan), p53 (1:300; M7001; Dako), and ki67 (1:200; ab16667; Abcam, Cambridge, UK) for 60 min at room temperature, and ki67 (1:200; ab16667; Abcam) primary antibodies at room temperature for 60 min. The samples were washed and treated with a secondary antibody (Histofine Simple Stain MAX-PO(MULTI), Nichirei Bioscience, Tokyo, Japan) for 30 min at room temperature and treated with 3,3′-diaminobenzidine tetrahydrochloride (DAB, Cell

Signaling Technology) for 5 min to detect antigen–antibody binding. Finally, hematoxylin was used to counterstain the nuclei.

2.3. Evaluation of p62, XPO1, p53, and ki67

The expression of p62, XPO1, p53, and ki67 in the mucosal epithelium of the stained sections was evaluated. Although both strong and weak expressions were observed, we judged the cells as either positive or negative. A staining intensity equivalent to that of the positive control was considered positive. Images of representative strong and weakly expressing cases of each marker are shown in Figure 1.

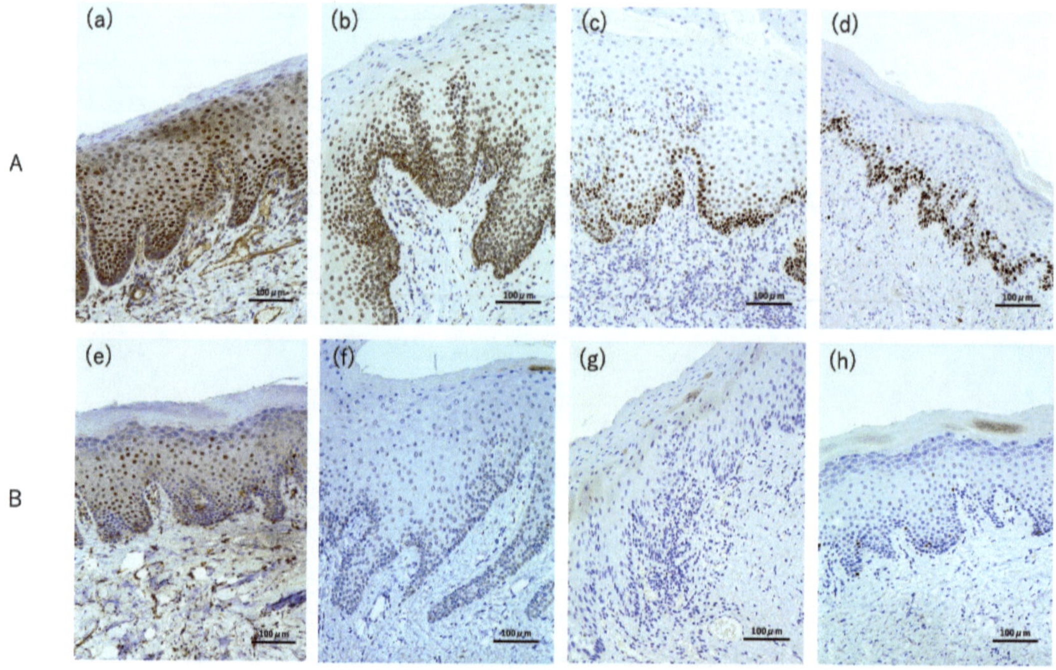

Figure 1. The result of p62, XPO1, p53, and ki67 immunohistochemical stains in OPMDs. (**A**) Representative sections showing strong expression of (**a**) p62, (**b**) XPO1, (**c**) p53, and (**d**) ki67. (**B**) Representative sections showing weak expression of (**e**) p62, (**f**) XPO1, (**g**) p53, and (**h**) ki67.

All immunostained markers were assessed using an optical microscope (BZ-X710; KEYENCE, Osaka, Japan). Regions showing representative staining in sections were selected at a low-magnification field of view (5×) and evaluated at a medium-magnification field of view (20× and 40×). Three oral surgeons (RT, FU, and ST) determined the number of all epithelial cells and the number of positive cells for each marker at random. The positive cell occupancy rate for each was calculated, considering each receiver operating characteristic.

Staining for p62 was evaluated separately for nucleus, cytoplasm, and aggregation. In accordance with our previous research, we defined a cell positive for p62 aggregation staining as one with at least one dot of accumulation in the cytoplasm in 20× and 40× field of view [18]. Representative images of p62 aggregation are shown in Figure 2. XPO1, p53, and ki67 were evaluated only for staining for nucleus.

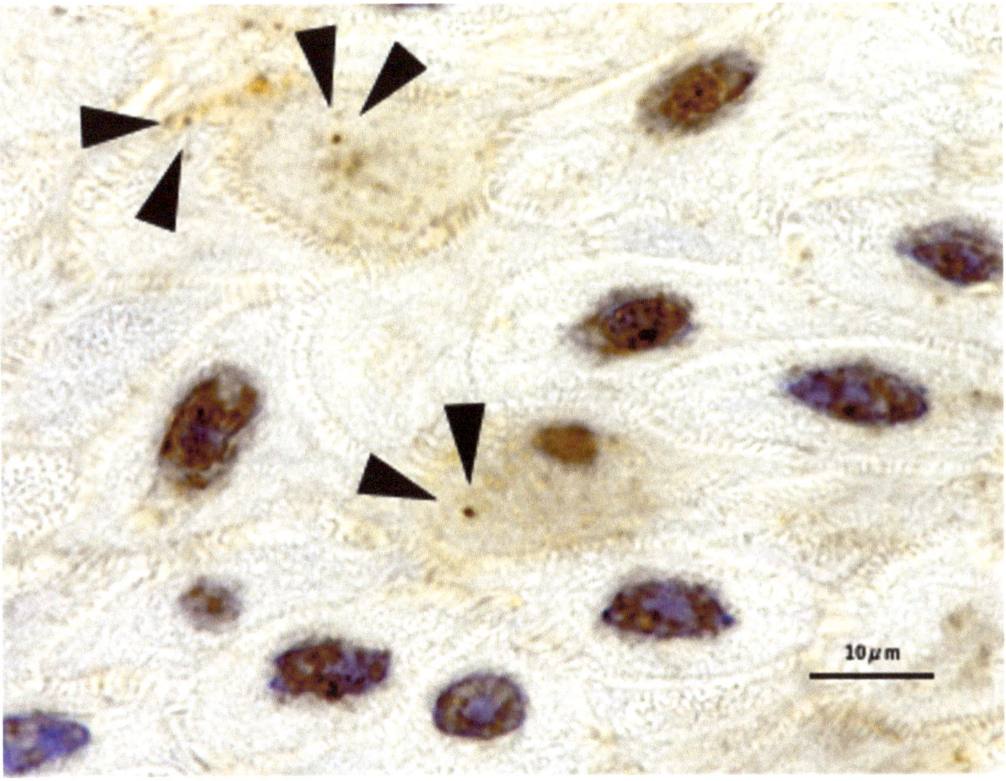

Figure 2. Representative example of p62 aggregation expression. p62 aggregation was defined as the presence of at least one dot indicating cytoplasmic accumulation. Arrows indicate representative p62 aggregation images.

2.4. Statistical Analysis

The correlation of protein expression in all OPMDs cases was evaluated using Spearman's correlation test and is presented as a scatter plot. In addition, univariate analysis was performed using Fisher's exact test with clinical characteristics and the Mann–Whitney U test for the association with each protein expression rate in cases of malignant transformation of OPMDs.

All p values less than 0.05 were considered statistically significant.

SPSS software ver. 28 (IBM Corp., Armonk, New York, NY, USA) was used for statistical analysis.

3. Results

3.1. Correlation of Each Protein Expression in OPMDs

The expression levels of p62, XPO1, p53, and ki67 in OPMDs, which were correlated using the Spearman's correlation test, are shown in Figure 3. The nuclear p62 expression was negatively correlated with ki67 expression (Figure 3a; $r = -0.321$; $p < 0.01$). The expression of p62 in the cytoplasm positively correlated with ki67 expression (Figure 3b; $r = 0.353$; $p < 0.01$) and XPO1 expression (Figure 3c; $r = 0.380$; $p < 0.01$). XPO1 expression positively correlated with ki67 expression (Figure 3d; $r = 0.258$; $p = 0.03$) and p53 expression (Figure 3e; $r = 0.262$; $p = 0.03$). The correlations for proteins other than these five were not significantly different.

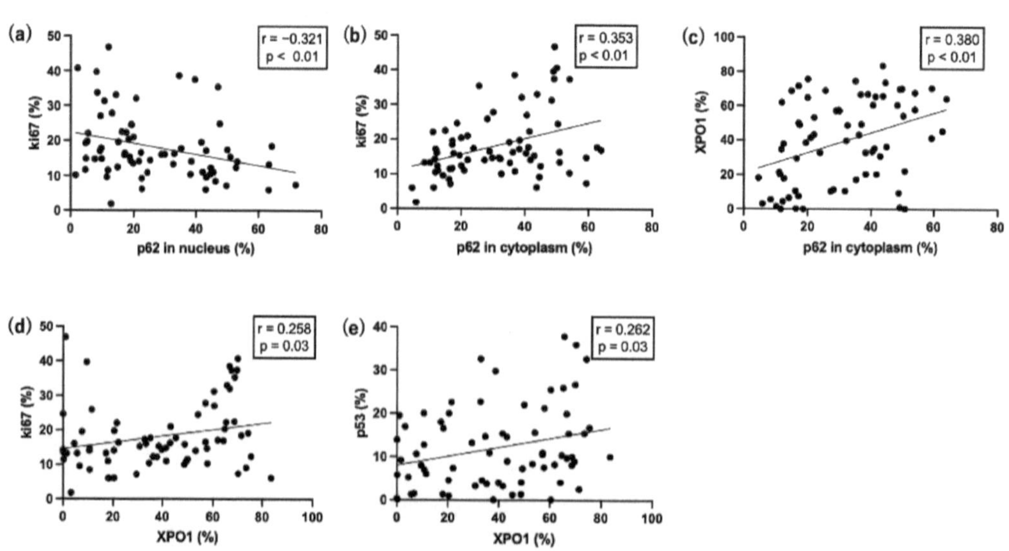

Figure 3. The correlation of each protein in OPMDs. In OPMDs, (**a**) p62 and ki67 were negatively correlated. (**b**) p62 in the cytoplasm and ki67, (**c**) p62 in the cytoplasm and XPO1; (**d**) XPO1 and ki67, and (**e**) XPO1 and p53 showed a positive correlation.

3.2. The Association between Malignant Transformation of OPMDs and Clinical Characteristics or the Expression of Each Protein

The results of univariate analysis of the association between the malignant transformation of OPMDs and the clinical characteristics or the expression rate of each protein are shown in Tables 2 and 3. No significant differences were found between malignant transformation and sex, age, drinking habits, smoking habits, or presence of epithelial dysplasia. Regarding protein expression, p62 in the nucleus ($p = 0.03$), p62 in cytoplasm ($p = 0.03$), and aggregation ($p < 0.03$) were associated with malignant transformation. Actually, the expression of p62 in the nucleus was significantly lower in the malignant transformation cases, whereas p62 expression in the cytoplasm and aggregation was significantly higher. In contrast, the expression levels of XPO1, p53, and ki67 were not associated with malignant transformation.

Table 2. Relationships between cancer development of OPMDs and clinical characteristics.

Characteristics		Developed Cancer Cases (%)	No Cancer Cases (%)	*p*
Sex	Male	3 (6.4)	44 (93.6)	0.39
	Female	3 (13.0)	20 (87.0)	
Age	65>	1 (2.7)	36 (97.3)	0.09
	65≤	5 (15.2)	28 (84.8)	
Drinking	Yes	4 (7.8)	47 (92.2)	0.66
	No	2 (10.5)	17 (89.5)	
Smoking	Yes	3 (7.0)	40 (93.0)	0.67
	No	3 (11.1)	24 (88.9)	

Table 2. *Cont.*

Characteristics		Developed Cancer Cases (%)	No Cancer Cases (%)	*p*
Location	Tongue	3 (12.5)	21 (87.5)	0.41
	Others	3 (7.0)	43 (91.5)	
Disorders	Leukoplakia	4 (8.0)	46 (92.0)	1.00
	OLP	2 (10.0)	18 (90.0)	
Epithelial dysplasia	Positive	2 (13.3)	13 (86.7)	0.60
	Negative	4 (7.3)	51 (92.7)	

Table 3. Relationship between cancer development of OPMDs and the expressions of each protein.

Parameter	Developed Cancer Expression (%)	No Cancer Expression (%)	*p*
p62 in nucleus	12.9 ± 5.9	29.7 ± 19.0	0.03 *
p62 in cytoplasm	45.4 ± 12.5	29.8 ± 15.5	0.03 *
p62 aggregation	6.5 ± 3.6	2.1 ± 1.9	<0.01 *
XPO1	41.6 ± 22.9	38.9 ± 25.0	0.78
p53	14.2 ± 8.5	11.9 ± 9.4	0.44
Ki67	20.3 ± 6.5	18.0 ± 9.7	0.20

* $p < 0.05$ significant.

4. Discussion

This study yields two major findings. First, in OPMDs, nuclear p62 expression was negatively correlated with cell proliferation, whereas cytoplasmic expression was positively correlated with cell proliferation. In addition, p62 cytoplasmic expression correlated with XPO1 expression, suggesting that p62 may be transported to the cytoplasm by XPO1. In the cases of OPMDs malignant transformation cases, the p62 expression was altered in the subcellular localization from the nucleus to the cytoplasm, and the p62 aggregation expression was also significantly higher, suggested that selective autophagy is abnormal.

Selective autophagy plays a role in maintaining cellular homeostasis but is known to play a dual role in malignant tumors, depending on the type, stage, and genetic context of the cancer. In the early stages of tumorigenesis, it prevents chronic tissue damage by maintaining the genomic stability and by preventing the accumulation of carcinogenic mutations and inhibits the accumulation of cancer inducers. This enabled it to function as a cancer suppressor; however, in the late stages of tumorigenesis, it strengthens cancer cell survival and stress resistance by preventing tumor cell damage and satisfying high metabolic cravings. This eventually promotes tumorigenesis and leads to the accumulation of therapeutic resistance, which contributes to the progression and metastasis of malignant tumors [21]. In other words, at some stages of tumorigenesis, there is an alteration in the function and role of selective autophagy.

p62 is an adaptor protein for selective autophagy, which is resolved by autophagy along with its target proteins; therefore, the accumulation of p62 is thought to signify autophagy impairment [22]. Although intracellular protein metabolism involves the ubiquitin–proteasome and autophagy systems, when autophagy is impaired, p62 binds non-selectively to ubiquitinated proteins and prevents their transport to the proteasome, leading to the accumulation of ubiquitinated proteins and p62 [23]. Komatsu et al. reported that ubiquitinated proteins and p62 accumulate in autophagy-deficient mice and ubiquitin–p62 aggregates are markedly formed [24]. In addition, p62 aggregates promote intracellular accumulation by reducing the rate of nuclear-cytoplasmic shuttling owing to the large size of the polymer [25]. Therefore, the dynamics of p62 may reflect various intracellular signals, such as selective autophagy and nucleocytoplasmic transport signaling.

Originally, p62 was a shuttle protein that continuously and rapidly moved between the nucleus and cytoplasm, and this movement was regulated by the phosphorylation and aggregation of p62 [25]. Various studies have investigated the subcellular localization of p62; however, its function in the nucleus remains unclear.

ki67 is a protein expressed in all cell cycles except quiescence (G0 phase) and is used as a cell proliferation marker [26]. ki67 is seldom detected in normal cells but is highly expressed in most malignant tumors, including OSCC [27,28]. In this study, we found that in OPMDs, the nuclear expression of p62 was negatively correlated with cell proliferation, whereas the cytoplasmic expression of p62 was positively correlated with cell proliferation. Thus, there may be some connection between p62 in cytoplasm and cell proliferation. It has been reported that p62 accumulates in the cytoplasm rather than the nuclear in various types of malignant tumors [29,30]. Malignant tumors are characterized by abnormal cell proliferation, and the results of this study are consistent with those of previous studies. In addition, Iwadate, et al. reported cytoplasmic p62 expression may be involved in tumor growth and tolerance to cellular stress [31], which is also consistent with our study. These results suggest that changes in the subcellular localization of p62 may be associated with cell differentiation and proliferation.

XPO1 binds to the nuclear export signal (NES) of its target protein, forms a complex, and is transported from the nucleus to the cytoplasm via the nuclear membrane [32]. XPO1 participates in the transport of approximately 220 proteins via the NES [33]. XPO1 is overexpressed in many types of malignancies, and tumor proteins such as YAP1, c-ABL, and SNAIL, and tumor suppressors such as p53, p27, and RB are transported by XPO1 [32]. Therefore, XPO1 is a potential therapeutic target for malignant tumors. A multidomain protein containing various types of protein–protein interdomains, p62 has an NES and moves from the nucleus to the cytoplasm [25]. Previous report has also shown that p62 migrates and accumulates in the nucleus when gastric and liver cancer cells are treated with KPT-8602, an XOP1 inhibitor [34]. Herein, the cytoplasmic expression of p62 correlated with XPO1 expression, indicating that p62 may be transported to the cytoplasm in association with XPO1. Therefore, it was predicted that XPO1 overexpression in malignant tumors causes p62 to move from the nucleus to the cytoplasm, resulting in a change in its subcellular localization.

In addition to autophagy, p62 is involved in the Keap1-Nrf2 pathway [11]. p62 has a Keap1-interaction region that binds to the Keap1 domain involved in its interaction with Nrf2. Thus, p62 overexpression inhibit the binding between Keap1 and Nrf2 and activated Nrf2. Activated Nrf2 translocates to the nucleus and induces the expression of several Nrf2 target genes, including p62. This activates a positive feedback mechanism for p62, and p62 accumulation progresses [35]. In addition, p62 is not metabolized in the presence of autophagy disorders and further accumulation occurs. A schematic representation of the relationship between p62, XPO1, and autophagy in OPMDs and a positive feedback mechanism of p62 is shown in Figure 4.

p53 acts as a tumor suppressor by inducing genes involved in cell cycle arrest, apoptosis, senescence, and repairing DNA [36]. p53 is the most frequently mutated protein in OSCC [37]. In addition, mutant p53 in patients are associated with decreased survival and resistance to radiotherapy and chemotherapy. Normally, p53 is degraded so rapidly that it cannot be detected by immunohistochemistry (IHC); however, mutant p53 has an extended half-life and can be detected by IHC [38]. XPO1 and ki67 are correlated with colorectal cancer. This suggests that the overexpression of XPO1 leads to uncontrolled cell division and tumor growth through subcellular localization changes in cell cycle inhibitory proteins, such as p21 and p53, and apoptotic proteins [39]. In this study, the correlations between XPO1 and p53 and between XPO1 and ki67 were shown in OPMDs, which correspond to this report.

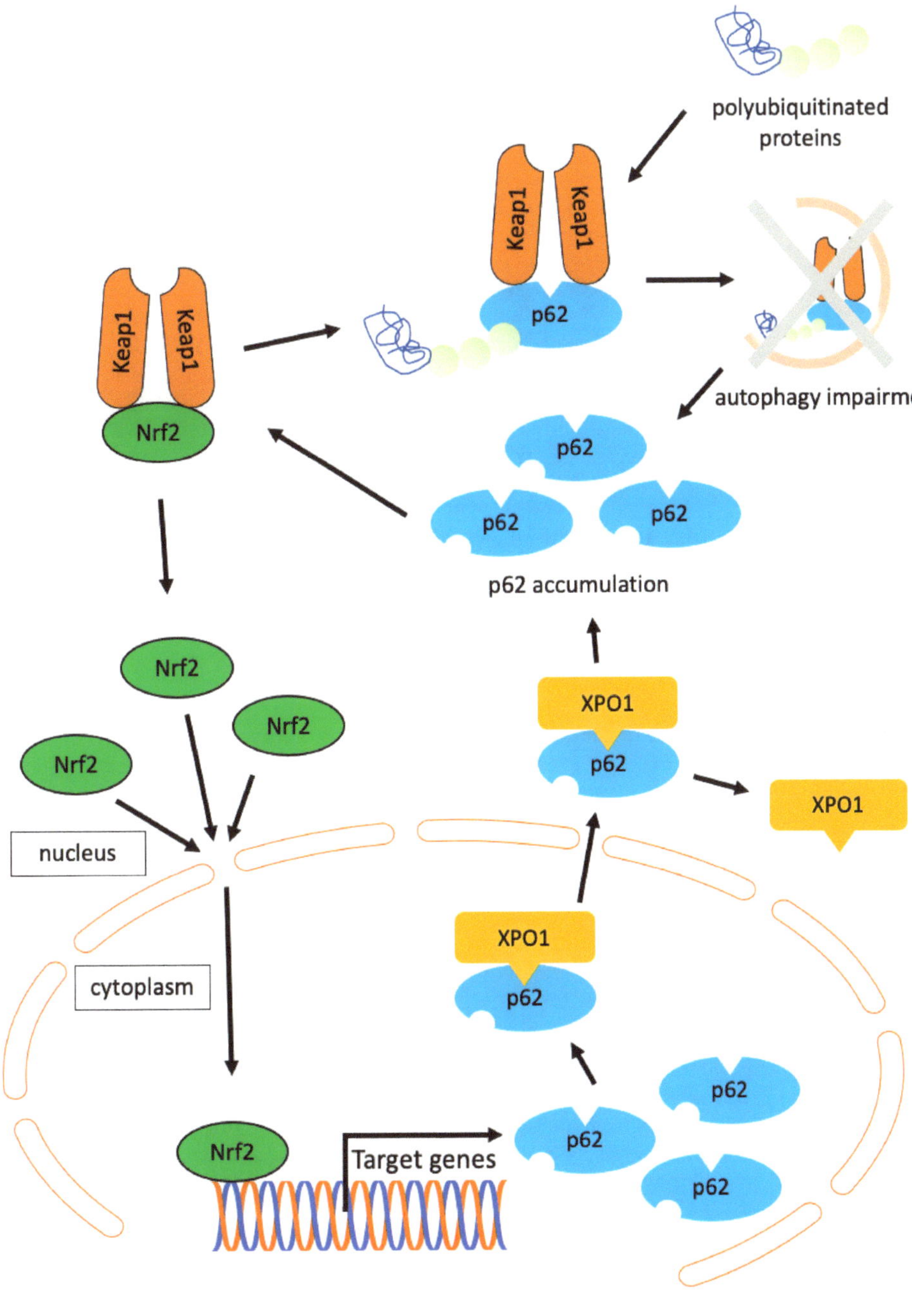

Figure 4. Diagram of the relationship between p62, XPO1, and autophagy in OPMDs as predicted by the results herein. The Keap1–Nrf2 complex separates in the presence of p62, Keap1 binds tightly to p62, and the activated Nrf2 translocate to the nucleus. In the nucleus, Nrf2 induces the expression of several target genes, including p62. Nuclear p62 binds XPO1 and translocate to the cytoplasm. Normally, polyubiquitinated proteins are digested together with p62 by selective autophagy, and p62 is thought to accumulate due to impaired autophagy.

Several studies have reported changes in p62 subcellular localization in normal tissues and malignant tumors. Lin et al. reported no difference in p62 nuclear expression but significantly increased the p62 cytoplasmic expression in OSCC compared to normal oral mucosa. The combination of the high p62 cytoplasmic expression and the low p62 nuclear expression in OSCC is also associated with a lower overall survival and disease-specific survival [19]. Kitamura et al. found that the p62 nuclear expression decreased and the cytoplasmic expression increased in malignant prostate tissues compared to benign prostatic tissues, suggesting that p62 migrates from the nucleus to the cytoplasm following tumorigenic transformation [30]. Interestingly, in cases of cancer development in OPMDs, p62 showed a change in subcellular localization from the nucleus to the cytoplasm compared with non-cancer development cases, which is similar to the results of these studies. In addition, no significant difference in the XPO1 expression was observed in cases of malignant transformation of OPMDs, suggesting that the nucleocytoplasmic transport system was not abnormal. p62 aggregation was also significantly increased, suggesting impaired autophagy. In other words, these results suggest that in cases of cancer development in OPMDs, some abnormalities in autophagy, increased cytoplasmic expression of p62, and decreased nuclear expression of p62 occur prior to the actual malignant transformation.

This study has several limitations. First, in this study, 6 cases (8.6%) were malignantly converted, which is similar to the malignant conversion rate reported previously [2]. However, this study was conducted at a single institution, which may have limited the sample size and selection of cases, leading to bias. Second, there are no uniform standards for immunostaining. To implement these biomarkers, further studies should be conducted using uniform criteria to obtain quantitative data. Third, regarding the correlation of expression proteins in OPMDs, the correlation coefficients ranged between 0.25 and 0.38, suggesting that the correlation is weak. However, in reality, various factors are expected to be involved in protein expression, and the correlation may have decreased possibility; therefore, a complex clarification of protein expression is a future issue. Finally, further studies are needed to clarify the diverse functions of p62, as this study focused on autophagy. For example, as mentioned in Figure 4, the intercellular behavior of p62 also involves the Nrf2-Keap1 pathway, which is an interesting topic. However, this is the first study to examine the association between the expression of p62, an autophagy-related protein, and cancer development in OPMDs. In particular, the subcellular localization of p62 may serve as a biomarker for assessing the risk of cancer development in OPMDs. Further investigation is needed to elucidate the mechanism of the malignant transformation of OPMDs and identify new biomarkers.

5. Conclusions

In this study, we found that the properties of p62 in OPMDs with respect to cell proliferative capacity were reversed between the nuclear and cytoplasmic expression. In addition, in cases of malignant transformation of OPMDs, the subcellular localization of p62 changes from nuclear to cytoplasmic, and p62 aggregation increases, suggesting that selective autophagy is impaired. This indicates that changes in the subcellular localization of p62 may be a potential biomarker for assessing the risk of malignant transformation in OPMDs.

Author Contributions: Conceptualization, R.T. and T.Y.; methodology, F.U. and E.W.; formal analysis, R.I., S.F. and N.I.-K.; investigation, R.T., F.U. and S.T.; writing—original draft preparation, R.T.; writing—review and editing, F.U.; visualization, K.Y.; project administration, H.B.; supervision, T.Y. All authors have read and agreed to the published version of the manuscript.

Funding: This research was supported by JSPS KAKNHI Grant Number JP22H03258.

Institutional Review Board Statement: This study was conducted in accordance with the guidelines of the Declaration of Helsinki and approved by the Institutional Review Board of the University of Tsukuba Hospital (ethical approval number: R04-040).

Informed Consent Statement: Informed consent was obtained from all participants involved in the study.

Data Availability Statement: The data presented in this study are available upon request from the corresponding author. The data are not publicly available because of the lack of conflicts of interest.

Conflicts of Interest: The authors declare no conflict of interest.

References

1. Reibel, J.; Gale, N.; Hille, J.; Hunt, J.L.; Lingen, M.; Muller, S.; Sloan, P.; Tilakaratne, W.M.; Westra, W.H.; Williams, M.D. Oral potentially malignant disorders and oral epithelial dysplasia. In *WHO Classification of Head and Neck Tumors*, 4th ed.; El-Naggar, A.K., Chan, J.K.C., Grandis, J.R., Takata, T., Slootweg, P., Eds.; iaRC: Lyon, France, 2017; pp. 112–115.
2. Iocca, O.; Sollecito, T.P.; Alawi, F.; Weinstein, G.S.; Newman, J.G.; De Virgilio, A.; Di Maio, P.; Spriano, G.; Pardiñas López, S.; Shanti, R.M. Potentially malignant disorders of the oral cavity and oral dysplasia: A systematic review and meta-analysis of malignant transformation rate by subtype. *Head Neck* **2020**, *42*, 539–555. [CrossRef]
3. Chaturvedi, A.K.; Udaltsova, N.; Engels, E.A.; Katzel, J.A.; Yanik, E.L.; Katki, H.A.; Lingen, M.W.; Silverberg, M.J. Oral leukoplakia and risk of progression to oral cancer: A population-based cohort study. *J. Natl. Cancer Inst.* **2020**, *112*, 1047–1054. [CrossRef]
4. Gupta, S.; Jawanda, M.K.; Madhushankari, G.S. Current challenges and the diagnostic pitfalls in the grading of epithelial dysplasia in oral potentially malignant disorders: A review. *J. Oral Biol. Craniofac. Res.* **2020**, *10*, 788–799. [CrossRef]
5. Kumari, P.; Debta, P.; Dixit, A. Oral potentially malignant disorders: Etiology, pathogenesis, and transformation into oral cancer. *Front. Pharmacol.* **2022**, *13*, 825266. [CrossRef]
6. Dionne, K.R.; Warnakulasuriya, S.; Zain, R.B.; Cheong, S.C. Potentially malignant disorders of the oral cavity: Current practice and future directions in the clinic and laboratory. *Int. J. Cancer* **2015**, *136*, 503–515. [CrossRef] [PubMed]
7. Kuribayashi, Y.; Tsushima, F.; Sato, M.; Morita, K.; Omura, K. Recurrence patterns of oral leukoplakia after curative surgical resection: Important factors that predict the risk of recurrence and malignancy. *J. Oral Pathol. Med.* **2012**, *41*, 682–688. [CrossRef] [PubMed]
8. Faruk, M.O.; Ichimura, Y.; Komatsu, M. Selective autophagy. *Cancer Sci.* **2021**, *112*, 3972–3978. [CrossRef]
9. Mizushima, N.; Komatsu, M. Autophagy: Renovation of cells and tissues. *Cell* **2011**, *147*, 728–741. [CrossRef] [PubMed]
10. Lamark, T.; Svenning, S.; Johansen, T. Regulation of selective autophagy: The p62/SQSTM1 paradigm. *Essays Biochem.* **2017**, *61*, 609–624. [CrossRef] [PubMed]
11. Sánchez-Martín, P.; Saito, T.; Komatsu, M. p62/SQSTM1: 'Jack of all trades' in health and cancer. *FEBS J.* **2019**, *286*, 8–23. [CrossRef]
12. Cuyler, J.; Murthy, P.; Spada, N.G.; McGuire, T.F.; Lotze, M.T.; Xie, X.Q. Sequestsome-1/p62-targeted small molecules for pancreatic cancer therapy. *Drug Discov. Today* **2022**, *27*, 362–370. [CrossRef]
13. Klionsky, D.J.; Petroni, G.; Amaravadi, R.K.; Baehrecke, E.H.; Ballabio, A.; Boya, P.; Bravo-San Pedro, J.M.; Cadwell, K.; Cecconi, F.; Choi, A.M.K.; et al. Autophagy in major human diseases. *EMBO J.* **2021**, *40*, e108863. [CrossRef]
14. Yin, X.; Hu, L.; Feng, X.; Wang, H.; Zhang, C.; Wang, S. Simultaneous activation of impaired autophagy and the mammalian target of rapamycin pathway in oral squamous cell carcinoma. *J. Oral Pathol. Med.* **2019**, *48*, 705–711. [CrossRef] [PubMed]
15. Wang, X.; Wu, W.K.K.; Gao, J.; Li, Z.; Dong, B.; Lin, X.; Li, Y.; Gong, J.; Qi, C.; Peng, Z.; et al. Autophagy inhibition enhances PD-L1 expression in gastric cancer. *J. Exp. Clin. Cancer Res.* **2019**, *38*, 140. [CrossRef]
16. Thompson, H.G.; Harris, J.W.; Wold, B.J.; Lin, F.; Brody, J.P. p62 overexpression in breast tumors and regulation by prostate-derived Ets factor in breast cancer cells. *Oncogene* **2003**, *22*, 2322–2333. [CrossRef] [PubMed]
17. Terabe, T.; Uchida, F.; Nagai, H.; Omori, S.; Ishibashi-Kanno, N.; Hasegawa, S.; Yamagata, K.; Gosho, M.; Yanagawa, T.; Bukawa, H. Expression of autophagy-related markers at the surgical margin of oral squamous cell carcinoma correlates with poor prognosis and tumor recurrence. *Hum. Pathol.* **2018**, *73*, 156–163. [CrossRef] [PubMed]
18. Yoshida, T.; Terabe, T.; Nagai, H.; Uchida, F.; Hasegawa, S.; Nagao, T.; Miyabe, S.; Ishibashi-Kanno, N.; Yamagata, K.; Warabi, E.; et al. Association between p62 expression and clinicopathological characteristics in oral leukoplakia. *Clin. Exp. Dent. Res.* **2019**, *5*, 389–397. [CrossRef] [PubMed]
19. Liu, J.L.; Chen, F.F.; Lung, J.; Lo, C.H.; Lee, F.H.; Lu, Y.C.; Hung, C.H. Prognostic significance of p62/SQSTM1 subcellular localization and LC3B in oral squamous cell carcinoma. *Br. J. Cancer* **2014**, *111*, 944–954. [CrossRef]
20. Leemans, C.R.; Braakhuis, B.J.; Brakenhoff, R.H. The molecular biology of head and neck cancer. *Nat. Rev. Cancer* **2011**, *11*, 9–22. [CrossRef]
21. Li, X.; He, S.; Ma, B. Autophagy and autophagy-related proteins in cancer. *Mol. Cancer* **2020**, *19*, 12. [CrossRef]
22. Mathew, R.; Karp, C.M.; Beaudoin, B.; Vuong, N.; Chen, G.; Chen, H.Y.; Bray, K.; Reddy, A.; Bhanot, G.; Gelinas, C.; et al. Autophagy suppresses tumorigenesis through elimination of p62. *Cell* **2009**, *137*, 1062–1075. [CrossRef] [PubMed]
23. Wang, Y.; Le, W.D. Autophagy and Ubiquitin-Proteasome System. *Adv. Exp. Med. Biol.* **2019**, *1206*, 527–550. [CrossRef] [PubMed]
24. Komatsu, M.; Waguri, S.; Koike, M.; Sou, Y.S.; Ueno, T.; Hara, T.; Mizushima, N.; Iwata, J.; Ezaki, J.; Murata, S.; et al. Homeostatic levels of p62 control cytoplasmic inclusion body formation in autophagy-deficient mice. *Cell* **2007**, *131*, 1149–1163. [CrossRef] [PubMed]

25. Pankiv, S.; Lamark, T.; Bruun, J.A.; Øvervatn, A.; Bjørkøy, G.; Johansen, T. Nucleocytoplasmic shuttling of p62/SQSTM1 and its role in recruitment of nuclear polyubiquitinated proteins to promyelocytic leukemia bodies. *J. Biol. Chem.* **2010**, *285*, 5941–5953. [CrossRef]
26. Scholzen, T.; Gerdes, J. The Ki-67 protein: From the known and the unknown. *J. Cell Physiol.* **2000**, *182*, 311–322. [CrossRef]
27. Yang, C.; Zhang, J.; Ding, M.; Xu, K.; Li, L.; Mao, L.; Zheng, J. Ki67 targeted strategies for cancer therapy. *Clin. Transl. Oncol.* **2018**, *20*, 570–575. [CrossRef]
28. Torres-Rendon, A.; Roy, S.; Craig, G.T.; Speight, P.M. Expression of Mcm2, geminin and Ki67 in normal oral mucosa, oral epithelial dysplasias and their corresponding squamous-cell carcinomas. *Br. J. Cancer* **2009**, *100*, 1128–1134. [CrossRef]
29. Inoue, D.; Suzuki, T.; Mitsuishi, Y.; Miki, Y.; Suzuki, S.; Sugawara, S.; Watanabe, M.; Sakurada, A.; Endo, C.; Uruno, A.; et al. Accumulation of p62/SQSTM1 is associated with poor prognosis in patients with lung adenocarcinoma. *Cancer Sci.* **2012**, *103*, 760–766. [CrossRef]
30. Kitamura, H.; Torigoe, T.; Asanuma, H.; Hisasue, S.I.; Suzuki, K.; Tsukamoto, T.; Satoh, M.; Sato, N. Cytosolic overexpression of p62 sequestosome 1 in neoplastic prostate tissue. *Histopathology* **2006**, *48*, 157–161. [CrossRef]
31. Iwadate, R.; Inoue, J.; Tsuda, H.; Takano, M.; Furuya, K.; Hirasawa, A.; Inazawa, J. High expression of p62 protein is associated with poor prognosis and aggressive phenotypes in endometrial cancer. *Am. J. Pathol.* **2015**, *185*, 2523–2533. [CrossRef]
32. Azizian, N.G.; Li, Y. XPO1-dependent nuclear export as a target for cancer therapy. *J. Hematol. Oncol.* **2020**, *13*, 61. [CrossRef] [PubMed]
33. Fu, S.C.; Huang, H.C.; Horton, P.; Juan, H.F. ValidNESs: A database of validated leucine-rich nuclear export signals. *Nucleic Acids Res.* **2013**, *41*, D338–D343. [CrossRef] [PubMed]
34. Gruffaz, M.; Yuan, H.; Meng, W.; Liu, H.; Bae, S.; Kim, J.S.; Lu, C.; Huang, Y.; Gao, S.J. CRISPR-Cas9 Screening of Kaposi's Sarcoma-Associated Herpesvirus-Transformed Cells Identifies XPO1 as a Vulnerable Target of Cancer Cells. *mBio* **2019**, *10*, e00866-19. [CrossRef] [PubMed]
35. Jiang, T.; Harder, B.; Rojo de la Vega, M.; Wong, P.K.; Chapman, E.; Zhang, D.D. p62 links autophagy and Nrf2 signaling. *Free Radic. Biol. Med.* **2015**, *88*, 199–204. [CrossRef] [PubMed]
36. Duffy, M.J.; Synnott, N.C.; Crown, J. Mutant p53 as a target for cancer treatment. *Eur. J. Cancer* **2017**, *83*, 258–265. [CrossRef] [PubMed]
37. Lindemann, A.; Takahashi, H.; Patel, A.A.; Osman, A.A.; Myers, J.N. Targeting the DNA Damage Response in OSCC with TP53 Mutations. *J. Dent. Res.* **2018**, *97*, 635–644. [CrossRef]
38. Deshmukh, A.V.; Gupta, A.; Chaudhari, A.G.; Gangane, N.M. Correlation of p53 expression with Clinical Presentation and Prognosis of Oral Squamous Cell Carcinoma Patients: A Pilot Study. *Indian J. Otolaryngol. Head Neck Surg.* **2022**, *74*, 1836–1840. [CrossRef]
39. Aladhraei, M.; Kassem Al-Thobhani, A.; Poungvarin, N.; Suwannalert, P. Association of XPO1 Overexpression with NF-κB and Ki67 in Colorectal Cancer. *Asian Pac. J. Cancer Prev.* **2019**, *20*, 3747–3754. [CrossRef]

current issues in molecular biology

MDPI

Article

PRI-724 and IWP-O1 Wnt Signaling Pathway Inhibitors Modulate the Expression of Glycolytic Enzymes in Tongue Cancer Cell Lines

Robert Kleszcz *, Jarosław Paluszczak, Marta Belka and Violetta Krajka-Kuźniak

Chair and Department of Pharmaceutical Biochemistry, Poznan University of Medical Sciences, Rokietnicka 3, 60-806 Poznań, Poland; paluszcz@ump.edu.pl (J.P.); mbelka@ump.edu.pl (M.B.); vkrajka@ump.edu.pl (V.K.-K.)
* Correspondence: kleszcz@ump.edu.pl

Abstract: The dysregulation of energetic metabolism is one of the hallmarks of cancer cells. Indeed, the growth of head and neck squamous cell carcinoma (HNSCC) cells depends heavily on glycolytic activity, which can be considered a potential therapeutic target. Wnt signaling is one of the pathways that undergoes upregulation in HNSCC. Our previous studies have shown that Wnt signaling inhibitors—PRI-724 and IWP-O1—attenuate tongue SCC survival and reduce glucose uptake and lactate release. The aim of this research was to further evaluate the possible mechanisms of the previously observed effects. We assessed the effect of PRI-724 and IWP-O1 on the expression of selected glycolytic enzymes: phosphofructokinase M, pyruvate kinase M2, and lactate dehydrogenase. Relative transcript expression was assessed by real-time PCR, and protein levels by Western blot. Moreover, clinical data concerning mRNA and protein expression, gene promoter methylation, and HNSCC patients' survival time were analyzed by the UALCAN tool, and protein–protein interaction was assessed using the STRING database. Experimental and bioinformatic data confirmed the relation between Wnt signaling and glycolytic enzymes in tongue cancer cells and HNSCC clinical samples. Overall, the inhibition of glucose metabolism by Wnt signaling inhibitors is a promising mode of action against tongue cancer cells.

Keywords: Wnt signaling; the Warburg effect; aerobic glycolysis; head and neck cancer; tongue cancer; PRI-724; IWP-O1; β-catenin; pyruvate kinase; lactate dehydrogenase

Citation: Kleszcz, R.; Paluszczak, J.; Belka, M.; Krajka-Kuźniak, V. PRI-724 and IWP-O1 Wnt Signaling Pathway Inhibitors Modulate the Expression of Glycolytic Enzymes in Tongue Cancer Cell Lines. *Curr. Issues Mol. Biol.* **2023**, *45*, 9579–9592. https://doi.org/10.3390/cimb45120599

Academic Editors: Emma Adriana Ozon and Violeta Popovici

Received: 29 October 2023
Revised: 24 November 2023
Accepted: 27 November 2023
Published: 29 November 2023

1. Introduction

The development of cancer is a complex, multistep process. Tumor cells gain and present characteristic hallmarks. The foundations of cancer cells' characteristics include their self-sufficiency in growth signals with concomitant insensitivity to antigrowth signals, evasion of apoptosis with limitless proliferation, promotion of angiogenesis, local invasion, and systemic metastasis [1]. The first twenty years of the 21st century allowed significant progress in cancer biology. Currently, cancer cells are also described by genome instability and mutations co-existing with non-mutational epigenetic reprogramming, inflammatory microenvironments supporting the avoidance of the immune response, unlocked phenotypic plasticity, senescence of cells, abnormalities in residing microbiomes and neuronal signaling, and finally, dysregulated cellular metabolism [2,3].

Changes in energy metabolism in cancer cells are known as the Warburg effect or aerobic glycolysis. Over a hundred years since its discovery by Otto Warburg, the understanding of this hallmark of cancer has been improved and revised many times. The enhanced activity of the glycolytic pathway enables effective redirection of metabolites, including standard oxidative fates of pyruvate via entering the citric acid cycle, the production of lactate (and related extracellular acidification), or the use of glycolytic intermediates for the biosynthesis of molecules essential for cancer cell growth and division, e.g., via the pentose phosphate pathway. Beyond energy (ATP) synthesis, the reorganization of energy metabolism is engaged in the modulation of cancer cell signaling, e.g., by promoting the formation of reactive oxygen species (ROS), modifying epigenetic mechanisms of gene expression, and promoting the expression and activity of metabolism-related transcription factors, like c-Myc and Hif-1α [4,5]. Currently, experimental observations indicate that the Warburg effect is a consequence of cancer phenotype development, and an adaptation to the hypoxic microenvironment by tumor cells is crucial during cancer progression [5].

Head and neck squamous cell carcinoma (HNSCC) is one of the leading tumor types worldwide [6]. Its occurrence depends on both genetic and epigenetic abnormalities, and in a subset of cases, it is caused by human papillomavirus (HPV) infection. Importantly, HPV-negative HNSCC tumors present a worse prognosis and higher diversity in molecular changes, for instance, in the activity of pro-tumorigenic signaling pathways [7]. The significance of particular hallmarks of cancer among HNSCC differs, and not all are experimentally well-proven [8]. Regarding metabolic alterations, accelerated glycolysis was vital for HNSCC tumor progression and the avoidance of immune system responses [9]. Moreover, high glycolytic activity measured based on 18F-FDG positron-emission tomography was a negative prognostic factor for HNSCC patients [10]. Thus, metabolic pathways are now considered potential targets of HNSCC therapy [11,12].

Several molecular pathways were described to be connected with HNSCC development. Our previous research focusing on Wnt signaling and the Akt kinase revealed that Wnt pathway inhibitors—PRI-724 and IWP-O1—had minor to moderate influence on tongue cancer cell proliferation, cytotoxicity, and apoptosis, while Akt inhibitors led to stronger effects on cell survival. The combinations of Wnt signaling and Akt inhibitors partly improved the anticancer effects [13]. Akt kinase and Wnt pathways were shown to be engaged in controlling cellular energetics [14,15]. Still, their role in modulating glycolytic activity in HNSCC has not been comprehensively described. Previously, we showed that PRI-724 and IWP-O1 significantly reduced glucose uptake, and PRI-724 also reduced lactate release in tongue cancer cells. In contrast, Akt kinase inhibition did not show any substantial effects [13]. Similar advantageous effects were observed in experiments using combinations of Wnt signaling inhibitors (PRI-724 and IWP-O1) with direct glycolytic inhibitors (2-deoxyglucose and lonidamine), since individually applied glycolytic inhibitors had limited influence on glucose consumption and lactate release [16].

Based on our previous findings, this study aimed to gain more insight into the mechanism of the observed functional effects of Wnt inhibitors on glucose metabolism. For this purpose, we experimentally assessed the influence of PRI-724, IWP-O1, and Akt inhibitor on the expression of key glycolytic enzymes in CAL 27, SCC-25, and BICR 22 tongue squamous cell carcinoma cell lines. Additionally, bioinformatic analyses were performed in order to further validate the inter-relationship between the aberrations in Wnt signaling and glycolysis, and its significance for disease progression. This allowed the clinical context to be brought into the in vitro experimental results.

2. Materials and Methods

2.1. Cells and Culture Conditions

Commercially available tongue squamous cell carcinoma cell lines were used in the experiments: CAL 27 and SCC-25 cell lines (derived from primary tumors) were purchased from the American Type Culture Collection (ATCC, Manassas, VA, USA), while BICR 22 cells (derived from a tongue cancer lymph node metastasis) were purchased from the European Collection of Authenticated Cell Cultures (ECACC, Porton Down, Wiltshire, UK).

The CAL 27 and BICR 22 cells were grown in high-glucose DMEM medium (Biowest, Nuaillé, France), supplemented with 10% FBS (EURx, Gdańsk, Poland) and a 1% antibiotic solution (penicillin and streptomycin; Biowest, Nuaillé, France). SCC-25 cells were grown in a 1:1 mixture of DMEM medium with a F12 medium containing 1.2 g/L sodium bicarbonate, 2.5 mM L-glutamine, 15 mM HEPES, and 0.5 mM sodium pyruvate (Biowest, Nuaillé, France) supplemented with 10% FBS (EURx, Gdańsk, Poland), a 1% antibiotic solution (penicillin and streptomycin; Biowest, Nuaillé, France), and 400 ng/mL hydrocortisone

(Sigma-Aldrich, St. Louis, MO, USA). All cells were cultured under standard conditions (37 °C, 5% CO_2, 95% humidity) in an incubator (Memmert, Schwabach, Germany).

2.2. Inhibitors

Three small-molecule inhibitors were used in the experiments. PRI-724 (Selleck Chemicals, Pittsburgh, PA, USA) and IWP-O1 (Sigma-Aldrich, St. Louis, MO, USA) are Wnt signaling inhibitors, which target the interaction between β-catenin and CREB binding protein (CREBBP) or the activity of Porcupine (*O*-acyltransferase), respectively. Moreover, Akt inhibitor X (Sigma-Aldrich, St. Louis, MO, USA) was used to inhibit the Akt kinase. Stock solutions of the compounds were prepared in DMSO and stored in aliquots at −20 °C.

In all the experiments, IC25 concentrations of individual compounds were used, and combinations were composed of two compounds with equal potency (IC25 + IC25). The IC25 values were previously determined by the MTS viability assay [13], and amount to: PRI-724 = 2.6 μM (CAL 27 cells), 0.85 μM (SCC-25), and 3.0 μM (BICR 22 cells); IWP-O1 = 1.0 μM (CAL 27 cells), 10.0 μM (SCC-25 cells), and 3.0 μM (BICR 22 cells); Akt inhibitor = 5.0 μM (CAL 27 cells), 7.0 μM (SCC-25 cells), and 4.0 μM (BICR 22 cells).

2.3. Isolation of RNA, Reverse Transcription, and Quantitative Real-Time PCR

Total RNA was isolated from cells treated with the compounds for 48 h using the Universal RNA Purification Kit (EURx, Gdańsk, Poland), and samples were subjected to reverse transcription using the RevertAid First Strand cDNA Synthesis Kit (Thermo Scientific, Waltham, MA, USA), in accordance with the manufacturer's instructions.

For the qR-T PCR analyses, the SG qPCR Master Mix (EURx, Gdańsk, Poland) and LightCycler 96 (Roche, Basel, Switzerland) were used. The initial enzyme activation at 95 °C lasted 10 min, and was followed by 40 three-step cycles consisting of denaturation (95 °C for 15 s), primer annealing (56 °C for 30 s), elongation (72 °C for 30 s with fluorescence measurement), and subsequent melting curve analysis. The relative mRNA expression of phosphofructokinase M—*PFKM*, pyruvate kinase M2—*PKM2*, and lactate dehydrogenase A—*LDHA* was determined. The TATA-box-binding protein (*TBP*) expression was used to normalize the data, and the ΔΔCt method served for fold-change quantification. Three independent experiments were performed with three technical repeats for each sample during qR-T PCR. The sequences of the primers used in the research were previously published [17].

2.4. Isolation of Protein Extracts and Western Blot Assay

Total protein extracts were isolated from cells treated with the compounds for 48 h by lysis with Laemmli buffer, followed by immediate protein denaturation by heating at 96 °C for 15 min. Protein concentration was assessed using the Pierce BCA Protein Assay Kit (Thermo Scientific, Waltham, MA, USA), and the absorbance was read using an Infinite M200 multi-plate reader (Tecan, Grödig, Austria). Protein content was further analyzed by Western blot.

Lysates were separated onto 7.5%, 10%, or 12% SDS-PAGE slab gels. Proteins were transferred to the nitrocellulose Immobilon P membrane (Sigma-Aldrich, St. Louis, MO, USA). After blocking for 2 h with 10% skimmed milk, proteins were probed with mouse anti-PFKM, mouse anti-PKM, mouse anti-LDHA, and rabbit anti-β-actin primary antibodies (Santa Cruz Biotechnology, Dallas, TX, USA). The horseradish peroxidase HRP-conjugated anti-mouse IgG or anti-rabbit IgG secondary antibodies (Bosterbio, Pleasanton, CA, USA) were used in the staining reaction. Bands were visualized using the chemiluminescent HRP Substrate Clarity ECL Kit (BioRad Laboratories, Hercules, CA, USA). The amount of immunoreactive products in each lane was determined using the ChemiDoc Imaging System (BioRad Laboratories, Hercules, CA, USA). Values were calculated as relative absorbance units (RQ) per mg of protein and expressed as a percentage of the control.

2.5. Bioinformatic Analysis of Gene Expression in Clinical Samples

Data available from the Cancer Genome Atlas (TCGA) were analyzed using the UALCAN tool (https://ualcan.path.uab.edu; accessed on 4 October 2023) [18,19] in order to evaluate the differences in the level of expression of *PFKM*, *PKM2*, and *LDHA* between healthy controls and HNSCC patients, and also between HPV-positive and HPV-negative HNSCC cases. Moreover, the changes in the promoter methylation level of these genes were assessed. Additionally, the association between glycolysis-related gene expression and HNSCC patient survival was determined using Kaplan–Meier plots generated in the UALCAN tool.

Furthermore, the UALCAN tool was used to analyze the changes in the level of expression of genes encoding the components of the Wnt pathway, which constitute molecular targets of the small-molecule inhibitors used in this study, and Wnt pathway-dependent transcription factors.

2.6. Bioinformatic Analysis of Protein Level in Clinical Samples

Protein level data available from the National Cancer Institute's Clinical Proteomic Tumor Analysis Consortium (CPTAC) were analyzed using the UALCAN tool (https://ualcan.path.uab.edu; accessed on 4 October 2023) [18,19] to assess the differences in the level of expression of the enzymes related to the glycolytic pathway between healthy controls and HNSCC patients.

2.7. STRING Analysis

The STRING database (version 12.0) from the STRING CONSORTIUM 2023 (https://string-db.org; accessed on 5 October 2023) [20] was used to evaluate protein–protein interactions in order to show molecular inter-relationships between the Akt kinase and the Wnt/β-catenin pathway, as well as β-catenin and glycolysis.

2.8. Statistical Analysis

All data in this study were analyzed using GraphPad Instat, version 3.10 (GraphPad Software, San Diego, CA, USA) and are presented as mean ± SD. The differences between experimental groups were evaluated by a Student's *t*-test (two groups) and one-way ANOVA test with the Tukey post hoc test (multiple groups). $p < 0.05$ was considered statistically significant.

3. Results

3.1. Higher Expression of PFKM and PKM2 Genes among HNSCC Patients Is a Negative Prognostic Marker of Survival Time

Phosphofructokinase is a key enzyme controlling the early steps of glycolysis, while pyruvate kinase regulates the last step of pyruvate synthesis. Moreover, cancer cells using aerobic glycolysis require lactate dehydrogenase to produce energy and acidify the microenvironment. Based on the TCGA data, we checked the transcript expression level of isoenzymes which are important for cancer cells, i.e., PFKM, *PKM2*, and *LDHA*, among HNSCC patients (Figure 1A). For all three genes, the transcript level was significantly higher in HNSCC samples compared to normal tissues.

Head and neck cancers can develop independently of or due to HPV infection, which affects therapy outcomes. A higher level of the expression of *PKM2* and *LDHA* genes was shown in HPV-negative samples, which are more challenging to treat (Figure 1B). Thus, the Warburg effect may be an important factor in HPV-negative HNSCC progression.

One of the epigenetic mechanisms influencing gene expression is the DNA methylation of gene promoter regions. Although the basal methylation level of the analyzed genes was low (0.065–0.140), it was further significantly lowered in primary HNSCC tumor samples (Figure 1C).

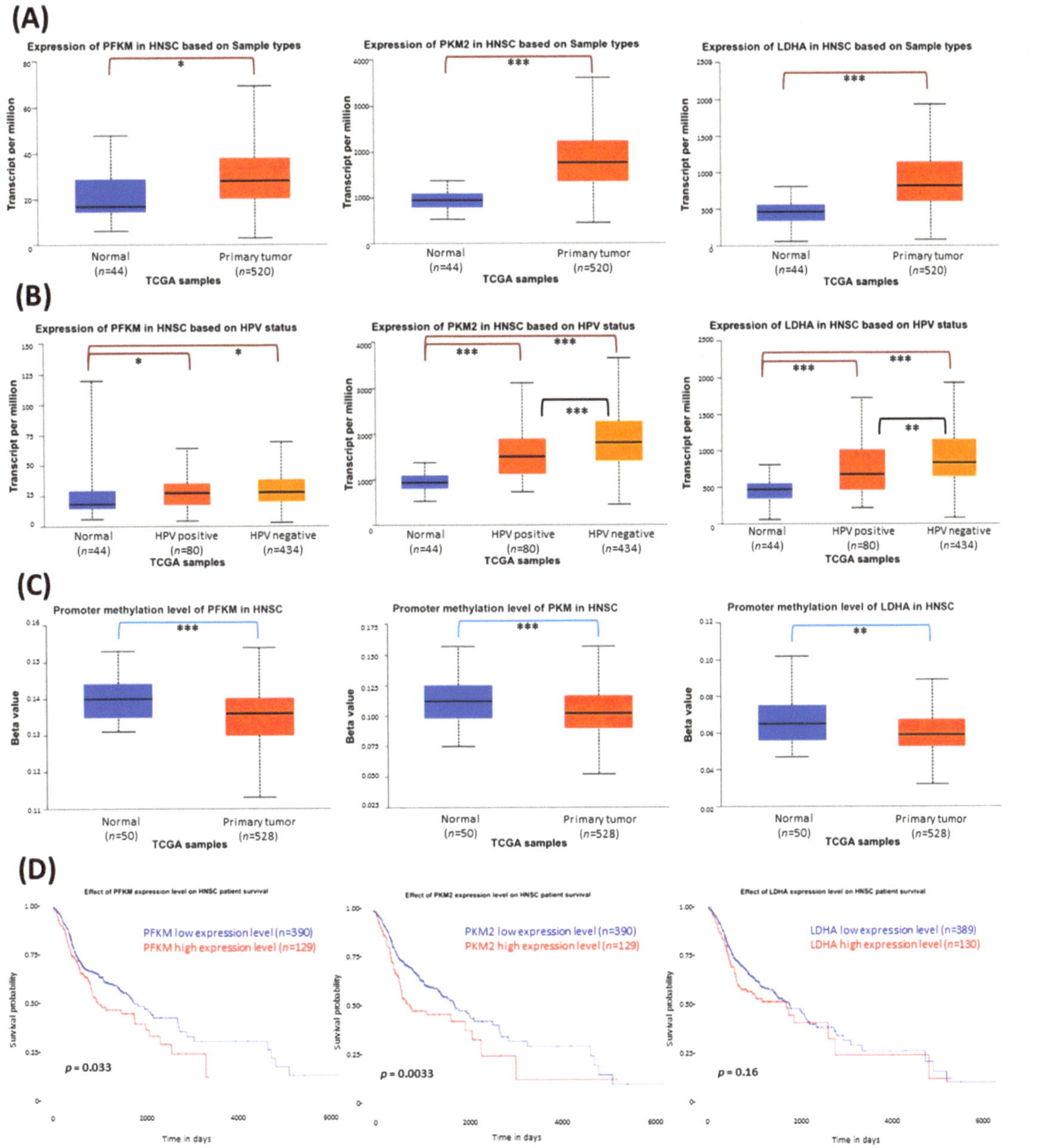

Figure 1. The results of the analysis of phosphofructokinase M (*PFKM*), pyruvate kinase M2 (*PKM2*), and lactate dehydrogenase A (*LDHA*) gene expression data available from the Cancer Genome Atlas using the UALCAN tool. (**A**) The differences in the level of gene expression between normal tissue and HNSCC samples. (**B**) The differences in the gene expression level depending on HPV status. (**C**) The differences in gene promoter methylation level between normal tissue and HNSCC samples. The Beta value indicates the level of DNA methylation ranging from 0 (unmethylated) to 1 (fully methylated). (**D**) Kaplan–Meier plots showing the association between high and low expression levels of glycolytic-related genes and survival time of HNSCC patients. The asterisk (*) denotes statistically significant changes, * $p < 0.05$, ** $p < 0.01$, *** $p < 0.001$.

Finally, Figure 1D presents Kaplan–Meier plots to compare the influence of high or low expression levels of particular glycolytic-related genes on the survival of HNSCC patients. The increased expression of *PFKM* ($p = 0.033$) and *PKM2* ($p = 0.0033$) negatively affects patient survival time. The impact of *LDHA* expression level was not statistically significant ($p = 0.16$), but an unfavorable trend can be observed.

3.2. The Protein Level of PKM and LDHA Is Increased in HNSCC Patient-Derived Cells

In the next step, we analyzed whether the changes in transcript level translate into the protein expression level. Figure 2 presents data generated based on the National Cancer Institute's Clinical Proteomic Tumor Analysis Consortium database. The expression of PKM and LDHA in HNSCC protein samples was indeed increased. In turn, the protein level of PFKM was significantly lower compared to normal samples, although its transcript level was upregulated (Figure 1A).

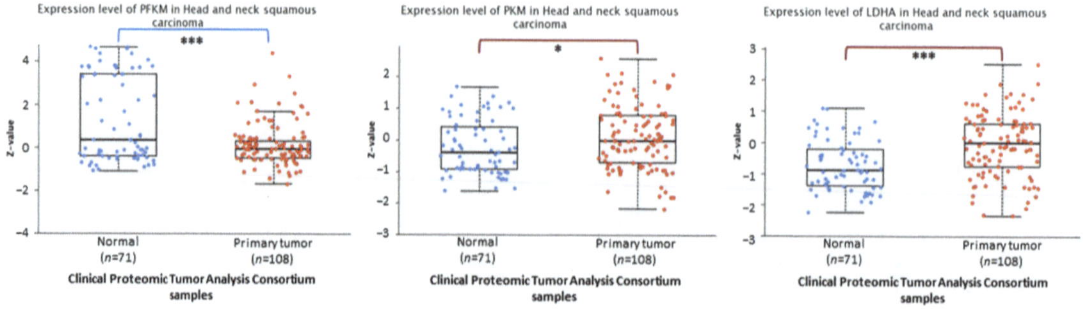

Figure 2. The results of the analysis of phosphofructokinase M (PFKM), pyruvate kinase M (PKM), and lactate dehydrogenase A (LDHA) protein expression data based on Clinical Proteomic Tumor Analysis Consortium (CPTAC) HNSCC samples using the UALCAN tool. Z-values represent standard deviations from the median across samples for the cancer type. Log2 Spectral count ratio values from CPTAC were first normalized within each sample profile and then normalized across samples. The asterisk (*) denotes statistically significant changes, * $p < 0.05$, *** $p < 0.001$.

3.3. Wnt Signaling Pathway Elements Are Functionally Related to Akt Kinase and Glycolytic Enzymes

Our previous research indicated the anticancer effects of Akt, Porcupine, β-catenin, and CREBBP inhibition in tongue squamous cell carcinoma cell lines [13]. Consistent with these results, the analysis of the TCGA data confirmed an increased level of the transcripts encoding these proteins (Figure 3A) in HNSCC clinical samples. Moreover, the higher mRNA levels of the four transcription factors (*TCF7*, *TCF7L1*, *TCF7L2*, *LEF1*) propagating the signals in the Wnt/β-catenin pathway (Figure 3B) also corroborate the pro-tumorigenic status of Wnt signaling in HNSCC.

In our previous studies, the inhibitors of the Wnt pathway (PRI-724 and IWP-O1) were able to modulate glucose utilization and lactate release in tongue cancer cells [13,16]. Moreover, the co-inhibition of Wnt signaling and the Akt kinase led to partly better effects than mono-treatment. Therefore, we wanted to evaluate the possible cross-talk between these signaling pathways (Wnt signaling and Akt kinase) and the glycolytic pathway by analyzing protein–protein interactions using the STRING database (Figure 3C). Many interactions were found for Akt kinase 1 and Wnt signaling elements: glycogen synthase kinase 3β (GSK3β), β-catenin (CTNNB1), and CREBBP. Thus, the simultaneous inhibition of the Akt kinase and Wnt pathway can potentially augment anticancer effects against HNSCC cells. In addition, CTNNB1 is directly related to PKM and LDHA, so it can potentially influence glycolytic flux. Thus, combining the Wnt pathway and Akt inhibitors could hypothetically enhance their suppressive effects against glycolytic enzyme expression.

Therefore, in the next step, we performed quantitative real-time PCR analyses of *PFKM*, *PKM2*, and *LDHA* gene expression in tongue cancer cells to validate these hypotheses.

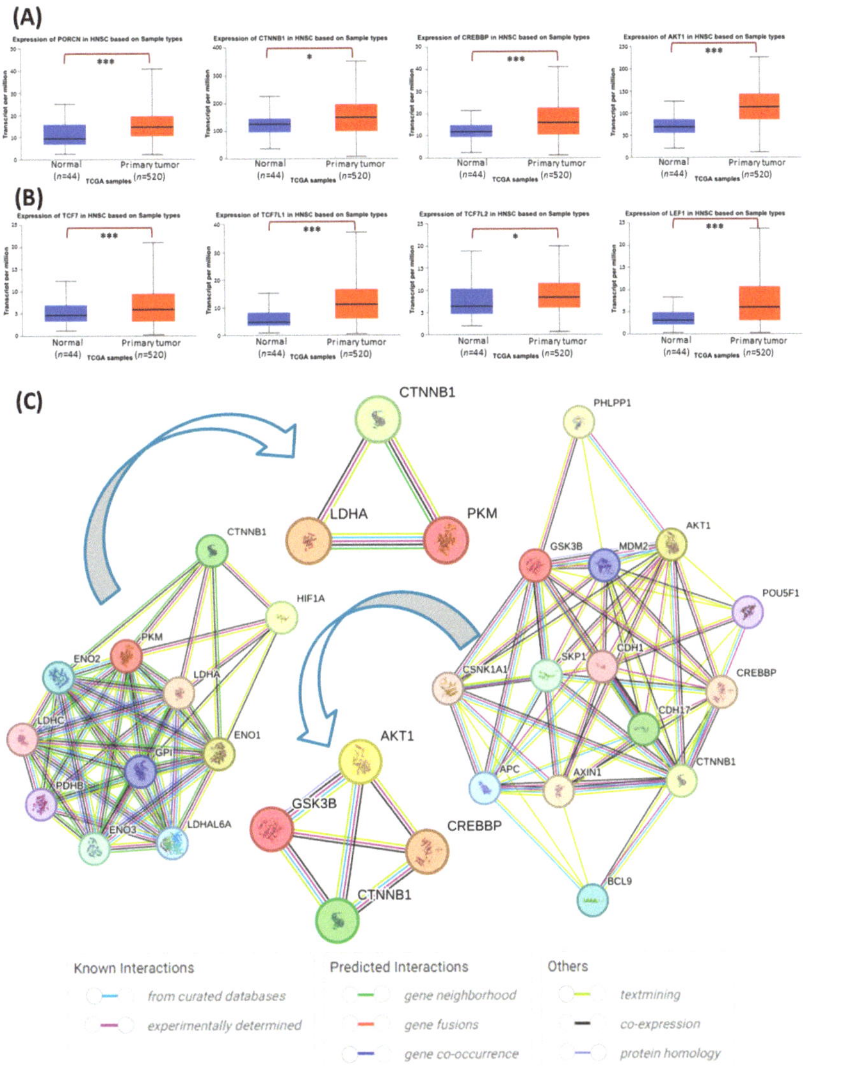

Figure 3. Analysis of Akt, Wnt signaling, and glycolytic pathway-related mRNA expression and functional protein–protein interactions. The results of the analysis based on data available from the Cancer Genome Atlas using the UALCAN tool showing the expression of: (**A**) genes being targets of small-molecule inhibitors used in the research, i.e., *PORCN*, *CTNNB1*, *CREBBP*, *AKT1*; (**B**) transcription factors responding to Wnt signaling activation, i.e., *TCF7*, *TCF7L1*, *TCFTL2*, *LEF1*. The asterisk (*) denotes statistically significant changes, * $p < 0.05$, *** $p < 0.001$. (**C**) The analyses of protein–protein interaction data available from the STRING database from the STRING CONSORTIUM 2023. The STRING graphs show the interaction of Wnt signaling elements with the Akt kinase and glycolysis proteins. Most important abbreviations: AKT1—Akt kinase 1, CREBBP—CREB binding protein, CTNNB1—β-catenin, GSK3B—glycogen synthase kinase 3β, LDHA—lactate dehydrogenase A, PKM—pyruvate kinase M, PORCN—porcupine.

3.4. Wnt Inhibitors Reduce the Transcript Level of PFKM, PKM2, and LDHA

The inhibition of Wnt signaling lowered the mRNA expression of *PFKM*, particularly by IWP-O1 in CAL 27 and BICR 22 cells and PRI-724 in SCC-25 cells (Figure 4A). Individually used Akt inhibitors had no influence. The addition of Akt inhibitors to active Wnt inhibitors presented no additional effect on the *PFKM* transcript level, except a mixture of PRI-724 and Akt inhibitors in CAL 27 cells.

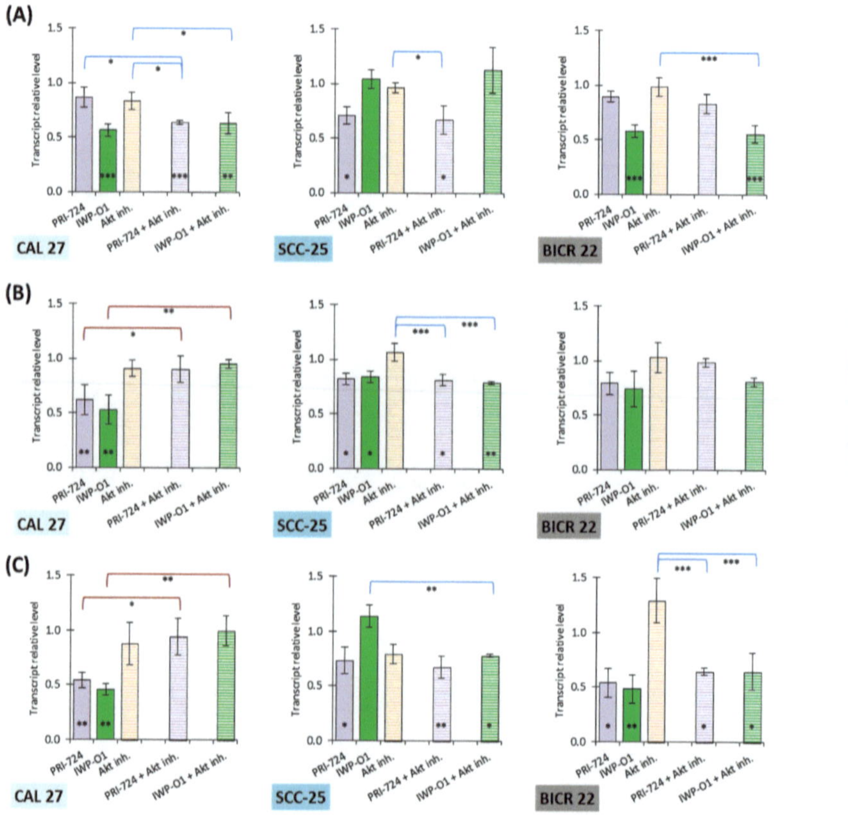

Figure 4. The effect of PRI-724, IWP-O1, and the Akt inhibitor on the relative transcript level of (**A**) phosphofructokinase M (*PFKM*), (**B**) pyruvate kinase M2 (*PKM2*), and (**C**) lactate dehydrogenase A (*LDHA*) in CAL 27, SCC-25, and BICR 22 cells. Mean values ± SD from three independent experiments with three replicates per qR-T PCR reaction are shown. The level of transcript in DMSO-treated cells was considered to be 1. The asterisk (*) inside the bar denotes statistically significant changes in comparison to the control, and above the bars denotes statistically significant changes in comparison to indicated experimental variants; * $p < 0.05$, ** $p < 0.01$, *** $p < 0.001$.

Both Wnt signaling inhibitors significantly reduced *PKM2* expression in CAL 27 and, to a lesser extent, SCC-25 cells (Figure 4B). Once again, the Akt inhibitor did not modulate the transcript level, and in combination, it eliminated the effects of Wnt pathway inhibitors (in CAL 27 cells) or had no influence (in SCC-25 cells). BICR 22 cells were resistant to all inhibitors.

Both Wnt signaling inhibitors importantly downregulated *LDHA* in CAL 27 and BICR 22 cells (Figure 4C). Moreover, PRI-724 slightly reduced the mRNA level in SCC-25 cells. The Akt inhibitor tended to change the expression of *LDHA*, but the results were insignificant. The combinations of PRI-724 or IWP-O1 with the Akt inhibitor showed no augmentation of effect or even worsened the effect (CAL 27 cells).

3.5. Wnt Signaling and Akt Kinase Inhibitors Strongly Downregulated LDHA Protein Expression

In the next step, we analyzed whether the changes in transcript expression levels elicited by the studied chemicals were followed by changes in the level of respective proteins (Figure 5). For individual compounds, the protein level of PFKM was significantly reduced only after the exposure of SCC-25 cells to the IWP-O1 inhibitor. On the contrary, the Akt inhibitor increased the expression of this enzyme in CAL 27 and BICR 22 cells. Interestingly, in CAL 27 cells, the combination of IWP-O1 and the Akt inhibitor reduced the PFKM protein level by ~32%.

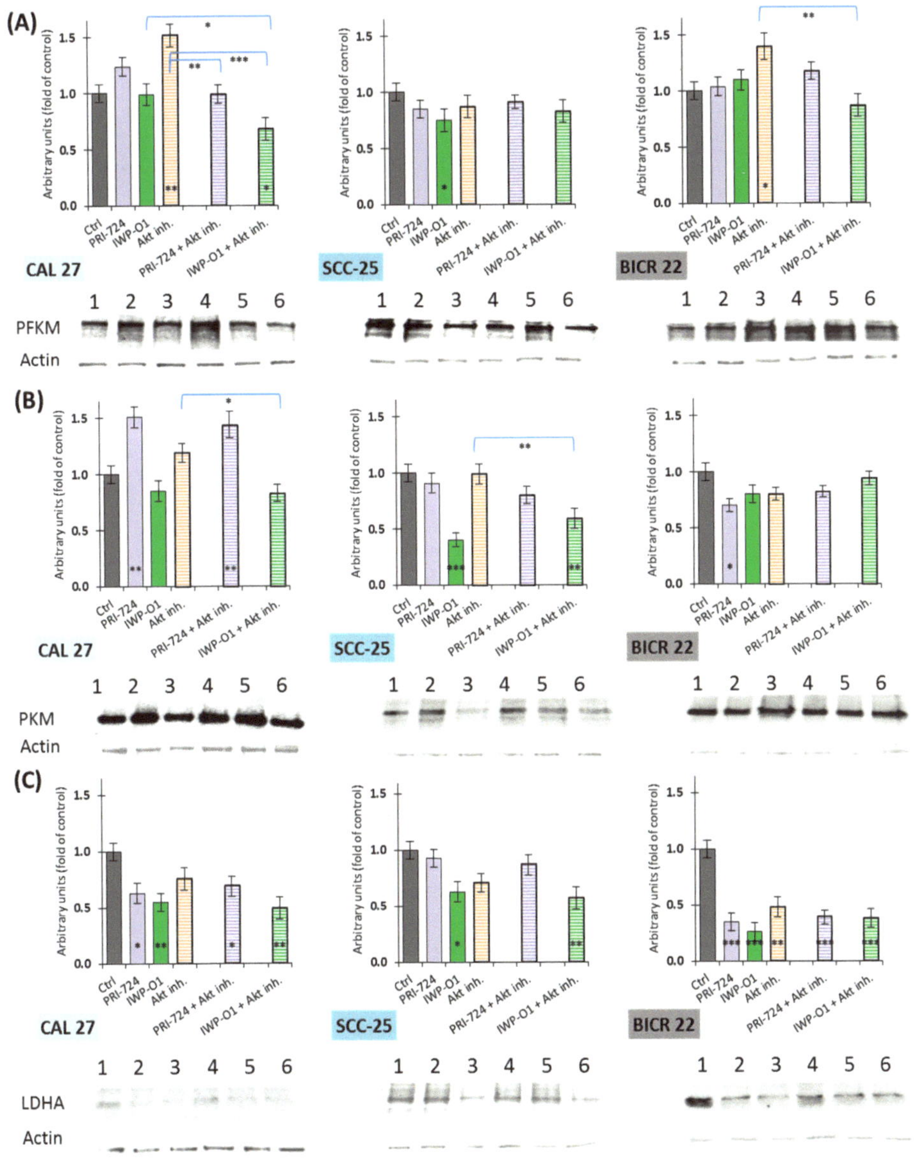

Figure 5. The effect of PRI-724, IWP-O1, and the Akt inhibitor on the relative protein level of (**A**) phosphofructokinase M (PFKM), (**B**) pyruvate kinase M2 (PKM2), and (**C**) lactate dehydrogenase

A (LDHA) in CAL 27, SCC-25, and BICR 22 cells. Mean values ± SD from two independent experiments are shown. The level of protein in DMSO-treated cells was considered to be 1. The asterisk (*) inside the bar denotes statistically significant changes in comparison to control, and above the bars denotes statistically significant changes in comparison to indicated experimental variants; * $p < 0.05$, ** $p < 0.01$, *** $p < 0.001$. Exemplary immunoblots are presented below the plots. 1—Ctrl, 2—PRI-724, 3—IWP-O1, 4—Akt inhibitor, 5—PRI-724 + Akt inhibitor, 6—IWP-O1 + Akt inhibitor.

Porcupine inhibitor IWP-O1 significantly reduced the expression of PKM in SCC-25 cells. In parallel, PRI-724 increased the PKM protein level in CAL 27 cells, was neutral in SCC-25 cells, and lowered its expression in BICR 22 cells. Co-treatment of cells with Wnt inhibitors and the Akt inhibitor did not show beneficial effects, while in SCC-25 cells, the expression level of PKM for IWP-O1 and the Akt inhibitor mixture was higher compared to individual IWP-O1 use. CAL 27 cells exposed to PRI-724 and the Akt inhibitor presented comparably high protein levels compared to PRI-724 alone.

Lactate dehydrogenase A was most susceptible to downregulation. In BICR 22 cells, each experimental variant significantly decreased the LDHA protein level. Furthermore, in CAL 27 cells, only the Akt inhibitor had an insignificant influence on LDHA expression. In turn, SCC-25 cells were affected by IWP-O1 and its combination with the Akt inhibitor. We did not detect notably better results of combinatorial treatments in comparison to exposure to single inhibitors in any of the cell lines.

4. Discussion

Changes in energy metabolism, known as the Warburg effect, are part of the characteristics of HNSCC cells. Indeed, glucose was identified as the dominant energy source for this type of tumor [21]. Thus, balancing glycolytic activity comprises a valid molecular target for HNSCC therapy. Previously, we described the anti-glycolytic effects of 2-deoxyglucose and Hif-1α and c-Myc inhibitors in hypopharyngeal FaDu cells [17]. In other studies, we detected a significant influence of PRI-724 and IWP-O1 Wnt signaling inhibitors on glucose utilization by tongue cancer cells [13,16]. Therefore, in this study, we further explored the association between the inhibition of Wnt signaling and/or the Akt kinase and the expression of crucial glycolysis-related genes in tongue cancer cells, in order to better recognize the interconnections between these processes.

Phosphofructokinase M, pyruvate kinase M2, and lactate dehydrogenase A were selected as pivotal isoenzymes in the glucose flux during aerobic glycolysis. Based on the TCGA database, we showed that all of them are overexpressed at the transcript level among HNSCC patients. Although the promoter methylation percentage was generally low, its value was slightly decreased in HNSCC tissues, which can favor gene expression. In addition, *PKM2* and *LDHA* genes present higher transcript copy numbers per million in HPV-negative cases, suggesting a greater metabolic disturbance in this group of difficult-to-treat tumors. Moreover, these molecular events have clinical importance, since high *PFKM* and *PKM2* mRNA expression correlates with worse prognosis, i.e., shorter survival of patients.

Activated Wnt signaling has been implicated in the enhancement of glycolytic activity in cancer cells, e.g., in nasopharyngeal carcinoma via the upregulation of pyruvate dehydrogenase kinase 1 [22] or by increasing glucose uptake in hepatocellular carcinoma [23]. The influence of Wnt1-inducible signaling protein 1 (WISP1) on glucose uptake, glycolytic flux, and lactate production was additionally observed in laryngeal squamous cell carcinoma [24]. Recently, Huang et al. (2023) pointed to DEP domain containing 1 (DEPDC1) as the glycolysis-related biomarker associated with oral squamous cell carcinoma progression. Such an effect is mediated by the activation of the Wnt/β-catenin signaling pathway [25].

Based on TCGA database analysis, we showed that *PORCN*, *CTNNB1*, and *CREBBP* genes are overexpressed in HNSCC, similarly to *AKT1* and four transcription factors acting downstream of the Wnt/β-catenin signaling pathway, which further corroborates the aberrant activation of Wnt signaling. Next, we performed the bioinformatic analysis of protein–protein interactions between glycolytic enzymes and Wnt pathway proteins. The STRING plots showed a three-sided relation between β-catenin, PKM, and LDHA, and the interaction between the Akt kinase and Wnt signaling elements, i.e., GSK-3β, β-catenin, and CREBBP. Indeed, the Akt/Wnt/β-catenin axis seems vital to the development of HNSCC [26–28]. Such data theoretically corroborate the possibility that Wnt inhibitors could decrease glycolytic activity in HNSCC cells, and that the combination of the Wnt pathway and Akt kinase inhibitors could improve the therapeutic effects of individual inhibitors.

We used two Wnt pathway inhibitors with distinct mechanisms of action to experimentally validate the influence of Wnt inhibition on the expression of glycolytic genes. PRI-724 blocks the interaction between β-catenin and CREBBP in the nucleus, while IWP-O1 targets porcupine responsible for the maturation of Wnt ligands. These small-molecule inhibitors of Wnt signaling were able to downregulate *PFKM*, *PKM2*, and *LDHA* gene expression, which is in line with the interaction between the Wnt pathway (β-catenin) and glycolytic genes. Conversely, individually used Akt inhibitors had no meaningful influence on transcript levels. Almost no additional reduction was observed in combinations with a Wnt signaling inhibitor, except for a mixture with PRI-724 in CAL 27 cells. Therefore, the assumption of the enhancement of anti-glycolytic effects by simultaneous targeting of Akt and Wnt signaling was not confirmed by our experimental results. We also performed protein level analysis of PFKM, PKM2, and LDHA. IWP-O1 showed a higher ability to decrease the levels of proteins of interest. Furthermore, in CAL 27 cells, the combination with Akt inhibitor enhanced its activity and reversed the direction of Akt inhibitor influence. However, the protein expression analysis did not confirm the activity of the PRI-724 inhibitor in most cases, and even the direction of effect was the opposite. On the other hand, the PRI-724 inhibitor of the β-catenin–CREBBP interaction was the most effective in reducing glucose uptake and lactate release in the same tongue cancer cell lines [13]. We suppose that additional molecular mechanisms, like posttranslational modifications, can orchestrate the final effect on glycolysis. For instance, the phosphorylation of Tyrosine 105 in PKM2 was described as a frequent molecular event in HNSCC cells correlated with high glycolytic activity [29].

Aberrant Wnt signaling modulates glycolysis, favoring lactate production, and other metabolic processes, including glutaminolysis and lipogenesis. By altering metabolism, which leads to extracellular fluid acidification, the Wnt pathway can actively regulate the tumor microenvironment, negatively modify the immune response, and promote tumor development [30]. Interestingly, β-catenin induces the expression of activating transcription factor 3 (ATF3) and inhibits the transcription of C-C motif chemokine ligand 4 (CCL4), which results in the impaired infiltration and activation of ATF3-related CD103+ dendritic cells, reduced CD8+ effector T cell priming and infiltration, and finally, a lack of proper responses to immune checkpoint blockade [31]. Argentiero et al. (2019) compared in silico RNA-seq data of 64 lymph node-positive and 79 lymph node-negative pancreatic ductal adenocarcinoma (PDAC) patient-derived samples [32]. Notably, the altered expression of a cluster covering the Wnt signaling pathway-related genes was associated with pleiotropic effects on immune response, epithelial modeling, and cytokine regulation in lymph node-positive PDAC cases. Lymph node-positive PDAC patients showed an increased number of activated dendritic cells and M2 macrophages with a concomitant decrease in effector T cells. Further experiments using PDAC lymph node-positive PANC-1 cells, PDAC lymph node-negative MIAPaCa-2 cells, and XAV-939 Wnt signaling inhibitor demonstrated, e.g., increased cytotoxic and antiproliferative effects of the Wnt inhibitor when PDAC cells were co-cultured with peripheral blood mononuclear cells. In addition, the inhibitor of Wnt signaling restored the expression of the macrophage stimulating 1 (MST-1) protein related to cancer suppression via the production of reactive oxygen species and inhibited angiopoietin-like 4 (ANGPTL4) expression related to angiogenesis, tumor migration, and regulation of immune homeostasis [32]. More information about Wnt pathway connections with immune cells, immunotherapy, and tumor microenvironments can be found, e.g., in [33–37]. The influence of Wnt signaling on metabolic and immune changes is an excellent target for future research on anti-cancer therapy.

The experiments described in this article confirmed that Wnt signaling inhibitors have a significant role in the attenuation of glycolysis in tongue squamous cancer cells. The inhibitory influence on PFKM, PKM2, and LDHA expression generally supports the metabolic effects observed in our previous work, in which we have shown the changes in glucose consumption, lactate release, and cellular ATP level [13]. Conversely, the inhibition of the Akt kinase may have a lower effect on glucose metabolism in tongue tumor cells than initially thought. This study was limited to assessing the expression of three crucial glycolytic genes. As the conclusions derived from our in vitro model are limited, future studies will thus need to include cell lines from different HNSCC localizations, patient-derived primary cells, and three-dimensional organoids in co-culture with other cell types present in the tumor microenvironment. Moreover, we have not evaluated the long-term effects of the studied inhibitors, or the potential resistance mechanisms that can affect the treatment outcome, and these factors will need to be addressed in the future to optimize the anti-cancer effects of the chemicals. Nevertheless, the observed correlation between Wnt signaling and energy metabolism suggests a more comprehensive effect of drugs targeting this pathway on the hallmarks of HNSCC cancer cells.

Author Contributions: Conceptualization, R.K.; methodology, R.K. and V.K.-K.; software, R.K. and V.K.-K.; validation, R.K. and V.K.-K.; formal analysis, R.K.; investigation, R.K., J.P., M.B. and V.K.-K.; resources, R.K., J.P. and V.K.-K.; data curation, R.K. and V.K.-K.; writing—original draft preparation, R.K. and V.K.-K.; writing—review and editing, J.P. and R.K.; visualization, R.K.; supervision, R.K.; project administration, R.K.; funding acquisition, R.K. and V.K.-K. All authors have read and agreed to the published version of the manuscript.

Funding: This research was funded by the Poznan University of Medical Sciences.

Institutional Review Board Statement: Not applicable.

Informed Consent Statement: Not applicable.

Data Availability Statement: Data are contained within the article.

Conflicts of Interest: The authors declare no conflict of interest.

References

1. Hanahan, D.; Weinberg, R.A. The hallmarks of cancer. *Cell* **2000**, *100*, 57–70. [CrossRef] [PubMed]
2. Senga, S.S.; Grose, R.P. Hallmarks of cancer—The new testament. *Open Biol.* **2021**, *11*, 200358. [CrossRef] [PubMed]
3. Hanahan, D. Hallmarks of Cancer: New Dimensions. *Cancer Discov.* **2022**, *12*, 31–46. [CrossRef] [PubMed]
4. Liberti, M.V.; Locasale, J.W. The Warburg Effect: How Does it Benefit Cancer Cells? *Trends Biochem. Sci.* **2016**, *41*, 211–218. [CrossRef]
5. Pascale, R.M.; Calvisi, D.F.; Simile, M.M.; Feo, C.F.; Feo, F. The Warburg Effect 97 Years after Its Discovery. *Cancers* **2020**, *12*, 2819. [CrossRef]
6. Sung, H.; Ferlay, J.; Siegel, R.L.; Laversanne, M.; Soerjomataram, I.; Jemal, A.; Bray, F. Global Cancer Statistics 2020: GLOBOCAN Estimates of Incidence and Mortality Worldwide for 36 Cancers in 185 Countries. *CA Cancer J. Clin.* **2021**, *71*, 209–249. [CrossRef]
7. Kleszcz, R. Advantages of the Combinatorial Molecular Targeted Therapy of Head and Neck Cancer—A Step before Anakoinosis-Based Personalized Treatment. *Cancers* **2023**, *15*, 4247. [CrossRef]
8. González-Moles, M.Á.; Warnakulasuriya, S.; López-Ansio, M.; Ramos-García, P. Hallmarks of Cancer Applied to Oral and Oropharyngeal Carcinogenesis: A Scoping Review of the Evidence Gaps Found in Published Systematic Reviews. *Cancers* **2022**, *14*, 3834. [CrossRef]
9. Takahashi, H.; Kawabata-Iwakawa, R.; Ida, S.; Mito, I.; Tada, H.; Chikamatsu, K. Upregulated glycolysis correlates with tumor progression and immune evasion in head and neck squamous cell carcinoma. *Sci. Rep.* **2021**, *11*, 17789. [CrossRef]
10. Rijo-Cedeño, J.; Mucientes, J.; Álvarez, O.; Royuela, A.; Seijas Marcos, S.; Romero, J.; García-Berrocal, J.R. Metabolic tumor volume and total lesion glycolysis as prognostic factors in head and neck cancer: Systematic review and meta-analysis. *Head Neck* **2020**, *42*, 3744–3754. [CrossRef]
11. Yamamoto, M.; Inohara, H.; Nakagawa, T. Targeting metabolic pathways for head and neck cancers therapeutics. *Cancer Metastasis Rev.* **2017**, *36*, 503–514. [CrossRef] [PubMed]
12. Lin, X.; Zhou, W.; Liu, Z.; Cao, W.; Lin, C. Targeting cellular metabolism in head and neck cancer precision medicine era: A promising strategy to overcome therapy resistance. *Oral Dis.* 2022, *online ahead of print*. [CrossRef]

13. Kleszcz, R.; Paluszczak, J. The combinatorial inhibition of Wnt signaling and Akt kinase is beneficial for reducing the survival and glycolytic activity of tongue cancer cells. *J. Oral Pathol. Med.* **2022**, *51*, 231–239. [CrossRef] [PubMed]
14. Hoxhaj, G.; Manning, B.D. The PI3K-AKT network at the interface of oncogenic signalling and cancer metabolism. *Nat. Rev. Cancer* **2020**, *20*, 74–88. [CrossRef] [PubMed]
15. Lee, M.; Chen, G.T.; Puttock, E.; Wang, K.; Edwards, R.A.; Waterman, M.L.; Lowengrub, J. Mathematical modeling links Wnt signaling to emergent patterns of metabolism in colon cancer. *Mol. Syst. Biol.* **2017**, *13*, 912. [CrossRef] [PubMed]
16. Kleszcz, R.; Paluszczak, J. The Wnt Signaling Pathway Inhibitors Improve the Therapeutic Activity of Glycolysis Modulators against Tongue Cancer Cells. *Int. J. Mol. Sci.* **2022**, *23*, 1248. [CrossRef] [PubMed]
17. Kleszcz, R.; Paluszczak, J.; Krajka-Kuźniak, V.; Baer-Dubowska, W. The inhibition of c-MYC transcription factor modulates the expression of glycolytic and glutaminolytic enzymes in FaDu hypopharyngeal carcinoma cells. *Adv. Clin. Exp. Med.* **2018**, *27*, 735–742. [CrossRef] [PubMed]
18. Chandrashekar, D.S.; Bashel, B.; Balasubramanya, S.A.H.; Creighton, C.J.; Ponce-Rodriguez, I.; Chakravarthi, B.V.S.K.; Varambally, S. UALCAN: A Portal for Facilitating Tumor Subgroup Gene Expression and Survival Analyses. *Neoplasia* **2017**, *19*, 649–658. [CrossRef]
19. Chandrashekar, D.S.; Karthikeyan, S.K.; Korla, P.K.; Patel, H.; Shovon, A.R.; Athar, M.; Netto, G.J.; Qin, Z.S.; Kumar, S.; Manne, U.; et al. UALCAN: An update to the integrated cancer data analysis platform. *Neoplasia* **2022**, *25*, 18–27. [CrossRef]
20. Szklarczyk, D.; Kirsch, R.; Koutrouli, M.; Nastou, K.; Mehryary, F.; Hachilif, R.; Gable, A.L.; Fang, T.; Doncheva, N.T.; Pyysalo, S.; et al. The STRING database in 2023: Protein-protein association networks and functional enrichment analyses for any sequenced genome of interest. *Nucleic Acids Res.* **2023**, *51*, D638–D646. [CrossRef]
21. Sandulache, V.C.; Ow, T.J.; Pickering, C.R.; Frederick, M.J.; Zhou, G.; Fokt, I.; Davis-Malesevich, M.; Priebe, W.; Myers, J.N. Glucose, not glutamine, is the dominant energy source required for proliferation and survival of head and neck squamous carcinoma cells. *Cancer* **2011**, *117*, 2926–2938. [CrossRef] [PubMed]
22. Cai, C.F.; Ye, G.D.; Shen, D.Y.; Zhang, W.; Chen, M.L.; Chen, X.X.; Han, D.X.; Mi, Y.J.; Luo, Q.C.; Cai, W.Y.; et al. Chibby suppresses aerobic glycolysis and proliferation of nasopharyngeal carcinoma via the Wnt/β-catenin-Lin28/let7-PDK1 cascade. *J. Exp. Clin. Cancer Res.* **2018**, *37*, 104. [CrossRef] [PubMed]
23. Xu, Z.; Shao, J.; Zheng, C.; Cai, J.; Li, B.; Peng, X.; Chen, L.; Liu, T. The E3 ubiquitin ligase RBCK1 promotes the invasion and metastasis of hepatocellular carcinoma by destroying the PPARγ/PGC1α complex. *Am. J. Cancer Res.* **2022**, *12*, 1372–1392. [PubMed]
24. Wang, L.; Sun, J.; Gao, P.; Su, K.; Wu, H.; Li, J.; Lou, W. Wnt1-inducible signaling protein 1 regulates laryngeal squamous cell carcinoma glycolysis and chemoresistance via the YAP1/TEAD1/GLUT1 pathway. *J. Cell. Physiol.* **2019**, *234*, 15941–15950. [CrossRef] [PubMed]
25. Huang, G.; Chen, S.; Washio, J.; Paka Lubamba, G.; Takahashi, N.; Li, C. Glycolysis-Related Gene Analyses Indicate That DEPDC1 Promotes the Malignant Progression of Oral Squamous Cell Carcinoma via the WNT/β-Catenin Signaling Pathway. *Int. J. Mol. Sci.* **2023**, *24*, 1992. [CrossRef] [PubMed]
26. Lee, K.; Chang, J.W.; Oh, C.; Liu, L.; Jung, S.N.; Won, H.R.; Kim, Y.I.; Rha, K.S.; Koo, B.S. HOXB5 acts as an oncogenic driver in head and neck squamous cell carcinoma via EGFR/Akt/Wnt/β-catenin signaling axis. *Eur. J. Surg. Oncol.* **2020**, *46*, 1066–1073. [CrossRef]
27. Zhao, R.; Wang, S.; Liu, J.; Xu, C.; Zhang, S.; Shao, Y.; Duan, X. KLK11 acts as a tumor-inhibitor in laryngeal squamous cell carcinoma through the inactivation of Akt/Wnt/β-catenin signaling. *J. Bioenerg. Biomembr.* **2021**, *53*, 85–96. [CrossRef]
28. Zhang, P.; Tian, Q.; Gao, H.; Zhao, A.; Shao, Y.; Yang, J. Inhibition of MAC30 exerts antitumor effects in nasopharyngeal carcinoma via affecting the Akt/GSK-3β/β-catenin pathway. *J. Biochem. Mol. Toxicol.* **2022**, *36*, e23061. [CrossRef]
29. Boschert, V.; Teusch, J.; Müller-Richter, U.D.A.; Brands, R.C.; Hartmann, S. PKM2 Modulation in Head and Neck Squamous Cell Carcinoma. *Int. J. Mol. Sci.* **2022**, *23*, 775. [CrossRef]
30. El-Sahli, S.; Xie, Y.; Wang, L.; Liu, S. Wnt Signaling in Cancer Metabolism and Immunity. *Cancers* **2019**, *11*, 904. [CrossRef]
31. Li, X.; Xiang, Y.; Li, F.; Yin, C.; Li, B.; Ke, X. WNT/β-Catenin Signaling Pathway Regulating T Cell-Inflammation in the Tumor Microenvironment. *Front. Immunol.* **2019**, *10*, 2293. [CrossRef] [PubMed]
32. Argentiero, A.; De Summa, S.; Di Fonte, R.; Iacobazzi, R.M.; Porcelli, L.; Da Vià, M.; Brunetti, O.; Azzariti, A.; Silvestris, N.; Solimando, A.G. Gene Expression Comparison between the Lymph Node-Positive and -Negative Reveals a Peculiar Immune Microenvironment Signature and a Theranostic Role for WNT Targeting in Pancreatic Ductal Adenocarcinoma: A Pilot Study. *Cancers* **2019**, *11*, 942. [CrossRef] [PubMed]
33. Haseeb, M.; Pirzada, R.H.; Ain, Q.U.; Choi, S. Wnt Signaling in the Regulation of Immune Cell and Cancer Therapeutics. *Cells* **2019**, *8*, 1380. [CrossRef] [PubMed]
34. Suryawanshi, A.; Hussein, M.S.; Prasad, P.D.; Manicassamy, S. Wnt Signaling Cascade in Dendritic Cells and Regulation of Anti-tumor Immunity. *Front. Immunol.* **2020**, *11*, 122. [CrossRef]
35. Goldsberry, W.N.; Meza-Perez, S.; Londoño, A.I.; Katre, A.A.; Mott, B.T.; Roane, B.M.; Goel, N.; Wall, J.A.; Cooper, S.J.; Norian, L.A.; et al. Inhibiting WNT Ligand Production for Improved Immune Recognition in the Ovarian Tumor Microenvironment. *Cancers* **2020**, *12*, 766. [CrossRef]

36. Zhou, Y.; Xu, J.; Luo, H.; Meng, X.; Chen, M.; Zhu, D. Wnt signaling pathway in cancer immunotherapy. *Cancer Lett.* **2022**, *525*, 84–96. [CrossRef]
37. Pundkar, C.; Antony, F.; Kang, X.; Mishra, A.; Babu, R.J.; Chen, P.; Li, F.; Suryawanshi, A. Targeting Wnt/β-catenin signaling using XAV939 nanoparticles in tumor microenvironment-conditioned macrophages promote immunogenicity. *Heliyon* **2023**, *9*, e16688. [CrossRef]

current issues in molecular biology

MDPI

Article

The Potential MicroRNA Diagnostic Biomarkers in Oral Squamous Cell Carcinoma of the Tongue

Young-Nam Park [1,†], Jae-Ki Ryu [2,†] and Yeongdon Ju [2,*]

[1] Department of Dental Hygiene, Gimcheon University, Gimcheon 39528, Republic of Korea; ivy9797@empas.com
[2] Department of Biomedical Laboratory Science, Gimcheon University, Gimcheon 39528, Republic of Korea; rs0429@daum.net
* Correspondence: lrdrlr@naver.com; Tel.: +82-54-420-4157
† These authors contributed equally to this work.

Abstract: Oral squamous cell carcinoma (OSCC) of the tongue is a common type of head and neck malignancy with a poor prognosis, underscoring the urgency for early detection. MicroRNAs (miRNAs) have remarkable stability and are easily measurable. Thus, miRNAs may be a promising biomarker candidate among biomarkers in cancer diagnosis. Biomarkers have the potential to facilitate personalized medicine approaches by guiding treatment decisions and optimizing therapy regimens for individual patients. Utilizing data from The Cancer Genome Atlas, we identified 13 differentially expressed upregulated miRNAs in OSCC of the tongue. Differentially expressed miRNAs were analyzed by enrichment analysis to reveal underlying biological processes, pathways, or functions. Furthermore, we identified miRNAs associated with the progression of OSCC of the tongue, utilizing receiver operating characteristic analysis to evaluate their potential as diagnostic biomarkers. A total of 13 upregulated miRNAs were identified as differentially expressed in OSCC of the tongue. Five of these miRNAs had high diagnostic power. In particular, miR-196b has the potential to serve as one of the most effective diagnostic biomarkers. Then, functional enrichment analysis for the target gene of miR-196b was performed, and a protein–protein interaction network was constructed. This study assessed an effective approach for identifying miRNAs as early diagnostic markers for OSCC of the tongue.

Keywords: oral squamous cell carcinoma; microRNA; diagnosis; biomarker

Citation: Park, Y.-N.; Ryu, J.-K.; Ju, Y. The Potential MicroRNA Diagnostic Biomarkers in Oral Squamous Cell Carcinoma of the Tongue. *Curr. Issues Mol. Biol.* **2024**, *46*, 6746–6756. https://doi.org/10.3390/cimb46070402

Academic Editors: Violeta Popovici and Emma Adriana Ozon

Received: 15 May 2024
Revised: 22 June 2024
Accepted: 25 June 2024
Published: 1 July 2024

1. Introduction

Oral squamous cell carcinoma (OSCC) of the tongue is the most common cancer in the squamous cells of the tongue and is marked by an insidious nature [1]. It represents a significant portion of head and neck squamous cell carcinoma (HNSCC) cases and accounts for approximately 31.9% of oral cavity cancers [2,3]. The main risk factors associated with the development of tongue cancer are tobacco smoking and excessive alcohol consumption. The early symptoms of OSCC are foreign body sensation or swallowing pain [4]. The treatment of tongue cancer is contingent upon the staging and anatomical site of the tumor, typically necessitating an approach encompassing surgery, radiation therapy, and chemotherapy. Persistent investigation into early detection modalities and precision therapies is imperative for enhancing the therapeutic outcomes for individuals afflicted with tongue cancer. OSCC is one of the most diagnosed malignancies and a prominent contributor to mortality attributed to head and neck cancers [5]. OSCC has a poor prognosis because of the lack of a strong barrier for preventing tumor propagation [6]. Anticipating the prognosis of individuals diagnosed with OSCC holds importance in devising treatment strategies. In the early stage of OSCC, which is marked by a favorable prognosis, cancer-related mortality affects approximately 19% of patients [7]. The early detection of cancer correlates with elevated survival rates and diminished healthcare expenditures among patients, owing to decreased dependence on aggressive therapeutic modalities [8]. Early detection and treatment can greatly improve the prognosis for patients with OSCC [9].

MicroRNAs (miRNAs) are short noncoding RNAs that play crucial roles in the regulation of gene expression and comprise approximately 22 nucleotides [10,11]. miRNAs operate by binding to messenger RNA molecules, thereby either inhibiting translation initiation or degrading mRNA [12]. miRNAs exhibit differential expression, being either upregulated or downregulated and are associated with the status and progression of tumors [13]. Several miRNAs have enhanced potential as biomarkers for diagnosing or monitoring specific cancers because they demonstrate cancer-specific expression patterns [14]. miRNA plays a crucial role in understanding the biological processes of tumors and developing treatment strategies.

The Cancer Genome Atlas (TCGA) resource serves as a platform for the diagnosis, prognosis, and immunotherapy of cancer, including the exploration of potential miRNA-based biomarkers [15–17]. Graphical representations such as heatmaps and volcano plots facilitate the visualization of miRNA expression patterns and distinguish differential expression profiles. The objective of our study was to identify certain miRNAs essential for diagnosing OSCC by employing a methodology based on TCGA data. This approach, which includes diagnostic tools, discriminates between malignant and adjacent nontumorous tissues with remarkable sensitivity and specificity. Among the myriad candidates scrutinized, five miRNAs emerged as promising contenders for OSCC diagnosis, emphasizing their potential as biomarkers to augment the clinical management of tongue cancer, particularly in the early stages. Our investigation prioritized the exploration of biomarkers aimed at discerning OSCC. Utilizing bioinformatics, we performed a comprehensive analysis of OSCC transcriptome data, aiming to elucidate the molecular mechanisms underlying OSCC development and to identify specific molecules critical for its progression.

2. Materials and Methods

2.1. Acquisition of TCGA Data

The miRNA expression profiles of OSCCs were downloaded from the TCGA data portal (http://firebrowse.org/, accessed on 3 May 2024), comprising 128 tumor samples and 13 adjacent nontumorous tissue samples. OSCC expression profiles were extracted from the TCGA HNSCC dataset, with the primary tumor site filtered to include only those of the tongue within the HNSCC dataset. Each sample comprised 1046 miRNA expression values obtained through the Illumina HiSeq platform. The miRNA expression profiles were generated using miRNA sequencing data, which represented a record of the reads per million miRNAs that were mapped. Clinical information regarding OSCC was sourced from the TCGA data portal. Table 1 presents comprehensive characteristics of the study, including genders, ages, pathologic stages, and tumor-node-metastasis classifications.

Table 1. Clinical data of patients with oral squamous cell carcinoma of the tongue from The Cancer Genome Atlas.

Characteristic	Overall
Gender, *n* (%)	
Male	82 (64.1)
Female	46 (35.9)
Age (years)	
Mean ± SD	58.17 ± 13.27
NA	1
Pathologic stage, *n* (%)	
Stage I	15 (11.72)
Stage II	22 (17.19)
Stage III	30 (23.44)
Stage IV	51 (39.84)
Not Available	10 (7.81)

Table 1. *Cont.*

Characteristic	Overall
Pathological, T, *n* (%)	
T1	22 (17.19)
T2	45 (35.16)
T3	34 (26.56)
T4	19 (14.84)
TX	6 (4.69)
Not Available	2 (1.56)
Pathological, N, *n* (%)	
N0	53 (41.41)
N1	18 (14.06)
N2	44 (34.38)
NX	11 (8.59)
Not Available	2 (1.56)
Pathological, M, *n* (%)	
M0	43 (33.59)
MX	13 (10.16)
Not Available	72 (56.25)

Data are expressed as either frequency with percentages or means ± standard deviations; T: tumor; N: node; M: metastasis.

2.2. Selection of Candidate Diagnostic miRNAs

Identifying miRNAs as candidate biomarkers provides an important approach for advancing cancer diagnosis. Due to their stability and specificity, miRNAs are excellent diagnostic tools. Their utilization can lead to the development of accurate and sensitive diagnostic assays. By identifying cancer-specific miRNAs, researchers can improve early detection and treatment strategies for individual cancer patients. Heatmap analysis and volcano plots serve as valuable tools in elucidating discernible patterns of miRNA expression [18,19]. The receiver operating characteristic (ROC) curves can provide evaluations of the accuracy of the predictive model selection of biomarkers [20]. The area under the curve (AUC) is a fundamental metric obtained from the ROC curve, offering a holistic evaluation of a diagnostic test's performance. It quantifies the ability of a diagnostic test to discriminate between disease cases and non-cases [21]. A higher AUC indicates better diagnostic performance [22]. In studies evaluating diagnostic value, an AUC exceeding 0.90 indicates excellent diagnostic performance of the test [23]. We selected differentially expressed miRNAs, which were analyzed using a heatmap and a volcano plot. Subsequently, ROC curves were generated, and the AUC was calculated with a 95% confidence interval (95% CI). ROC analysis identifies the optimal cutoff point, where sensitivity and specificity are maximized, for diagnostic biomarker values [24]. The point on the curve that maximizes the Youden Index is selected as the optimal threshold [24]. Diagnostic sensitivity and specificity were computed using GraphPad Prism software 6.0. miRNAs exhibiting an AUC > 0.9 were identified as potential diagnostic biomarkers.

2.3. Prediction of microRNA Targets

Predicting the target genes of miRNAs is crucial for regulating gene expression within cells and comprehending their involvement in the progress of cancers [25]. Studying miRNA–target interactions could reveal the biological function of certain miRNAs. We identified potential target mRNAs through the online miRNA prediction database miRDB (https://mirdb.org/, accessed on 8 May 2024). miRDB is a database for miRNA target prediction and functional annotation [26].

2.4. Functional Enrichment Analysis

Functional enrichment analysis is a bioinformatics computational method utilized for the interpretation of large-scale omics data, such as gene expression profiles or genomic datasets. Functional enrichment analysis assists researchers with uncovering the underlying biological mechanisms associated with observed experimental results. Gene ontology (GO)-based approaches utilize functional annotations to predict cancer driver genes [27]. The GO provides a hierarchical framework of terms organized into three main categories, including biological process, molecular function, and cellular component. Each term within GO represents a specific biological activity, contributing to a comprehensive understanding of gene functions and cellular organization. GO analysis was conducted using the DAVID online tool (https://david.ncifcrf.gov/, accessed on 10 May 2024). DAVID is a widely used bioinformatics resource system, comprising a web service for functional enrichment analysis and annotation [28]. The Kyoto Encyclopedia of Genes and Genomes (KEGG) is a comprehensive database and resource that links genomic information to higher-order functional information [29]. A $p < 0.05$ was considered to indicate statistical significance. KEGG provides a suite of tools dedicated to pathway enrichment analysis, facilitating the identify of biological pathways related to a set of genes or proteins.

2.5. Protein–Protein Interaction Network

Protein–protein interaction (PPI) networks facilitate the identification of diagnostic and prognostic biomarkers in cancer research by elucidating altered protein interactions within cancer cells [30]. PPI network exploration utilized the Search Tool for Retrieval of Interacting Genes/Proteins (STRING) database, a widely acknowledged resource for protein interaction data analysis (https://string-db.org/, accessed on 27 May 2024) [31]. Visualization of the network analysis was accomplished using Cytoscape software 3.9.1. CytoHubba was used to score node genes using the maximum clique centrality (MCC) algorithm [32].

2.6. Statistical Analyses

Data analysis and visualization were conducted using GraphPad Prism software 6.0 (GraphPad Software, San Diego, CA, USA). The independent samples t-test was used to determine statistical differences between groups. ROC curve analysis was employed, and the AUC was calculated to assess the diagnostic significance. A significance threshold of $p < 0.05$ was applied for statistical interpretation.

3. Results

3.1. Identification of Differentially Expressed miRNAs in OSCC Patients from TCGA Database

A differential expression analysis was performed to identify miRNA expression levels that exhibited significant differences between OSCC tissues and adjacent nontumorous tissues (Figure 1). TCGA data were utilized for this analysis, which contrasted miRNA expression patterns in OSCC patients with those in HNSCC patients. A total of 13 upregulated miRNAs were identified as differentially expressed through heatmap analysis, with visualization facilitated by MultiExperiment Viewer 4.9.0 (Figure S1). Differentially expressed miRNAs with a p-value less than 0.05 and significant log2 fold changes were identified and visualized on the volcano plot [33]. The volcano plot was created using GraphPad Prism 6.0. (Figure 2).

3.2. Evaluation of the Diagnostic Values of the Five Potential miRNAs

Among the thirteen upregulated miRNAs, the AUC values (with a 95% CI) of hsa-miR-196a-1, hsa-miR-196b, hsa-miR-450a-2, hsa-miR-503, and hsa-miR-877 were 0.9447 (0.8991–0.9903), 0.9838 (0.9661–1.001), 0.9303 (0.8588–1.002), 0.9549 (0.9153–0.9946), and 0.9285 (0.8791–0.9779), respectively (Figure S2). These AUC values indicate the discriminative ability of each miRNA in distinguishing between OSCC tissues and adjacent nontumorous tissues, with higher values indicating better diagnostic efficacy. Notably, miR-196b exhibited the highest diagnostic value among the miRNAs evaluated. Detailed diagnostic values for five differentially expressed miRNAs are presented in Table 2.

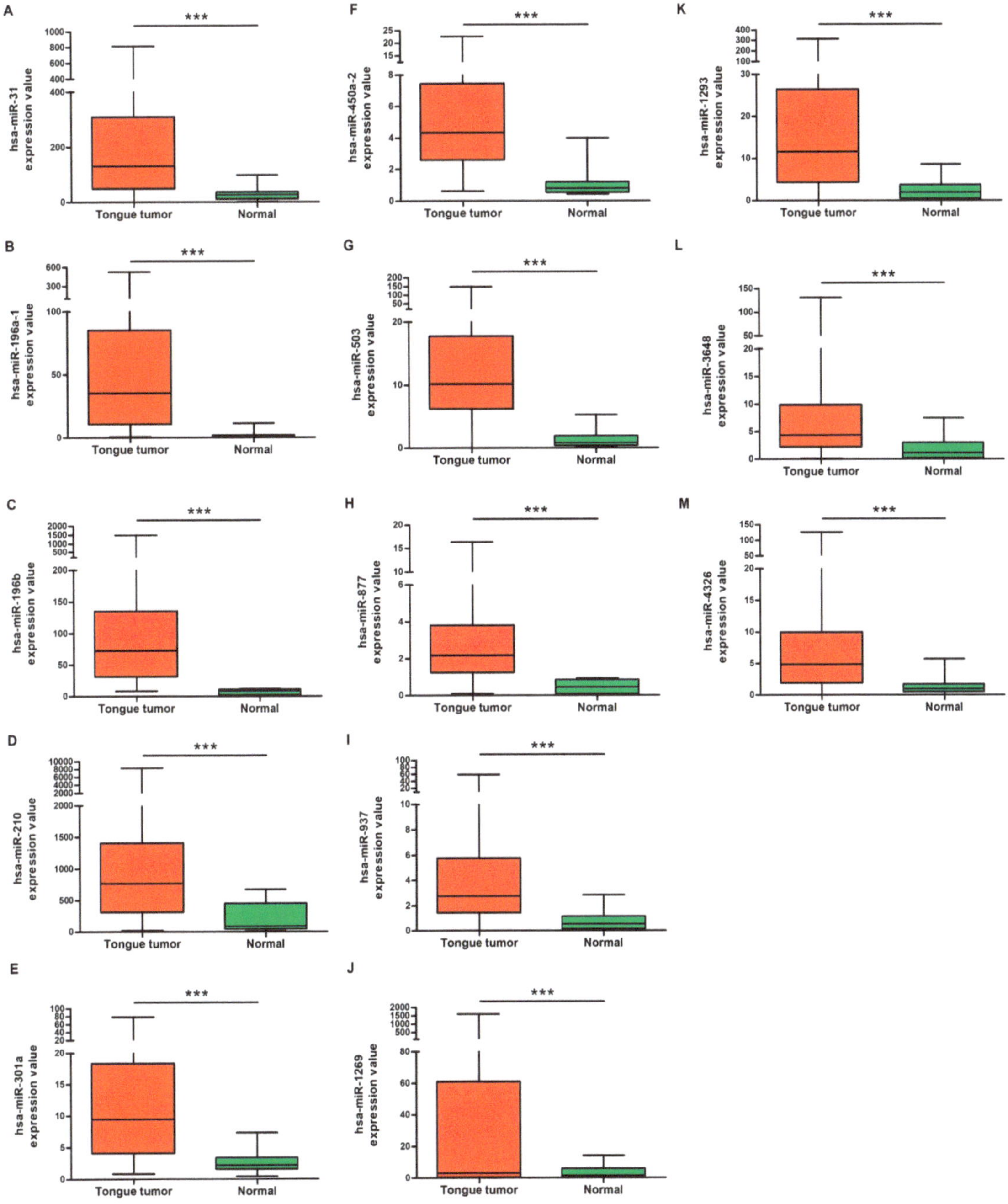

Figure 1. Relative expression levels of 13 differentially expressed miRNAs in oral squamous cell carcinoma of the tongue tissue and non-tumor tissue. (**A**) miR-31; (**B**) miR-196a-1; (**C**) miR-196b; (**D**) miR-210; (**E**) miR-301a; (**F**) miR-450a-2; (**G**) miR-503; (**H**) miR-877; (**I**) miR-937; (**J**) miR-1269; (**K**) miR-1293; (**L**) miR-3648; (**M**) miR-4326. *** $p < 0.001$.

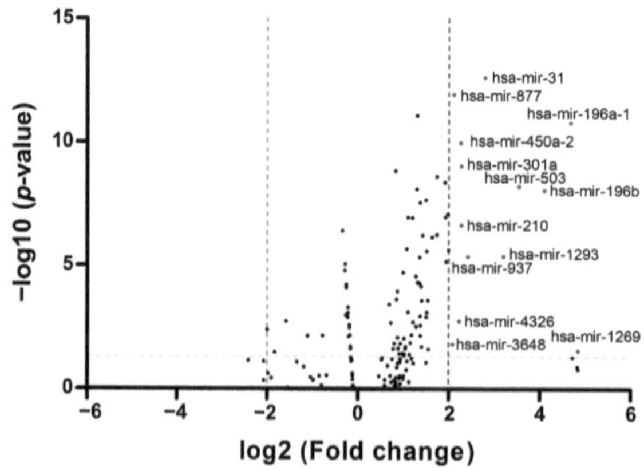

Figure 2. Volcano plot for the identification of differentially expressed miRNAs. A volcano plot was used to indicate the log2 fold change in 13 differentially expressed miRNAs. Differentially expressed miRNAs are ranked based on fold change and p-value. Adjustment for multiple testing was performed using Bonferroni correction, with a significance threshold set at $p < 0.01$.

Table 2. The areas under the receiver operating characteristic curve and diagnostic values of five differentially expressed miRNAs.

miRNAs	AUC	95% CI	Cutoff	Sensitivity (%)	Specificity (%)	p-Value
miR-196a-1	0.9447	0.8991–0.9903	<2.319	84.62	92.97	<0.0001
miR-196b	0.9838	0.9661–1.001	<11.47	92.31	95.31	<0.0001
miR-450a-2	0.9303	0.8588–1.002	<1.284	84.62	93.75	<0.0001
miR-503	0.9549	0.9153–0.9946	<2.26	84.62	93.75	<0.0001
miR-877	0.9285	0.8791–0.9779	<0.8552	84.62	83.59	<0.0001

3.3. Gene Ontology and KEGG Pathway Enrichment Analysis

We conducted a custom prediction search for the 369 target genes of miR-196b-5p and the 48 target genes of miR-196b-3p using miRDB. The results of the target genes for miR-196b-5p and miR-196b-3p are presented in Table S1. Furthermore, we utilized the DAVID database to identify potential biological functions of the target genes of miR-196b-5p and miR-196b-3p through GO enrichment analyses. The biological process group of the target genes of miR-196b-5p consisted of positive regulation of transcription from RNA polymerase II (GO:0045944), embryonic skeletal system morphogenesis (GO:0048704), and anterior/posterior pattern specification (GO:0009952). The biological process group of the target genes of miR-196b-3p included RNA 3′ uridylation (GO:0071076), positive regulation of 3′-UTR-mediated mRNA stabilization (GO:1905870), and a defense response to Gram-negative bacteria (GO:0050829). The cellular component group of the target genes of miR-196b-5p comprised nucleoplasm (GO:0005654), nucleus (GO:0005634), and cytoplasmic stress granule (GO:0010494). The cellular component group of the target genes of miR-196b-3p included cytoplasm (GO:0005737), cilium (GO:0005929), and cytosol (GO:0005829). The molecular function group of the target genes of miR-196b-5p included transcriptional activator activity, RNA polymerase II transcription (GO:0001228), DNA binding (GO:0003677), and chromatin binding (GO:0003682). The molecular function group of the target genes of miR-196b-5p consisted of calcium ion binding (GO:0005509) and lipopolysaccharide binding (GO:0001530). GO analysis of target genes is shown in Table S2. The KEGG pathway was mainly enriched in axon guidance (hsa04360), spinocerebellar ataxia (hsa05017), the MAPK signaling pathway (hsa04010), the Ras signaling pathway (hsa04014), the mRNA surveillance pathway (hsa03015), mitophagy (hsa04137), and protein digestion and absorption (hsa04974). The results of KEGG enrichment analysis are shown in Table S3.

3.4. Identification of the Hub Genes

The identification of hub genes in cancer is essential for elucidating the intricate molecular mechanisms that drive cancer development, progression, and responses to treatment. Hub genes serve as nodes within gene regulatory networks and signaling pathways implicated in cancer pathogenesis. The identification of hub genes plays a critical role in advancing our understanding of cancer biology. The STRING database was utilized to construct and analyze a PPI network, with the minimum correlation coefficient threshold set at 0.400. The top five hub genes of miR-196-5p in the PPI network were identified using the MCC algorithm implemented in CytoHubba. The identified hub genes include Homeobox A5 (HOXA5), Homeobox A7 (HOXA7), Homeobox B6 (HOXB6), Homeobox B7 (HOXB7), and PBX Homeobox 1 (PBX1) (Figure 3A). The top five hub genes of miR-196-3p were SMAD specific E3 ubiquitin protein ligase 1 (SMURF1), cullin 1 (CUL1), exportin 1 (XPO1), cytochrome P450 family 26 subfamily B member 1 (CYP26B1), and short-chain dehydrogenase/reductase family 9C member 7 (SDR9C7) (Figure 3B).

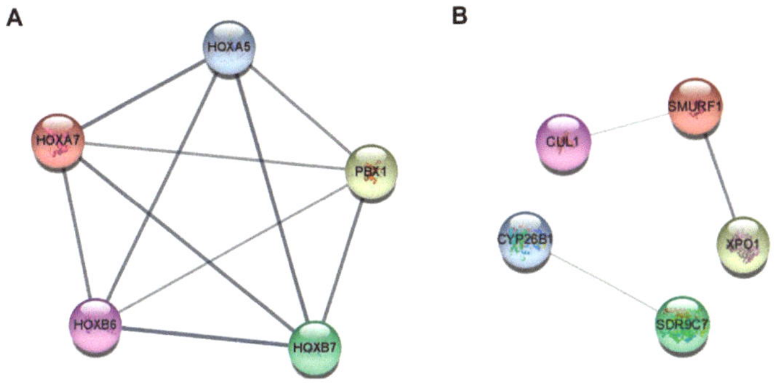

Figure 3. A protein–protein interaction network of the top five hub genes. (**A**) The top five hub genes of miR-196b-5p. (**B**) The top five hub genes of miR-196b-3p.

4. Discussion

Biomarkers represent a promising avenue for advancing personalized medicine and optimizing therapy tailored to individual patients, while concurrently facilitating the exploration of novel drug targets and the development of novel treatment strategies [34]. Diagnostic biomarkers play a crucial role in either detecting or confirming the presence of specific diseases or conditions [35]. In clinical applications, the effective utilization of miRNAs as diagnostic biomarkers necessitates adherence to stringent criteria, notably high sensitivity and specificity. Additionally, an ideal biomarker for a specific cancer type should exhibit significant differential expression [36]. Ultimately, biomarkers hold the potential to elevate standards of patient care and drive precision medicine initiatives forward [37]. miRNA profiling has garnered considerable attention as a valuable tool for tumor classification, early detection, disease prognosis, and therapeutic decision making. However, despite notable advancements, the field of miRNA research encounters persistent technical hurdles. The costs associated with miRNA profiling are still high and serve as a barrier to adoption. However, the utilization of miRNAs as biomarkers offers several advantages, including their manageable dimensionality, ease of testing, and clinical significance in disease diagnosis [38]. Detection of a select few miRNAs provides deeper insights into tumor developmental lineage and differentiation compared to profiling numerous mRNAs [39].

Diagnostic biomarkers for tongue cancer include various molecular entities such as long intergenic non-coding RNA, miRNAs, and metabolites. These biomarkers show promise for the early detection and monitoring of oral cancer. In previous research, the miRNA expression profiles of oral squamous cell carcinoma (OSCC) were investigated [40,41]. Salivary LINC00657 and miRNA-106a have been identified as potential diagnostic markers for OSCC [42]. Additionally, N-acetyl-D-glucosamine, L-pipecolic acid, and L-carnitine have been investigated as the signature diagnostic biomarkers for OSCC [43]. In this study, we identified 13 miRNAs that were differentially expressed in OSCC patients. The findings from this research demonstrate a notable upregulation in the expression of hsa-miR-31, hsa-miR-196a-1, hsa-miR-196b, hsa-miR-210, hsa-miR-301a, hsa-miR-450a-2, hsa-miR-503, hsa-miR-877, hsa-miR-937, hsa-miR-1293, hsa-miR-3648, and hsa-miR-4326 within tissue samples obtained from patients diagnosed with OSCC in comparison to control samples. Subsequently, through ROC analysis, we identified these selected miRNAs as diagnostic biomarkers of OSCC. Among these, five miRNAs demonstrated an AUC value of 0.9 or higher, with miR-196b emerging as particularly noteworthy in terms of diagnostic significance. The expression levels of miR-196a-1 are known to be markedly upregulated in tissue and plasma samples derived from individuals diagnosed with colorectal cancer compared with controls [44]. miR-450a-2 exhibited downregulation in both gastric cancer cells and tissue [45]. The downregulation of miR-503 leads to an inhibition in the progression in OSCC [46]. miR-877-3p targets vascular endothelial growth factor A (VEGFA), and the positive expression of VEGFA has been associated with a significantly poor prognosis In cases of OSCC [47]. miR-196b has been proposed as a potential biomarker for the management of oral cancer, and its overexpression has been associated with enhanced oral cancer cell migration, invasion, and lymph node metastasis [48]. Elevated expression levels of miR-196a and miR-196b have been observed in saliva samples from patients diagnosed with HNSCC. These findings highlight the potential of these miRNAs as diagnostic biomarkers for detecting HNSCC at an early stage [49]. Additionally, the combined determination of plasma miR-196a and miR-196b may be used as a diagnostic biomarker for the early detection of oral cancer [50].

Identifying target genes can aid in the development of personalized therapies based on genetic profiles. Predicted target genes of miR-196b were categorized based on their biological processes, cellular components, and molecular functions using GO analysis. Furthermore, analysis of enriched KEGG pathways has been conducted to elucidate the involvement of target genes of miR-196b in oral cancer. The expression of axon guidance genes is related to clinical features like pain and nodal status in oral cancer [51]. Clinical studies have reported dysphagia in patients with spinocerebellar ataxia type 2, type 3, type 6, and type 7 [52]. Dysphagia in oral cancer is commonly attributed to extensive tissue destruction, limited excursion of the remaining tissue, and sensory paralysis of the tongue [53]. The MAPK signaling pathway plays a crucial role in OSCC, where it interacts extensively with miRNAs to regulate cellular processes involved in OSCC development and progression [54]. The Ras signaling pathway is intricately linked to the pathogenesis of OSCC and contributes to key oncogenic processes such as cell proliferation, invasion, and metastasis [55]. The role of mRNA surveillance pathways, notably the nonsense-mediated mRNA decay (NMD) pathway, have been implicated in the pathogenesis of cancer. Additionally, the NMD pathway functions as a post-transcriptional regulator [55]. Mitophagy is a cellular process that selectively degrades dysfunctional mitochondria through autophagy [56]. Targeting mitochondria may be a promising way to treat OSCC [57]. Protein digestion and absorption are associated with oral cancer due to their role in nutrient uptake and cellular metabolism [58]. The top five hub genes of miR-196b-5p were screened using the MCC algorithm. These genes include HOXA5, HOXA7, HOXB6, HOXB7, and PBX1. Altered expression patterns of homeobox-containing HOX genes have been implicated in oral cancer [59]. The expression of HOXA5 was upregulated in OSCC samples compared to non-tumor tissue and was associated with survival rates [60]. Expression of HOXA7 in OSCC exhibited a substantial increase at both the mRNA and protein levels [61]. H3OXB6 was hypermethylated in OSCC cell lines (SCC4 and SCC9) derived from a human HNSCC [62]. HOXB7 is implicated in abnormal proliferation in oral carcinogenesis [63]. Expression analysis demonstrated the expression of PBX1 mRNA and protein within OSCC cells [64]. The top five hub genes of miR-196b-3p were SMURF1, CUL1, XPO1, CYP26B1, and SDR9C7. While this study presents promising findings, it is important to acknowledge several limitations. Conducting a larger cohort study is essential to further validate these findings. Additionally, investigating the functional roles of miRNAs through in vitro and in vivo studies in future research would significantly deepen our understanding of their biological mechanisms and signaling pathways. Nevertheless, our findings reveal a significant upregulation of miR-196b in tissues from oral cancer patients, suggesting its potential utility as a diagnostic biomarker.

5. Conclusions

A current method of diagnosing and screening OSCC of the tongue is the scalpel biopsy, which is time consuming and requires considerable expertise. While advanced imaging modalities such as computed tomography (CT) and magnetic resonance imaging (MRI) technologies have advanced significantly in recent decades, CT scans can only detect the presence of masses [65], thus underscoring the need for supplementary diagnostic tools. The differential expression patterns of miRNAs in cancerous tissue samples and adjacent nontumorous tissue samples serve as valuable biomarkers for diagnostic purposes [66]. In conclusion, this study elucidated a five-miRNA diagnostic model associated with patients diagnosed with OSCC of the tongue. Our findings highlighted the remarkable diagnostic potential of miR-196b. Additionally, five hub genes were selected among the target genes of miR-196b. Ensuring the stability and reproducibility of biomarkers is crucial for augmenting their clinical applicability. Further research will prioritize the standardization of biomarker measurement methodologies and validation of biomarker performance.

Supplementary Materials: The following supporting information can be downloaded at https://www.mdpi.com/article/10.3390/cimb46070402/s1: Table S1: Prediction of target genes of differentially expressed miR-196b; Table S2: GO analysis of target genes of miR-196b; Table S3: KEGG enrichment analysis of target genes of miR-196b. Figure S1: Heatmap for the identification of differentially expressed miRNAs. Figure S2: Receiver operating characteristic curve analysis of 13 differentially expressed miRNAs in oral squamous cell carcinoma of the tongue.

Author Contributions: Y.J. and Y.-N.P. performed the studies, prepared the figures, and wrote the first draft of the manuscript; Y.J. and J.-K.R. designed the project and edited the manuscript. All authors have read and agreed to the published version of the manuscript.

Funding: This research received no external funding.

Institutional Review Board Statement: Not applicable.

Informed Consent Statement: Not applicable.

Data Availability Statement: Data will be made available on request.

Acknowledgments: This study benefited from The Cancer Genome Atlas (TCGA).

Conflicts of Interest: The authors declare no conflicts of interest.

References

1. Liang, L.; Li, Y.; Ying, B.; Huang, X.; Liao, S.; Yang, J.; Liao, G. Mutation-associated transcripts reconstruct the prognostic features of oral tongue squamous cell carcinoma. *Int. J. Oral Sci.* **2023**, *15*, 1. [CrossRef]
2. Wang, S.; Li, K.; Zhao, T.; Sun, Y.; Zeng, T.; Wu, Y.; Ding, L.; Huang, X.; Celentano, A.; Yang, X.; et al. Oral tongue squamous cell carcinoma diagnosis from tissue metabolic profiling. *Oral Dis.* **2023**, *30*, 2158–2165. [CrossRef] [PubMed]
3. Yang, G.; Xiao, Z.; Ren, J.; Xia, R.H.; Wu, Y.; Yuan, Y.; Tao, X. Machine learning based on magnetic resonance imaging and clinical parameters helps predict mesenchymal-epithelial transition factor expression in oral tongue squamous cell carcinoma: A pilot study. *Oral Surg. Oral Med. Oral Pathol. Oral Radiol.* **2024**, *137*, 421–430. [CrossRef] [PubMed]
4. Luo, J.; Zhang, L.; Guo, L.; Yang, S. PKM2 regulates proliferation and apoptosis through the Hippo pathway in oral tongue squamous cell carcinoma. *Oncol Lett.* **2021**, *21*, 461. [CrossRef]
5. Lin, Y.W.; Kang, W.P.; Hong, C.Q.; Huang, B.L.; Qiu, Z.H.; Liu, C.T.; Chu, L.Y.; Xu, Y.W.; Guo, F.C. Nutritional and immune-related indicators-based Nomogram for predicting overall survival of surgical oral tongue squamous cell carcinoma. *Sci. Rep.* **2023**, *13*, 8525. [CrossRef]
6. Tang, W.; Wang, Y.; Yuan, Y.; Tao, X. Assessment of tumor depth in oral tongue squamous cell carcinoma with multiparametric MRI: Correlation with pathology. *Eur. Radiol.* **2022**, *32*, 254–261. [CrossRef]
7. Almangush, A.; Heikkinen, I.A.; Mäkitie, A.; Coletta, R.D.; Läärä, E.; Leivo, I.; Salo, T. Prognostic biomarkers for oral tongue squamous cell carcinoma: A systematic review and meta-analysis. *Br. J. Cancer* **2017**, *117*, 856–866. [CrossRef]

8. Crosby, D.; Bhatia, S.; Brindle, K.M.; Coussens, L.M.; Dive, C.; Emberton, M.; Esener, S.; Fitzgerald, R.C.; Gambhir, S.S.; Kuhn, P.; et al. Early Detection of Cancer. *Science* **2022**, *375*, eaay9040. [CrossRef] [PubMed]
9. Myers, D.; Allen, E.; Essa, A.; Gbadamosi-Akindele, M. Rapidly Growing Squamous Cell Carcinoma of the Tongue. *Cureus* **2020**, *12*, e7164. [CrossRef] [PubMed]
10. O'Brien, J.; Hayder, H.; Zayed, Y.; Peng, C. Overview of MicroRNA Biogenesis, Mechanisms of Actions, and Circulation. *Front. Endocrinol.* **2018**, *9*, 402. [CrossRef]
11. Hammond, S.M. An overview of microRNAs. *Adv. Drug Deliv. Rev.* **2015**, *87*, 3–14. [CrossRef] [PubMed]
12. Naeli, P.; Winter, T.; Hackett, A.P.; Alboushi, L.; Jafarnejad, S.M. The intricate balance between microRNA-induced mRNA decay and translational repression. *FEBS J.* **2023**, *290*, 2508–2524. [CrossRef] [PubMed]
13. Dillhoff, M.; Wojcik, S.E.; Bloomston, M. MicroRNAs in Solid Tumors. *J. Surg. Res.* **2009**, *154*, 349–354. [CrossRef] [PubMed]
14. Kim, T.; Croce, C.M.; Bloomston, M. MicroRNA: Trends in clinical trials of cancer diagnosis and therapy strategies. *Exp. Mol. Med.* **2023**, *55*, 1314–1321. [CrossRef] [PubMed]
15. Boichard, A.; Wagner, M.J.; Kurzrock, R. Angiosarcoma heterogeneity and potential therapeutic vulnerability to immune checkpoint blockade: Insights from genomic sequencing. *Genome Med.* **2020**, *12*, 61. [CrossRef]
16. Chen, G.; Luo, D.; Zhong, N.; Li, D.; Zheng, J.; Liao, H.; Li, Z.; Lin, X.; Chen, Q.; Zhang, C.; et al. GPC2 Is a Potential Diagnostic, Immunological, and Prognostic Biomarker in Pan-Cancer. *Front. Immunol.* **2022**, *13*, 857308. [CrossRef] [PubMed]
17. Sempere, L.F.; Azmi, A.S.; Moore, A. microRNA-based diagnostic and therapeutic applications in cancer medicine. *Genome Med.* **2021**, *12*, e1662. [CrossRef] [PubMed]
18. Qin, L.X. An Integrative Analysis of microRNA and mRNA Expression—A Case Study. *Cancer Inform.* **2008**, *6*, 369–379. [CrossRef] [PubMed]
19. Liu, Q.; Zheng, C.; Shen, H.; Zhou, Z.; Lei, Y. MicroRNAs-mRNAs Expression Profile and Their Potential Role in Malignant Transformation of Human Bronchial Epithelial Cells Induced by Cadmium. *Biomed. Res. Int.* **2015**, *2015*, 902025. [CrossRef]
20. Zou, K.H.; O'Malley, A.J.; Mauri, L. Receiver-operating characteristic analysis for evaluating diagnostic tests and predictive models. *Circulation* **2007**, *115*, 654–657. [CrossRef]
21. Walter, S. The partial area under the summary ROC curve. *Stat. Med.* **2005**, *24*, 2025–2040. [CrossRef] [PubMed]
22. Mandrekar, J.N. Receiver Operating Characteristic Curve in Diagnostic Test Assessment. *J. Thorac. Oncol.* **2010**, *5*, 1315–1316. [CrossRef] [PubMed]
23. Çorbacıoğlu, S.K.; Aksel, G. Receiver operating characteristic curve analysis in diagnostic accuracy studies: A guide to interpreting the area under the curve value. *Turk. J. Emerg. Med.* **2023**, *23*, 195–198. [CrossRef] [PubMed]
24. Unal, I. Defining an Optimal Cut-Point Value in ROC Analysis: An Alternative Approach. *Comput. Math Methods Med.* **2017**, *2017*, 3762651. [CrossRef]
25. Jorge, A.L.; Pereira, E.R.; Oliveira, C.S.; Ferreira, E.D.S.; Menon, E.T.N.; Diniz, S.N.; Pezuk, J.A. MicroRNAs: Understanding their role in gene expression and cancer. *Einstein* **2021**, *19*, eRB5996. [CrossRef]
26. Liu, X.J.; Yin, H.L.; Li, Y.; Hao, H.; Liu, Y.; Zhao, Q.L. The Construction and Analysis of a ceRNA Network Related to Salt-Sensitivity Hypertensives. *Biomed. Res. Int.* **2022**, *2022*, 8258351. [CrossRef] [PubMed]
27. Althubaiti, S.; Karwath, A.; Dallol, A.; Noor, A.; Alkhayyat, S.S.; Alwassia, R.; Mineta, K.; Gojobori, T.; Beggs, A.D.; Schofield, P.N.; et al. Ontology-based prediction of cancer driver genes. *Sci. Rep.* **2019**, *9*, 17405. [CrossRef] [PubMed]
28. Sherman, B.T.; Hao, M.; Qiu, J.; Jiao, X.; Baseler, M.W.; Lane, H.C.; Imamichi, T.; Chang, W. DAVID: A web server for functional enrichment analysis and functional annotation of gene lists (2021 update). *Nucleic Acid Res.* **2022**, *50*, W216–W221. [CrossRef]
29. Kanehisa, M.; Goto, S. KEGG: Kyoto Encyclopedia of Genes and Genomes. *Nucleic Acids Res.* **2000**, *28*, 27–30. [CrossRef] [PubMed]
30. Ruffalo, M.; Bar-Joseph, Z. Protein interaction disruption in cancer. *BMC Cancer* **2019**, *19*, 370. [CrossRef]
31. Szklarczyk, D.; Kirsch, R.; Koutrouli, M.; Nastou, K.; Mehryary, F.; Hachilif, R.; Gable, A.L.; Fang, T.; Doncheva, N.T.; Pyysalo, S.; et al. The STRING database in 2023: Protein-protein association networks and functional enrichment analyses for any sequenced genome of interest. *Nucleic Acids Res.* **2023**, *51*, D638–D646. [CrossRef] [PubMed]
32. Kalia, M. Feature selection with the Fisher score followed by the Maximal Clique Centrality algorithm can accurately identify the hub genes of hepatocellular carcinoma. *Sci. Rep.* **2019**, *9*, 17283.
33. Ding, W.; Xin, J.; Jiang, L.; Zhou, Q.; Wu, T.; Shi, D.; Lin, B.; Li, L.; Li, J. Characterisation of peripheral blood mononuclear cell microRNA in hepatitis B-related acute-on-chronic liver failure. *Sci. Rep.* **2015**, *5*, 13098. [CrossRef] [PubMed]
34. Kalia, M. Biomarkers for personalized oncology: Recent advances and future challenges. *Metabolism* **2015**, *64*, S16–S21. [CrossRef] [PubMed]
35. Califf, R.M. Biomarker definitions and their applications. *Exp. Biol. Med.* **2018**, *243*, 213–221. [CrossRef] [PubMed]
36. Condrat, C.E.; Thompson, D.C.; Barbu, M.G.; Bugnar, O.L.; Boboc, A.; Cretoiu, D.; Suciu, N.; Cretoiu, S.M.; Voinea, S.C. miRNAs as Biomarkers in Disease: Latest Findings Regarding Their Role in Diagnosis and Prognosis. *Cells* **2020**, *9*, 276. [CrossRef] [PubMed]
37. Hartl, D.; de Luca, V.D.; Kostikova, A.; Laramin, J.; Kennedy, S.; Ferrero, E.; Siegel, R.; Fink, M.; Ahmed, S.; Millholland, J.; et al. Translational precision medicine: An industry perspective. *J. Transl. Med.* **2021**, *19*, 245. [CrossRef] [PubMed]
38. Lussier, Y.A.; Stadler, W.M.; Chen, J.L. Advantages of genomic complexity: Bioinformatics opportunities in microrna cancer signatures. *J. Am. Med. Inform. Assoc.* **2012**, *19*, 156–160. [CrossRef]
39. Wang, J.; Chen, J.; Sen, S. MicroRNA as Biomarkers and Diagnostics. *J. Cell. Physiol.* **2016**, *231*, 25–30. [CrossRef]
40. Soga, D.; Yoshiba, S.; Shiogama, S.; Miyazaki, H.; Kondo, S.; Shintani, S. microRNA expression profiles in oral squamous cell carcinoma. *Oncol. Rep.* **2013**, *30*, 579–583. [CrossRef]

41. Gombos, K.; Horváth, R.; Szele, E.; Juhász, K.; Gocze, K.; Somlai, K.; Pajkos, G.; Ember, I.; Olasz, L. miRNA expression profiles of oral squamous cell carcinomas. *Anticancer Res.* **2013**, *33*, 1511–1517. [PubMed]
42. Tarrad, N.A.F.; Hassan, S.; Shaker, O.G.; AbdelKawy, M. "Salivary LINC00657 and miRNA-106a as diagnostic biomarkers for oral squamous cell carcinoma, an observational diagnostic study". *BMC Oral Health* **2023**, *23*, 994. [CrossRef] [PubMed]
43. Vimal, J.; George, N.A.; Kumar, R.R.; Kattoor, J.; Kannan, S. Identification of salivary metabolic biomarker signatures for oral tongue squamous cell carcinoma. *Arch. Oral Biol.* **2023**, *155*, 105780. [CrossRef]
44. Mehrjoei, B.; Haghnazari, L.; Bashiri, H.; Rezvani, N. The diagnostic potential of miR-196a-1 in colorectal cancer. *BMC Cancer* **2024**, *24*, 162. [CrossRef] [PubMed]
45. Raei, N.; Safaralizadeh, R.; Hesseinpourfeizi, M.; Yazdanbod, A.; Pourfarzi, F.; Latifi-Navid, S. Crosstalk between lncRNAs and miRNAs in gastrointestinal cancer drug resistance. *Life Sci.* **2021**, *284*, 119933. [CrossRef] [PubMed]
46. Han, L.; Cheng, J.; Li, A. hsa_circ_0072387 Suppresses Proliferation, Metastasis, and Glycolysis of Oral Squamous Cell Carcinoma Cells by Downregulating miR-503-5p. *Cancer Biother. Radiopharm.* **2021**, *36*, 84–94. [CrossRef]
47. Shao, Y.; Song, Y.; Xu, S.; Li, S.; Zhou, H. Expression Profile of Circular RNAs in Oral Squamous Cell Carcinoma. *Front. Oncol.* **2020**, *10*, 533616. [CrossRef] [PubMed]
48. Rajan, C.; Roshan, V.G.D.; Khan, I.; Manasa, V.G.; Himal, I.; Kattoor, J.; Thomas, S.; Kondaiah, P.; Kannan, S. MiRNA expression profiling and emergence of new prognostic signature for oral squamous cell carcinoma. *Sci. Rep.* **2021**, *11*, 7298. [CrossRef]
49. Álvarez-Teijeiro, S.; Menéndez, S.T.; Villaronga, M.; Rodrigo, J.P.; Manterola, L.; de Villalaín, L.; de Vicente, J.C.; Alonso-Durán, L.; Fernández, M.P.; Lawrie, C.H.; et al. Dysregulation of Mir-196b in Head and Neck Cancers Leads to Pleiotropic Effects in the Tumor Cells and Surrounding Stromal Fibroblasts. *Sci. Rep.* **2017**, *7*, 17785. [CrossRef]
50. Lu, Y.C.; Chang, J.T.C.; Huang, Y.C.; Huang, C.C.; Chen, W.H.; Lee, L.Y.; Huang, B.S.; Chen, Y.J.; Li, H.F.; Cheng, A.J. Combined determination of circulating miR-196a and miR-196b levels produces high sensitivity and specificity for early detection of oral cancer. *Clin. Biochem.* **2015**, *48*, 115–121. [CrossRef]
51. Bhattacharya, A.; Janal, M.N.; Veeramachaneni, R.; Dolgalev, I.; Dubeykovskaya, Z.; Tu, N.H.; Kim, H.; Zhang, S.; Wu, A.K.; Hagiwara, M.; et al. Oncogenes overexpressed in metastatic oral cancers from patients with pain: Potential pain mediators released in exosomes. *Sci. Rep.* **2020**, *10*, 14724. [CrossRef] [PubMed]
52. Rub, U.; Brunt, E.R.; Petrasch-Parwez, E.; Schols, L.; Theegarten, D.; Auburger, G.; Seidel, K.; Schultz, C.; Gierga, K.; Paulson, H.; et al. Degeneration of ingestion-related brainstem nuclei in spinocerebellar ataxia type 2, 3, 6 and 7. *Neuropathol. Appl. Neurobiol.* **2006**, *32*, 635–649. [CrossRef] [PubMed]
53. Son, Y.R.; Choi, K.H.; Kim, T.G. Dysphagia in Tongue Cancer Patients. *Ann. Rehabil. Med.* **2015**, *39*, 210–217. [CrossRef] [PubMed]
54. Cheng, Y.; Chen, J.; Shi, Y.; Fang, X.; Tang, Z. MAPK Signaling Pathway in Oral Squamous Cell Carcinoma: Biological Function and Targeted Therapy. *Cancers* **2022**, *14*, 4625. [CrossRef] [PubMed]
55. Bongiorno, R.; Colombo, M.P.; Lecis, D. Deciphering the nonsense-mediated mRNA decay pathway to identify cancer cell vulnerabilities for effective cancer therapy. *J. Exp. Clin. Cancer Res.* **2021**, *40*, 376. [CrossRef] [PubMed]
56. Youle, R.J.; Narendra, D.P. Mechanisms of mitophagy. *Nat. Rev. Mol. Cell Biol.* **2011**, *12*, 9–14. [CrossRef]
57. Bai, J.; Wu, L.; Wang, X.; Wang, Y.; Shang, Z.; Jiang, E.; Shao, Z. Roles of mitochondria in oral squamous cell carcinoma therapy: Friend or foe? *Cancers* **2022**, *14*, 5723. [CrossRef]
58. Rodríguez-Molinero, J.; Migueláñez-Medrán, B.D.C.; Puente-Gutiérrez, C.; Delgado-Somolinos, E.; Martín Carreras-Presas, C.; Fernández-Farhall, J.; López-Sánchez, A.F. Association between Oral Cancer and Diet: An Update. *Nutrients* **2021**, *13*, 1299. [CrossRef]
59. Padam, K.S.R.; Morgan, R.; Hunter, K.; Chakrabarty, S.; Kumar, N.A.N.; Radhakrishnan, R. Identification of HOX signatures contributing to oral cancer phenotype. *Sci. Rep.* **2022**, *12*, 10123. [CrossRef]
60. Rodini, C.O.; Xavier, F.C.; Paiva, K.B.; De Souza Setúbal Destro, M.F.; Moyses, R.A.; Michaluarte, P.; Carvalho, M.B.; Fukuyama, E.E.; Head and Neck Genome Project Gencapo; Tajara, E.H.; et al. Head and Neck Genome Project Gencapo. *Int. J. Oncol.* **2012**, *40*, 1180–1188. [CrossRef]
61. Duan, X.; Chen, H.; Ma, H.; Song, Y. The expression and significance of the HOXA7 gene in oral squamous cell carcinoma. *J. Oral Sci.* **2017**, *59*, 329–335. [CrossRef] [PubMed]
62. Xavier, F.C.; Destro, M.F.; Duarte, C.M.; Nunes, F.D. Epigenetic repression of HOXB cluster in oral cancer cell lines. *Arch. Oral Biol.* **2014**, *59*, 783–789. [CrossRef] [PubMed]
63. Marcinkiewicz, K.M.; Gudas, L.J. Altered epigenetic regulation of homeobox genes in human oral squamous cell carcinoma cells. *Exp. Cell Res.* **2014**, *320*, 128–143. [CrossRef] [PubMed]
64. Platais, C.; Radhakrishnan, R.; Ebensberger, S.N.; Morgan, R.; Lambert, D.W.; Hunter, K.D. Targeting HOX-PBX interactions causes death in oral potentially malignant and squamous carcinoma cells but not normal oral keratinocytes. *BMC Cancer* **2018**, *18*, 723. [CrossRef] [PubMed]
65. Wang, Q.; Gao, P.; Wang, X.; Duan, Y. The early diagnosis and monitoring of squamous cell carcinoma via saliva metabolomics. *Sci. Rep.* **2014**, *4*, 6802. [CrossRef]
66. Hu, Y.; Dingerdissen, H.; Gupta, S.; Kahsay, R.; Shanker, V.; Wan, Q.; Yan, C.; Mazumder, R. Identification of key differentially expressed MicroRNAs in cancer patients through pan-cancer analysis. *Comput. Biol. Med.* **2018**, *103*, 183–197. [CrossRef]

current issues in molecular biology

MDPI

Article

A Tissue Engineered 3D Model of Cancer Cell Invasion for Human Head and Neck Squamous-Cell Carcinoma

Manuel Stöth [1], Anna Teresa Mineif [2], Fabian Sauer [2], Till Jasper Meyer [1], Flurin Mueller-Diesing [1], Lukas Haug [3], Agmal Scherzad [1], Maria Steinke [1,2,4], Angela Rossi [4,†] and Stephan Hackenberg [1,*,†]

1 Department of Otorhinolaryngology, Plastic, Aesthetic and Reconstructive Head and Neck Surgery, University Hospital Würzburg, 97080 Würzburg, Germany; stoeth_m@ukw.de (M.S.); meyer_t2@ukw.de (T.J.M.); muellerdie_f@ukw.de (F.M.-D.); scherzad_a@ukw.de (A.S.); maria.steinke@isc.fraunhofer.de (M.S.)

2 Chair of Tissue Engineering and Regenerative Medicine, University Hospital Würzburg, 97070 Würzburg, Germany; anna@mineif-online.de (A.T.M.)

3 Institute of Pathology, University of Würzburg, 97080 Würzburg, Germany; lukas.haug@uni-wuerzburg.de

4 Fraunhofer Institute for Silicate Research ISC, 97082 Würzburg, Germany; angela.rossi@isc.fraunhofer.de

* Correspondence: hackenberg_s@ukw.de

† These authors contributed equally to this work.

Abstract: Head and neck squamous-cell carcinoma (HNSCC) is associated with aggressive local invasiveness, being a main reason for its poor prognosis. The exact mechanisms underlying the strong invasive abilities of HNSCC remain to be elucidated. Therefore, there is a need for in vitro models to study the interplay between cancer cells and normal adjacent tissue at the invasive tumor front. To generate oral mucosa tissue models (OMM), primary keratinocytes and fibroblasts from human oral mucosa were isolated and seeded onto a biological scaffold derived from porcine small intestinal submucosa with preserved mucosa. Thereafter, we tested different methods (single tumor cells, tumor cell spots, spheroids) to integrate the human cancer cell line FaDu to generate an invasive three-dimensional model of HNSCC. All models were subjected to morphological analysis by histology and immunohistochemistry. We successfully built OMM tissue models with high in vivo–in vitro correlation. The integration of FaDu cell spots and spheroids into the OMM failed. However, with the integration of single FaDu cells into the OMM, invasive tumor cell clusters developed. Between segments of regular epithelial differentiation of the OMM, these clusters showed a basal membrane penetration and lamina propria infiltration. Primary human fibroblasts and keratinocytes seeded onto a porcine carrier structure are suitable to build an OMM. The HNSCC model with integrated FaDu cells could enable subsequent investigations into cancer cell invasiveness.

Keywords: head and neck squamous-cell carcinoma; tissue engineering; oral mucosa; 3D tissue model

Citation: Stöth, M.; Mineif, A.T.; Sauer, F.; Meyer, T.J.; Mueller-Diesing, F.; Haug, L.; Scherzad, A.; Steinke, M.; Rossi, A.; Hackenberg, S. A Tissue Engineered 3D Model of Cancer Cell Invasion for Human Head and Neck Squamous-Cell Carcinoma. *Curr. Issues Mol. Biol.* **2024**, *46*, 4049–4062. https://doi.org/10.3390/cimb46050250

Academic Editors: Emma Adriana Ozon and Violeta Popovici

Received: 13 March 2024
Revised: 23 April 2024
Accepted: 25 April 2024
Published: 28 April 2024

1. Introduction

Head and neck squamous-cell carcinoma (HNSCC) occurs in the cell lining of the oral cavity, pharynx, and larynx. According to estimates from the Global Cancer Observatory (GLOBOCAN), HNSCC is the sixth most prevalent cancer globally, with approximately 890,000 new cases and 450,000 deaths each year, posing a significant public health challenge [1]. Major risk factors for HNSCC include tobacco and alcohol consumption [2], as well as human papillomavirus (HPV) and Ebstein–Barr virus (EBV) infection. Betel quid chewing [3], poor oral hygiene [4], and chronic traumatism [5] are additional factors that contribute to an increased risk of developing this cancer. Treatment for HNSCC primarily involves a combination of surgery, radiotherapy, and chemotherapy, based on the tumor stage and pathological diagnosis. Further treatment options include immunotherapy and targeted therapy. The local recurrence rate of HNSCC varies widely, depending on several factors such as the location of the tumor, the stage at diagnosis, and the treatment methods used. Despite the use of multimodal therapy, the local–regional recurrence rate for advanced HNSCC can be as high as 50% [6]. Notably, recurrent and metastatic disease emerges as a major cause of mortality in HNSCC [7]. Despite remarkable advances in surgery, radiation, and medical therapy over the past few decades, the prognosis for HNSCC remains poor, with a 5-year overall survival rate ranging from 25% for hypopharyngeal cancer to 59% for laryngeal cancer [8].

An in-depth understanding of HNSCC biology is critical to improve the prognosis. One challenge in achieving this is the development of suitable cancer models. So far, several in vitro cancer studies have been carried out using two-dimensional (2D) monolayer cell cultures. Although such models are easy to handle and reproducible, they fail to include the complex interplay between cancer cells, non-malignant cells, and acellular components within the tumor microenvironment (TME) [9]. For example, gene expression differs between 2D monolayer culture and three-dimensional (3D) cancer models, emphasizing the role of cell–cell interaction in cancer [10]. Furthermore, 3D cancer models have been shown to be more resistant to anticancer drugs compared to 2D models [11]. In addition, there is compelling evidence highlighting the importance of non-malignant cells in tumor progression [12]. Additionally, acellular components, including molecules of the extracellular matrix (ECM), have been shown to influence the phenotype of cancer cells [13]. Organoids represent an advanced 3D culture technology to mimic the complex physiology within human organs and tumors. However, organoids fail to integrate the whole TME [14] and have limitations in terms of studying cancer cell invasiveness and metastasis [15]. Conventional and patient-derived xenograft animal models are acknowledged as sophisticated tools for cancer research. They allow for detailed study of cancer metastasis [16] and the TME [17,18], and enable in vivo drug testing [19–21]. However, their informative value is also limited: in addition to ethical issues, costs, and availability [22], there are genetic [23], immunological [24], and stromal [25] differences between human, grafted, and experimental animal tumors. All the mentioned models make valuable contributions to preclinical cancer research. Nevertheless, all platforms have specific disadvantages that limit their use as cancer models. Thus, there is an unmet need for further cancer models to study HNSCC.

Compared to 2D in vitro models, tissue-engineered 3D models represent a complex culture format more closely resembling the physiological conditions of human tumors [26]. With the integration of TME components, cancer cells can be subjected to cellular, stromal, and biophysical stimuli. A suitable cell carrier is crucial for the generation of such 3D tissue models. This carrier represents a scaffold of ECM, which ensures the 3D structure of the model and allows cell-to-ECM interactions. A distinction is made between natural and synthetic cell carriers [27]. Advantages of natural cell carriers include low antigenicity [27], abundance in structural and functional proteins, and the fact that they allow for better epithelial cell attachment compared to synthetic cell carriers [28]. An example of a natural scaffold is the acellular carrier structure of porcine small intestinal submucosa with preserved mucosa (SIS/MUC), which consists of cross-linked collagen and elastin fibers [29]. It has previously been used to study the drug responses of tumor cell lines in human 2D and 3D lung cancer models [30]. In addition to providing a biologically intact ECM, SIS-based scaffolds enable fibroblast migration, epithelial differentiation, and the formation of a basement membrane [31]. The latter is essential for generating an HNSCC model adequately able to study tumor cell invasiveness—a major driver of tumor metastasis and recurrence.

Such a model could provide an ideal basis to elucidate specific interactions within the TME and to gain more insights into the interplay between cancer cells and normal adjacent tissue at the invasive tumor front. Therefore, the aim of this study was to generate a 3D tissue model of invasive HNSCC in which tumor cells are integrated into a healthy oral mucosal model (OMM) consisting of epithelium, basement membrane, and connective tissue.

2. Materials and Methods

Patients and specimens: This study was conducted according to the guidelines of the Declaration of Helsinki and was approved by the local ethic committee (Institutional review board number: 182/10). After written informed consent was obtained from each subject, fresh oral mucosa was obtained exclusively from donors with intact oral mucosa who underwent elective surgery (e.g., oral surgical repositioning osteotomies, trauma surgery, or metal removal). Exclusion criteria included a history of radiotherapy to the head and neck, inflammatory or malignant oral diseases, smoking, and high-risk alcohol consumption. Samples were collected from eight participants aged between 18 and 73 years.

Primary cell isolation: Following intraoperative collection, oral mucosa tissue was cut into 2 mm slices and then incubated in dispase (Life-Technologies, Carlsbad, CA, USA) at 4 °C for 16 h. Thereafter, the epithelium and lamina propria were detached from one another. The epithelium was then processed into a single-cell suspension of keratinocytes by mechanical and brief enzymatic disaggregation with trypsin (Life-Technologies). Thereafter, keratinocytes were transferred into a culture flask. The remaining tissue pieces consisting of lamina propria were placed into a T25 cell culture flask, allowing fibroblasts to spread out of the tissue and further proliferate. Once keratinocytes and fibroblasts reached confluency, they were cryopreserved until further usage.

Cell culture: Fibroblasts were grown in DMEM (Thermo Fisher Scientific, Waltham, MA, USA) supplemented with 10% fetal calf serum (FCS) (Bio&Sell, Feucht, Germany) and keratinocytes in E1-Medium consisting of EpiLife®Medium (Life Technologies) supplemented with 60 μM calcium chloride, 5 mL human keratinocyte growth supplement (Life Technologies), 0.2% bovine pituitary extract, 1 μg/mL insuline-like growth factor-I, 0.18 μg/mL hydrocortison, 5 μg/mL transferrin, and 0.2 ng/mL epidermal growth factor. For co-culture experiments under submerged and air–liquid interface (ALI) culture, E2 and E3-Media were used. E2-Medium consisted of E1-Medium supplemented with 2.4 mL 300 mM calcium chloride, and E3-Medium was further supplemented with 500 μL keratinocyte growth factor (Sigma Aldrich, St. Louis, MS, USA) and 500 μL ascorbyl phosphate. The hypopharyngeal squamous cell cancer cell line FaDu [32] was obtained from the American Type Culture Collection. Cells were grown in RPMI (Life Technologies) supplemented with 10% FCS. All cell culture media contained 0.1 mg/mL streptomycin and 100 U/mL penicillin (Life Technologies). All cells were tested for mycoplasma using the Venor®GeM Classic Kit (Minerva Biolabs, Berlin, Germany) every 4 weeks to ensure that they were free of contamination. For the generation of multicellular tumor spheroids, 50 μL of 0.5% agarose was added to each well of a 96-well plate and incubated overnight at 4 °C. 5000 FaDu cells in 100 μL were seeded into each well and incubated for 4 d. For microscopic detection, FaDu cells were transduced lentivirally to constitutively express red fluorescent protein (RFP). Therefore, FaDu cells were incubated overnight with RFP lentivirus and polybrene (Sigma Aldrich). Next, 2 mL of cell-specific medium was added. On the third day, the medium along with viral particles was changed and rinsed, and the culture was continued until the twenty-sixth day. Next, selection of RFP expressing cells was performed by adding 0.25 μg/mL puromycin (InvivoGen, Toulouse, France), increasing the dose to 0.5 μg/mL on day 31 and 1 μg/mL on day 36. Successful transduction was confirmed by fluorescence microscopy with the detection of more than 90% RFP-expressing FaDu cells.

Three-dimensional human oral mucosal model: Animal experiments were performed after previous approval by the ethical committee of the local government (approval number: 55.2-2532-2-256). All procedures conformed to the guidelines from Directive 2010/63/EU of the European Parliament on the protection of animals used for scientific purposes. After obtaining porcine small intestine as the carrier structure for our HNSCC model, the scaffold was prepared according to methods outlined in prior studies [29]. As a 3D scaffold for tissue model generation, decellularized porcine SIS/MUC was used, as previously described [31]. Briefly, pieces of the scaffold were mounted into cell crowns 13 mm in diameter. After preincubation, 50,000 fibroblasts were seeded in E1-Medium from the apical side to the cell crown. After incubation for seven days, 25,000 keratinocytes from the same donor were seeded on the apical side as well. Fibroblasts were used at passage 4, and keratinocytes at passage 2. The cells were cultured under submerged conditions with E2-Medium for 24 h, and subsequently as an ALI culture with E3-Medium for 12 d (Figure 1a). Eleven OMMs from eight different donors were generated. For each model, three technical replicates were generated.

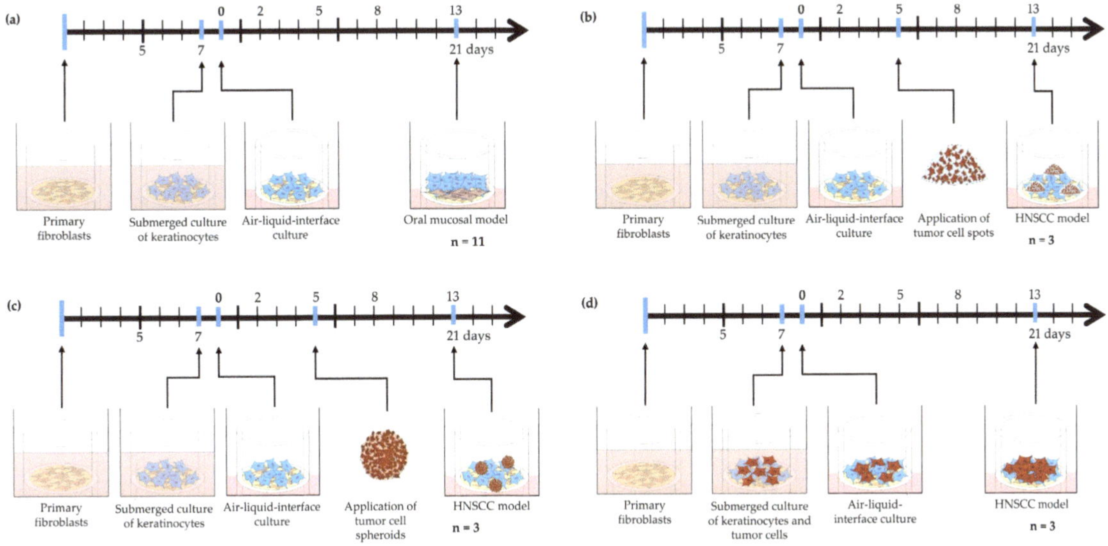

Figure 1. (**a**) Experimental design used to generate an oral mucosal model (OMM) and (**b–d**) different methods of integrating FaDu tumor cells into the OMM. (**b**) Integration of multicellular tumor cell spots. (**c**) Integration of multicellular tumor spheroids. (**d**) Integration of single tumor cells. Oral mucosal model (OMM), head and neck squamous-cell carcinoma (HNSCC).

Three-dimensional human head and neck squamous-cell carcinoma model: FaDu tumor cells were integrated into the human OMM in three different ways (Figure 1b–d). (i) Integration of multicellular tumor cell spots: On day 5 of ALI cultivation, medium was removed and 10,000 FaDu cells were seeded in 2 µL of RPMI medium from the apical side onto the OMM. After adding E3 media, ALI culture was continued for 9 more days. (ii) Integration of multicellular tumor spheroids: On day 5 of ALI cultivation, medium was removed and tumor cell spheroids were placed in the center of the apical side of the OMM. The culture was continued in the same way as in (i). For tumor cell detection, RFP-expressing FaDu cells were used. (iii) Integration of single tumor cells: On day 8 after incubation of fibroblasts within the SIS/MUC, a mixed suspension consisting of different ratios of FaDu cells and keratinocytes were seeded on the apical side into the cell crowns. For each method, three HNSCC models from two different donors were built. For each model, three technical replicates were generated. As in the OMM, the cells were cultured under submerged conditions with E2 medium for 24 h, and subsequently as an ALI culture with E3 Medium for 14 d.

Histology and immunohistochemistry: Immunohistochemistry (IHC) was performed as previously described [33]. Briefly, samples were fixed in 4% formalin for 3 h, then embedded in paraffin and sectioned at 5 µm thickness using a Leica SM2010 R Sliding Microtome (Leica, Wetzlar, Germany). Following antigen retrieval using sodium citrate buffer and blocking of endogenous peroxidases using 3% H_2O_2, primary antibodies were diluted with antibody dilution buffer (DCS, Hamburg, Germany) in the mixing ratio specified below, added following the provider's instructions, and incubated for 60 min at room temperature: Cytokeratin (CK) 10 1:100 (M7002, DAKO, Eching, Germany), CK14

1:2000 (HPA023040, Sigma-Aldrich, St. Louis, MO, USA), Ki67 1:100 (ab16667, Abcam, Cambridge, MA, USA), Collagen IV 1:500 (ab6586, Abcam), p53 1:50 (M7001, DAKO), and Vimentin 1:2000 (ab92547, Abcam). Primary antibody detection, signal enhancing, and chromogenic visualization were performed using the DCS Super Vision 2 HRP-Polymer-Kit (DAKO) according to the provider's instructions. Negative controls were utilized for each experiment by omission of primary antibodies. Mayer's hemalum solution (Merck, Darmstadt, Germany) was used for nuclear counterstain. H&E staining was performed using the following protocol. Tissue sections were stained for cell nuclei with Mayer's hemalum solution, followed by a ten-minute flush with tap water. Then, the sections were stained with eosin for three minutes and subsequently rinsed under running tap water for 30 s to eliminate excess dye. After staining, the tissue underwent dehydration through an ascending series of alcohol concentrations, followed by immersion in xylene. Sister slides were used for all histology and IHC staining. Images were acquired using the BZ-9000 BIOREVO System (Keyence, Neu-Isenburg, Germany) and processed with BZ-II Analyzer and BZ-II Viewer software (Version 2.1, Keyence). Figures were illustrated with Microsoft Paint (Redmond, WA, USA) and compiled with Microsoft PowerPoint (Version 2312).

3. Results

Generation of a reproducible 3D human oral mucosal model: The first step in establishing a human HNSCC model was the generation of a healthy OMM, which resembled the human oral mucosa. The aim was to evaluate the suitability of primary fibroblasts and keratinocytes from the oral cavity to generate such OMM. The growth pattern and properties of the OMM were assessed via H&E staining and IHC. A decellularized porcine SIS/MUC was used for the generation of the OMM. SIS-based scaffolds have already proved to be a suitable basis on which to build human cancer tissue models [30,31,34].

In all models, H&E staining showed a regular formed, keratinized, stratified squamous epithelium without any histological features of dysplasia. The epithelium of the OMM consisted of 10–15 cell layers with a thickness of around 100–120 μm and a regular stratification (Figure 2a). The thickness of the epithelium and lamina propria was comparable between different donors. However, in comparison to oral mucosa, the OMM exhibited a lower mucosal thickness. The overall thickness of the model, including the scaffold, was around 350 μm, with a diameter of 13 mm. IHC allowed us to further analyze the morphology and cell composition of the OMM (Figure 2b). In both the OMM and in oral mucosa, Vimentin-positive fibroblasts were exclusively found within the lamina propria, and collagen IV was expressed within the basal membrane and lamina propria. Furthermore, a continuously formed basal membrane was clearly identifiable by more intense collagen IV staining. Laminin V stained the basal membrane in oral mucosa. An identical Laminin V staining pattern was detectable in the OMM, providing further evidence for the formation of a basal membrane in the OMM. CK14 is a marker for keratinocytes of the basal layer [35]. Compared to oral mucosa, CK14 staining was more intense in OMM, but also decreased towards the apical surface. As in oral mucosa, suprabasal cells in the OMM stained positive for CK10. A matching staining was found in the OMM. All layers of oral mucosa stained moderately positive for CK5/6. Similarly, all layers of the OMM were positive for CK5/6, though with more intense staining. Proliferating Ki67 cells were found in the basal epithelial layer and occasionally within the lamina propria.

Overall, 11 OMM from eight different donors were generated and evaluated. The success in establishing the OMM indicates the suitability of primary oral fibroblasts and keratinocytes. Furthermore, the comparable properties between each OMM indicate the reproducibility of the model.

Failed integration of multicellular tumor cell spots and tumor spheroids into the 3D human oral mucosal model: Next, we aimed to use the OMM as a basis to generate human HNSCC models. Therefore, we tested different ways of integrating FaDu hypopharyngeal squamous cell carcinoma cells into the OMM. One way was the seeding of FaDu tumor cell spots onto the OMM (Figure 1b). The histological analysis of H&E staining revealed a regular formed, keratinized, stratified squamous epithelium, as seen in the OMM. There were no signs of either malignancy or tumor cells. A further characterization was performed by IHC, which revealed a morphology similar to the OMM (Figure 3a).

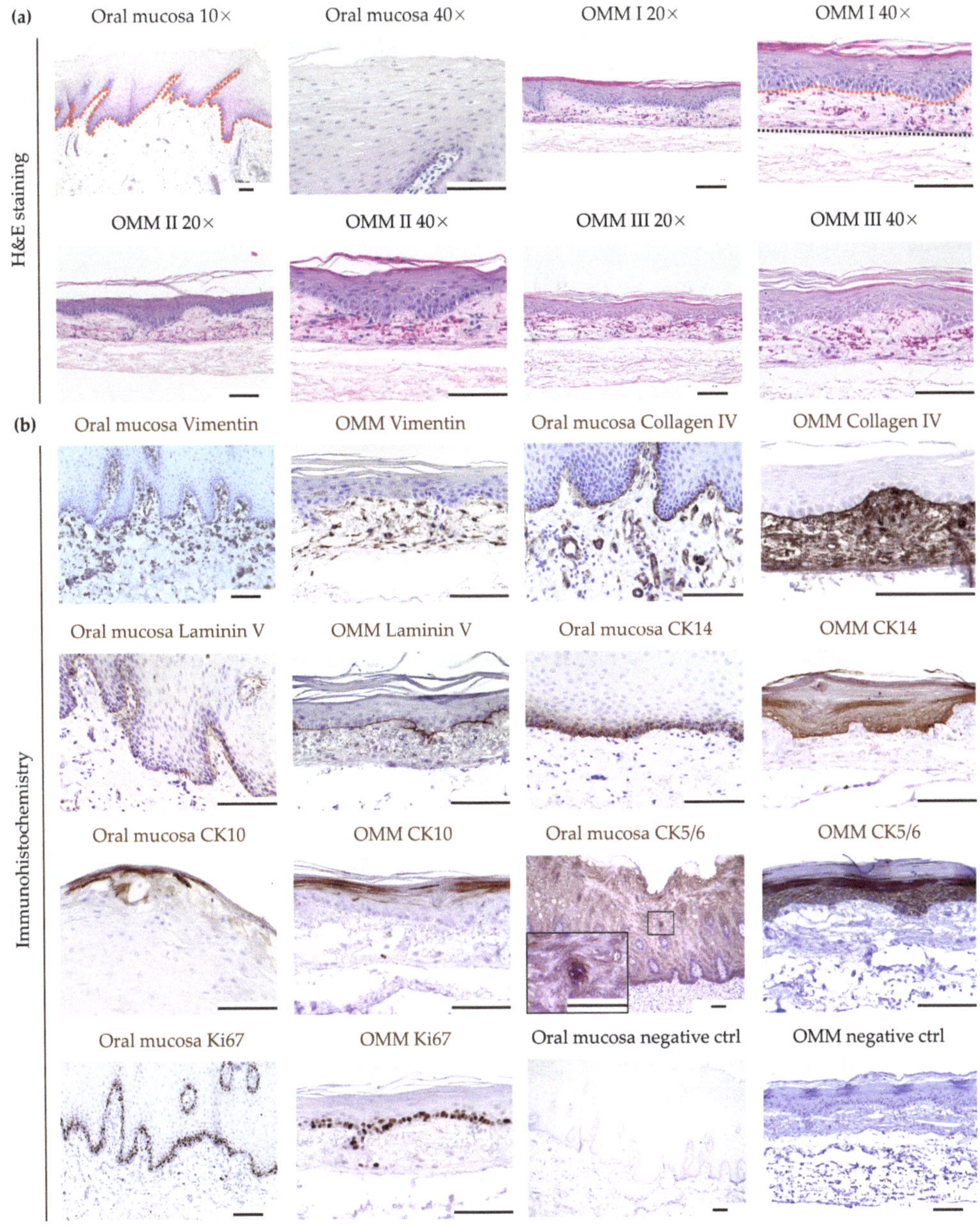

Figure 2. Generation of a 3D bioequivalent oral mucosa model (OMM). (**a**) Comparison of human oral mucosa with the OMM by H&E staining. The red dotted line separates the stratified squamous epithelium from the underlying lamina propria. The black dotted line separates the lamina propria from residual SIS/MUC. (**b**) Comparison of human oral mucosa with the OMM by immunohistochemistry. Scale bars represent 100 μm. Oral mucosa model (OMM), cytokeratin (CK), control (ctrl).

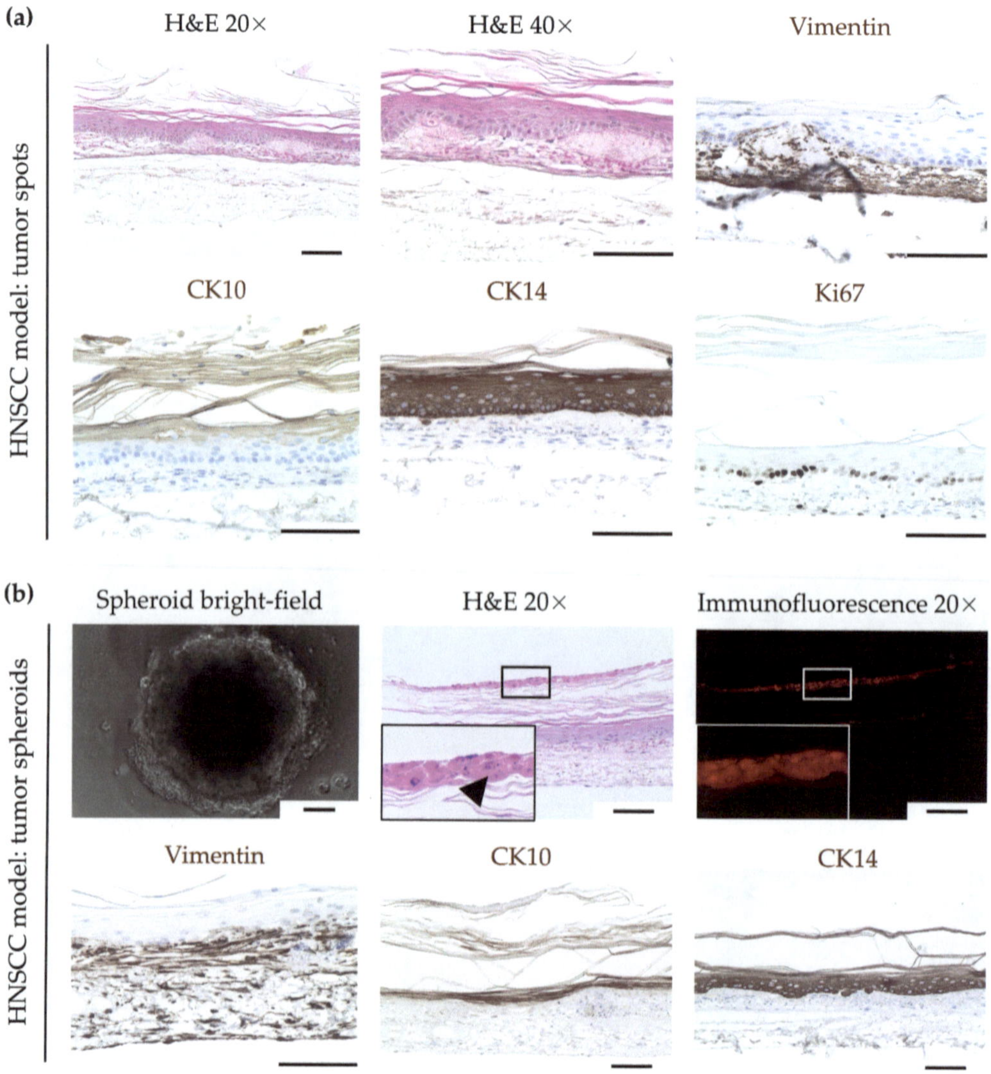

Figure 3. Failed integration of (**a**) FaDu tumor cell spots (**b**) and tumor spheroids into the oral mucosa model (OMM). FaDu cells were transduced lentivirally to constitutively express red fluorescent protein (RFP). Analysis by H&E staining and immunohistochemistry showed a keratinized stratified squamous epithelium, similar to the OMM. While, in (**a**), no tumor cells could be detected, in (**b**), RFP-positive tumor cells appeared on the stratum corneum as remnants of the spheroid. The arrowhead indicates tumor cells atop the stratum corneum, which are strongly eosinophilic with fragmented nuclei. Scale bars represent 100 μm. Head and neck squamous-cell carcinoma (HNSCC), cytokeratin (CK), red fluorescent protein (RFP).

Another method was the seeding of multicellular tumor spheroids (Figure 3b) onto the OMM on day 5 of ALI culture (Figure 1c). Histological analysis after nine more days showed tumor cells located on top of the stratum corneum, as seen with H&E staining and RFP-positive cells. The tumor cells were strongly eosinophilic with fragmented nuclei, possibly indicating apoptotic cells. All epithelial layers below, as well as the lamina propria, had regular morphology without any signs of malignancy, as seen in the OMM. This was further supported by IHC, showing a similar morphology compared to the OMM (Figure 3b).

Generation of a human head and neck squamous-cell carcinoma model through early integration of single tumor cells: Another way to transform the OMM into a human HNSCC model was the early integration of single tumor cells as a mixed suspension consisting of FaDu cells and keratinocytes (Figure 1d). For this purpose, different suspension ratios were seeded into the cell crowns eight days after the incubation of fibroblasts into the SIS/MUC. After a further 13 days in culture, the histological analysis revealed remarkable morphological differences between the cell ratios utilized compared to the previous models (Figure 4).

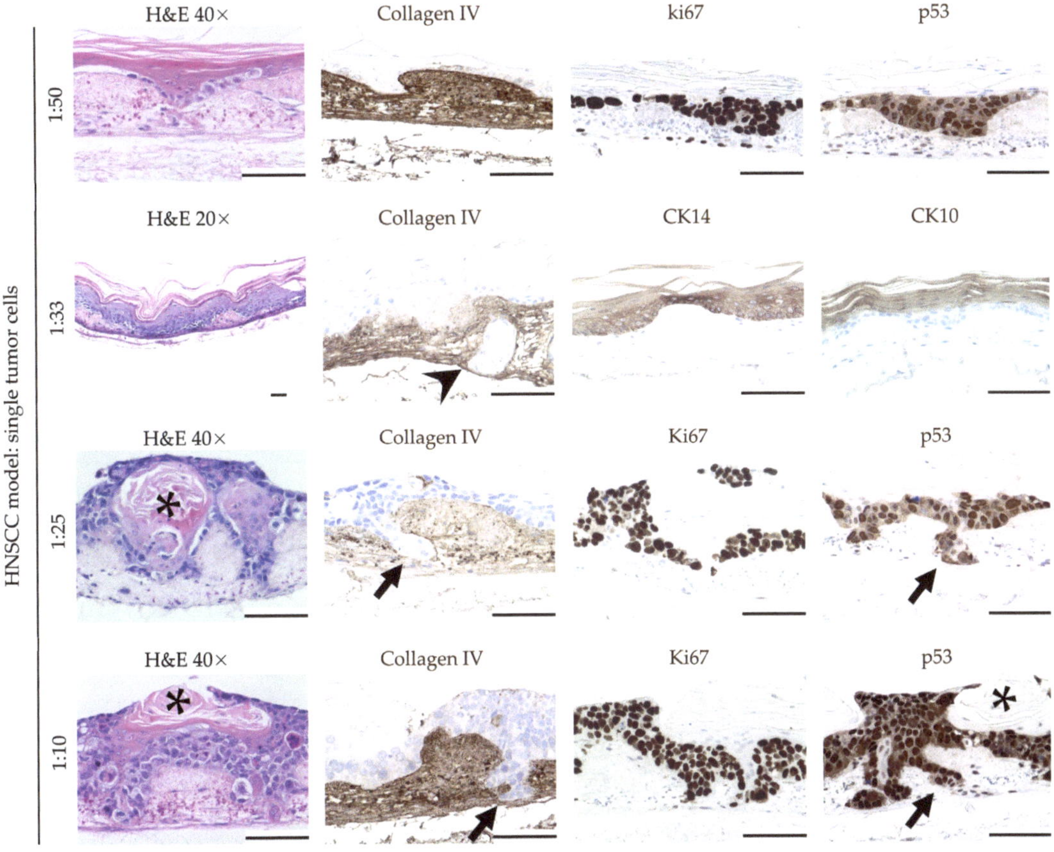

Figure 4. Integration of single tumor cells as a mixed suspension consisting of FaDu cells and keratinocytes. Depending on the mixing ratio, sections of normal stratification could still be detected at ratios up to 1:33. This was no longer the case at a ratio of 1:25. From a ratio of 1:33, there were clear signs of invasive tumor growth with basement membrane penetration and invasion into the lamina propria (arrowhead and arrows). Horn beads (asterisk), as another sign of malignancy, could be seen at ratios of 1:25 and 1:10. Scale bars represent 100 µm. Head and neck squamous-cell carcinoma (HNSCC), cytokeratin (CK).

When using a tumor cell-to-keratinocyte ratio of 1:50, we could still observe signs of stratification, as supported by regular upper cell layers and a continuously formed stratum corneum. The cells within the basal layers appeared to be irregular, with a nuclear pleomorphism and differences in size, indicating the formation of tumor cell clusters. This observation was further supported by local accumulations of Ki67- and p53-positive cells within these clusters. However, a penetration of the basal membrane with invasion into the lamina propria was not detected.

In contrast, at a ratio of 1:33, some segments of the model showed a disturbed epithelial architecture and invasive tumor growth. The arrowhead in Figure 4 highlights a segment of penetration of the basal membrane with invasion into the lamina propria. However, other segments of the model showed regular architecture, with organized stratification and a continuously formed stratum corneum. Basal membrane penetration was also detected at the ratios of 1:25 and 1:10. The arrows in Figure 4 show examples of tumor cell clusters invading into the lamina propria. In addition, most of the epithelial cells were positive for the proliferation marker Ki67 and the tumor suppressor p53. A regular epithelial architecture and an equal distribution of fibroblasts were no longer detectable. Instead, atypical cornification was found (Figure 4, asterisk).

In summary, integrating multicellular tumor spots and tumor spheroids into the OMM on day 5 of ALI culture failed. In contrast, early seeding of a mixture of keratinocytes and FaDu during the formation of the OMM enabled the establishment of a human HNSCC model. Depending on the ratio of keratinocytes and FaDu tumor cells, penetration of the basal membrane as well as a disrupted epithelial architecture could be detected. Table 1 provides a comparison of the results obtained from the different seeding approaches.

Table 1. Comparison of the epithelial characteristics and tumor formation observed from the different seeding approaches.

		Oral Mucosa Model	Tumor Formation
Tumor spots		Regularly formed epithelium	No signs of malignancy or tumor formation
Tumor spheroids		Regularly formed epithelium	Tumor cells located on top of stratum corneum
Single tumor cells	1:50	Continuously formed stratum corneum, cells in the basal layers appeared irregular with nuclear pleomorphism and differences in size	Formation of tumor cell clusters; no invasion into lamina propria
	1:33	Disturbed epithelial architecture	Invasive tumor growth
	1:25	No regular epithelial structure, atypical cornification	Invasive tumor growth
	1:10	No regular epithelial structure, atypical cornification	Invasive tumor growth

4. Discussion

The aim of the present study was to generate a 3D model of HNSCC that mimics the crosstalk at the invasive tumor front of human HNSCC by enabling cellular and stromal interactions. Our model is based on invasive cancer cells within an OMM which was established from primary human cells seeded on a porcine SIS/MUC. The focus of the study was set on morphological analysis based on histology and IHC. In summary, we showed that (1) an OMM could be generated from primary human fibroblasts and keratinocytes seeded onto a porcine SIS/MUC; (2) this OMM was reproducible, showed a regular pattern of differentiation, and therefore shared relevant properties with oral mucosa; and (3) HNSCC cells could be integrated into this model, which ultimately led to the generation of a 3D HNSCC model with features such as basal membrane penetration and cancer cell invasion into the lamina propria.

An OMM was generated as a first step to serve as a basis for the HNSCC model. In line with other studies, we were able to show that primary keratinocytes and fibroblasts were suitable for the generation of a reproducible OMM [36–38]. As in the oral mucosa, we detected a keratinized stratified squamous epithelium in the OMM, which was separated from the lamina propria by a continuously formed basal membrane. IHC for the basal membrane marker collagen IV stained the entire lamina propria. This was most likely due to the natural occurrence of collagen IV in the SIS/MUC [39]. However, a continuous basal membrane was clearly visible by IHC. This, as well as the regular distribution of fibroblasts within the lamina propria, indicate the bio-similarity of the OMM. In addition to a suitable cell carrier, the presence of fibroblasts is particularly important in the establishment of an OMM and HNSCC model. For example, fibroblasts were shown to be significantly involved in the deposition of bioactive molecules within the ECM and in the formation of a basal membrane [40]. In the OMM, fibroblasts were evenly distributed and arranged. Interestingly, this was not the case in the HNSCC model, indicating cross-talk between cancer cells and fibroblasts. In addition to directly promoting tumor cell motility [41], fibroblasts can further support tumor cell invasion by modifying the ECM [42]. Aptly, we showed tumor cell penetration of the basal membrane with invasion into the lamina propria in our HNSCC model, generated by application of a cell mixture of FaDu cells and primary keratinocytes. In contrast, other models of oral dysplasia [43] and squamous cell carcinoma [44] failed to demonstrate this feature. Therefore, our HNSCC model succeeds in mimicking these crucial processes of HNSCC biology. We further showed the formation of tumor cell clusters with an increased number of Ki67 and p53 cells, indicating ongoing cancer cell proliferation. Cancer cell distribution and epithelial differentiation were dependent upon the ratio of FaDu tumor cells to keratinocytes. In our opinion, the optimal ratio seems to be 1:33, as tumor cell clusters with basal membrane penetration can be detected along with segments of regular epithelial differentiation. This was not the case for the other ratios used in this study.

The integration of multicellular tumor spots and tumor spheroids into our OMM on day five of ALI culture did not succeed. However, Colley et al. were able to establish an oral carcinoma in situ model by applying FaDu spheroids onto a 3D OMM generated from normal oral keratinocytes and fibroblasts transferred onto a de-epidermized acellular dermis [38]. Compared to our study, Colley et al. had seeded spheroids on their tissue-engineered OMM already before raising the model to an ALI. In our OMM, a stratum corneum and, thus, a horny layer developed shortly after raising the model to an ALI. Therefore, it is likely that the tumor cells were not able to penetrate this horny layer, thus leading to a nutritional deprivation and cell death. This discrepancy indicates that tumor cells and tumor cell spheroids should be seeded onto the models before a stratum corneum is formed, thus enabling their integration. A potentially optimal time point might be the transition from a submerged to an ALI culture. Comprising tissue-specific primary human keratinocytes and fibroblasts, an OMM served as the basis for our HNSCC model. Importantly, the OMM closely approximated the in vivo situation from a histological and immunohistochemical point of view, enhancing the model's biological relevance and its suitability as a basis for our HNSCC model. This model stands out by replicating the stroma through the use of a bioactive 3D cell carrier which mimics the tissue-specific ECM composition and features a basal membrane. Furthermore, our SIS/MUC-based models gain significance from the considerable preservation of extracellular matrix (ECM) proteins across the evolutionary development of porcine and human organisms [45], suggesting that our porcine-derived scaffold is apt for developing an HNSCC model. Interestingly, it even becomes possible to implant porcine scaffold-based constructs into patients following the scaffold recellularization with human cells [46]. In summary, key aspects of our model are a biosimilar composition of the ECM as well as invasive tumor cell clusters with basal membrane penetration and lamina propria infiltration between segments of regular epithelial differentiation, based on an OMM from primary human fibroblasts and keratinocytes. This model might help to enable further studies on tumor cell invasion as well as on the cellular and stromal interactions at the invasive tumor front. However, limitations need to be addressed. Our study aimed to develop a cancer model to elucidate general aspects of HNSCC pathophysiology, utilizing the hypopharyngeal cancer cell line FaDu integrated into an OMM. This approach is supported by literature that identifies commonalities in the disease process across various sites within the head and neck region. However, it is also crucial to acknowledge that squamous cell carcinomas can exhibit significant variations depending on their anatomical origin [47,48]. Therefore, there is a need for subsequent studies to validate these findings using further cancer cell lines from other locations of the head and neck, aiming to enhance the translatability of the results to HNSCC. Our HNSCC model does not necessarily correspond to the in vivo conditions within the TME, as it lacks key features of the natural environment. Achieving full biological relevance is difficult due to the inherent complexity of the in vivo TME, containing diverse cell types and extracellular matrix components. Our model lacks important cell types of the TME such as inflammatory and further mesenchymal cells. However, as already demonstrated by others, additional cell types such as endothelial cells [49] can be integrated into ECM-based biological scaffolds. Cell lines that have been immortalized undergo manipulation to enable indefinite proliferation. An extended period in an in vitro setting can lead to alterations in the inherent characteristics of the cell population [50]. In addition to the integration of an immortalized cell line, as was performed in our work, further approaches such as the integration of patient-derived primary tumor explants proved to be feasible in another 3D model of head and neck cancer consisting of fibroblasts seeded on a viscose fiber fabric. In contrast to cell lines, tumor explants offer the benefit of maintaining key characteristics of the cancer lesion crucial for replicating the TME, such as leukocyte infiltration [51]. Isolating primasry tumor cells directly from patient samples would better replicate patient-specific conditions, potentially increasing the model's relevance in clinical settings. In the future, such approaches might enable testing treatment responses to various anticancer drugs, assessing the tumor's sensitivity to radiotherapy, or correlating the models to clinical histopathological characteristics like invasion and tumor growth. All in vitro models are exposed to certain microenvironmental factors, such as gradients of oxygen, nutrients, signaling molecules, and biophysical stimuli. The use of patient-derived serum, rather than calf serum, in the model could better mimic the natural tumor microenvironment, thereby providing more physiologically relevant data. However, these factors can only be controlled partially and do not exactly represent the in vivo environment of human tumors. Future efforts in HNSCC model development should focus on addressing these limitations by advancing models that incorporate primary HNSCC tumor tissue alongside various non-malignant cell types, aiming for a more comprehensive representation of the TME. By bridging the gap between basic in vitro conditions and the in vivo TME, such a model could additionally contribute to TME research and ultimately serve as platform for anti-cancer drug testing.

5. Conclusions

Primary human fibroblasts and keratinocytes seeded onto a porcine SIS/MUC formed a bioequivalent OMM. Cancer cells could be integrated into this OMM to generate an HNSCC model, which showed features of tumor cell invasiveness. Future work should focus on the integration of further components of the TME.

Author Contributions: Conceptualization, M.S. (Manuel Stöth), A.T.M., F.S., T.J.M., L.H., A.S., M.S. (Maria Steinke), A.R. and S.H.; formal analysis, M.S. (Manuel Stöth), A.T.M., F.S. and L.H.; investigation, A.T.M. and F.S.; methodology, A.T.M., F.S. and A.R.; project administration, A.R. and S.H.; resources, A.R. and S.H.; supervision, M.S. (Maria Steinke), A.R. and S.H.; visualization, M.S. (Manuel Stöth), A.T.M., F.S., T.J.M. and F.M.-D.; writing—original draft, M.S. (Manuel Stöth), T.J.M., A.S. and S.H.; writing—review and editing, M.S. (Manuel Stöth), A.T.M., T.J.M., F.M.-D., L.H., A.S., M.S. (Maria Steinke), A.R. and S.H. All authors have read and agreed to the published version of the manuscript.

Funding: This publication was supported by the Open Access Publication Fund of the University of Wuerzburg.

Institutional Review Board Statement: This study was conducted according to the guidelines of the Declaration of Helsinki and was approved by the local ethic committee (approval number: 182/10). Animal experiments were performed after previous approval by the ethical committee of the local government (approval number: 55.2-2532-2-256). All animal procedures conformed to the guidelines of Directive 2010/63/EU of the European Parliament on the protection of animals used for scientific purposes.

Informed Consent Statement: Informed consent was obtained from all subjects involved in the study.

Data Availability Statement: All data generated or analyzed during this study are included in this article. Further inquiries can be directed to the corresponding author.

Acknowledgments: We thank Heike Walles for her advice and insightful suggestions throughout the course of this research.

Conflicts of Interest: All authors have no conflicts of interest to declare that are relevant to the content of this article.

References

1. Barsouk, A.; Aluru, J.S.; Rawla, P.; Saginala, K.; Barsouk, A. Epidemiology, Risk Factors, and Prevention of Head and Neck Squamous Cell Carcinoma. *Med. Sci.* **2023**, *11*, 42. [CrossRef] [PubMed]
2. Pelucchi, C.; Gallus, S.; Garavello, W.; Bosetti, C.; La Vecchia, C. Cancer risk associated with alcohol and tobacco use: Focus on upper aero-digestive tract and liver. *Alcohol. Res. Health* **2006**, *29*, 193–198. [PubMed]
3. Warnakulasuriya, S.; Chen, T.H.H. Areca Nut and Oral Cancer: Evidence from Studies Conducted in Humans. *J. Dent. Res.* **2022**, *101*, 1139–1146. [CrossRef] [PubMed]
4. Hashim, D.; Sartori, S.; Brennan, P.; Curado, M.P.; Wunsch-Filho, V.; Divaris, K.; Olshan, A.F.; Zevallos, J.P.; Winn, D.M.; Franceschi, S.; et al. The role of oral hygiene in head and neck cancer: Results from International Head and Neck Cancer Epidemiology (INHANCE) consortium. *Ann. Oncol.* **2016**, *27*, 1619–1625. [CrossRef] [PubMed]
5. Lazos, J.P.; Piemonte, E.D.; Lanfranchi, H.E.; Brunotto, M.N. Characterization of Chronic Mechanical Irritation in Oral Cancer. *Int. J. Dent.* **2017**, *2017*, 6784526. [CrossRef] [PubMed]
6. Borsetto, D.; Sethi, M.; Polesel, J.; Tomasoni, M.; Deganello, A.; Nicolai, P.; Bossi, P.; Fabbris, C.; Molteni, G.; Marchioni, D.; et al. The risk of recurrence in surgically treated head and neck squamous cell carcinomas: A conditional probability approach. *Acta Oncol.* **2021**, *60*, 942–947. [CrossRef]
7. Faraji, F.; Cohen, E.E.W.; Guo, T.W. Evolving treatment paradigms in recurrent and metastatic head and neck squamous cell carcinoma: The emergence of immunotherapy. *Transl. Cancer Res.* **2023**, *12*, 1353–1358. [CrossRef]
8. Gatta, G.; Botta, L.; Sanchez, M.J.; Anderson, L.A.; Pierannunzio, D.; Licitra, L.; Group, E.W. Prognoses and improvement for head and neck cancers diagnosed in Europe in early 2000s: The EUROCARE-5 population-based study. *Eur. J. Cancer* **2015**, *51*, 2130–2143. [CrossRef] [PubMed]
9. Kapalczynska, M.; Kolenda, T.; Przybyla, W.; Zajaczkowska, M.; Teresiak, A.; Filas, V.; Ibbs, M.; Blizniak, R.; Luczewski, L.; Lamperska, K. 2D and 3D cell cultures—A comparison of different types of cancer cell cultures. *Arch. Med. Sci.* **2018**, *14*, 910–919. [CrossRef]
10. Melissaridou, S.; Wiechec, E.; Magan, M.; Jain, M.V.; Chung, M.K.; Farnebo, L.; Roberg, K. The effect of 2D and 3D cell cultures on treatment response, EMT profile and stem cell features in head and neck cancer. *Cancer Cell Int.* **2019**, *19*, 16. [CrossRef]
11. Loessner, D.; Stok, K.S.; Lutolf, M.P.; Hutmacher, D.W.; Clements, J.A.; Rizzi, S.C. Bioengineered 3D platform to explore cell-ECM interactions and drug resistance of epithelial ovarian cancer cells. *Biomaterials* **2010**, *31*, 8494–8506. [CrossRef]
12. Chen, Y.Q.; Kuo, J.C.; Wei, M.T.; Wu, M.C.; Yang, M.H.; Chiou, A. Fibroblast Promotes Head and Neck Squamous Cell Carcinoma Cell Invasion through Mechanical Barriers in 3D Collagen Microenvironments. *ACS Appl. Bio Mater.* **2020**, *3*, 6419–6429. [CrossRef] [PubMed]
13. Pickup, M.W.; Mouw, J.K.; Weaver, V.M. The extracellular matrix modulates the hallmarks of cancer. *EMBO Rep.* **2014**, *15*, 1243–1253. [CrossRef]
14. Fang, Z.; Li, P.; Du, F.; Shang, L.; Li, L. The role of organoids in cancer research. *Exp. Hematol. Oncol.* **2023**, *12*, 69. [CrossRef]
15. Buskin, A.; Scott, E.; Nelson, R.; Gaughan, L.; Robson, C.N.; Heer, R.; Hepburn, A.C. Engineering prostate cancer in vitro: What does it take? *Oncogene* **2023**, *42*, 2417–2427. [CrossRef]
16. Mermod, M.; Hiou-Feige, A.; Bovay, E.; Roh, V.; Sponarova, J.; Bongiovanni, M.; Vermeer, D.W.; Lee, J.H.; Petrova, T.V.; Rivals, J.P.; et al. Mouse model of postsurgical primary tumor recurrence and regional lymph node metastasis progression in HPV-related head and neck cancer. *Int. J. Cancer* **2018**, *142*, 2518–2528. [CrossRef] [PubMed]
17. Fu, Y.; Tian, G.; Li, J.; Zhang, Z.; Xu, K. An HNSCC syngeneic mouse model for tumor immunology research and preclinical evaluation. *Int. J. Mol. Med.* **2020**, *46*, 1501–1513. [CrossRef] [PubMed]
18. Brand, M.; Laban, S.; Theodoraki, M.N.; Doescher, J.; Hoffmann, T.K.; Schuler, P.J.; Brunner, C. Characterization and Differentiation of the Tumor Microenvironment (TME) of Orthotopic and Subcutaneously Grown Head and Neck Squamous Cell Carcinoma (HNSCC) in Immunocompetent Mice. *Int. J. Mol. Sci.* **2020**, *22*, 247. [CrossRef]
19. Ozawa, H.; Ranaweera, R.S.; Izumchenko, E.; Makarev, E.; Zhavoronkov, A.; Fertig, E.J.; Howard, J.D.; Markovic, A.; Bedi, A.; Ravi, R.; et al. SMAD4 Loss Is Associated with Cetuximab Resistance and Induction of MAPK/JNK Activation in Head and Neck Cancer Cells. *Clin. Cancer Res.* **2017**, *23*, 5162–5175. [CrossRef]

20. Wang, Y.; Zhu, Y.; Wang, Q.; Hu, H.; Li, Z.; Wang, D.; Zhang, W.; Qi, B.; Ye, J.; Wu, H.; et al. The histone demethylase LSD1 is a novel oncogene and therapeutic target in oral cancer. *Cancer Lett.* **2016**, *374*, 12–21. [CrossRef]
21. Feng, X.; Luo, Q.; Zhang, H.; Wang, H.; Chen, W.; Meng, G.; Chen, F. The role of NLRP3 inflammasome in 5-fluorouracil resistance of oral squamous cell carcinoma. *J. Exp. Clin. Cancer Res.* **2017**, *36*, 81. [CrossRef] [PubMed]
22. Mak, I.W.; Evaniew, N.; Ghert, M. Lost in translation: Animal models and clinical trials in cancer treatment. *Am. J. Transl. Res.* **2014**, *6*, 114–118. [PubMed]
23. Ben-David, U.; Ha, G.; Tseng, Y.Y.; Greenwald, N.F.; Oh, C.; Shih, J.; McFarland, J.M.; Wong, B.; Boehm, J.S.; Beroukhim, R.; et al. Patient-derived xenografts undergo mouse-specific tumor evolution. *Nat. Genet.* **2017**, *49*, 1567–1575. [CrossRef] [PubMed]
24. Jin, J.; Yoshimura, K.; Sewastjanow-Silva, M.; Song, S.; Ajani, J.A. Challenges and Prospects of Patient-Derived Xenografts for Cancer Research. *Cancers* **2023**, *15*, 4352. [CrossRef]
25. Peng, S.; Creighton, C.J.; Zhang, Y.; Sen, B.; Mazumdar, T.; Myers, J.N.; Lai, S.Y.; Woolfson, A.; Lorenzi, M.V.; Bell, D.; et al. Tumor grafts derived from patients with head and neck squamous carcinoma authentically maintain the molecular and histologic characteristics of human cancers. *J. Transl. Med.* **2013**, *11*, 198. [CrossRef]
26. Breslin, S.; O'Driscoll, L. Three-dimensional cell culture: The missing link in drug discovery. *Drug Discov. Today* **2013**, *18*, 240–249. [CrossRef]
27. Moharamzadeh, K.; Brook, I.M.; Van Noort, R.; Scutt, A.M.; Thornhill, M.H. Tissue-engineered oral mucosa: A review of the scientific literature. *J. Dent. Res.* **2007**, *86*, 115–124. [CrossRef]
28. Beckstead, B.L.; Pan, S.; Bhrany, A.D.; Bratt-Leal, A.M.; Ratner, B.D.; Giachelli, C.M. Esophageal epithelial cell interaction with synthetic and natural scaffolds for tissue engineering. *Biomaterials* **2005**, *26*, 6217–6228. [CrossRef]
29. Mertsching, H.; Walles, T.; Hofmann, M.; Schanz, J.; Knapp, W.H. Engineering of a vascularized scaffold for artificial tissue and organ generation. *Biomaterials* **2005**, *26*, 6610–6617. [CrossRef]
30. Stratmann, A.T.; Fecher, D.; Wangorsch, G.; Gottlich, C.; Walles, T.; Walles, H.; Dandekar, T.; Dandekar, G.; Nietzer, S.L. Establishment of a human 3D lung cancer model based on a biological tissue matrix combined with a Boolean in silico model. *Mol. Oncol.* **2014**, *8*, 351–365. [CrossRef]
31. Steinke, M.; Gross, R.; Walles, H.; Gangnus, R.; Schutze, K.; Walles, T. An engineered 3D human airway mucosa model based on an SIS scaffold. *Biomaterials* **2014**, *35*, 7355–7362. [CrossRef] [PubMed]
32. Rangan, S.R. A new human cell line (FaDu) from a hypopharyngeal carcinoma. *Cancer* **1972**, *29*, 117–121. [CrossRef] [PubMed]
33. Fecher, D.; Hofmann, E.; Buck, A.; Bundschuh, R.; Nietzer, S.; Dandekar, G.; Walles, T.; Walles, H.; Luckerath, K.; Steinke, M. Human Organotypic Lung Tumor Models: Suitable For Preclinical 18F-FDG PET-Imaging. *PLoS ONE* **2016**, *11*, e0160282. [CrossRef] [PubMed]
34. Sivarajan, R.; Kessie, D.K.; Oberwinkler, H.; Pallmann, N.; Walles, T.; Scherzad, A.; Hackenberg, S.; Steinke, M. Susceptibility of Human Airway Tissue Models Derived From Different Anatomical Sites to Bordetella pertussis and Its Virulence Factor Adenylate Cyclase Toxin. *Front. Cell Infect. Microbiol.* **2021**, *11*, 797491. [CrossRef]
35. Bonan, P.R.; Kaminagakura, E.; Pires, F.R.; Vargas, P.A.; Almeida, O.P. Cytokeratin expression in initial oral mucositis of head and neck irradiated patients. *Oral Surg. Oral Med. Oral Pathol. Oral Radiol. Endod.* **2006**, *101*, 205–211. [CrossRef] [PubMed]
36. Bhargava, S.; Chapple, C.R.; Bullock, A.J.; Layton, C.; MacNeil, S. Tissue-engineered buccal mucosa for substitution urethroplasty. *BJU Int.* **2004**, *93*, 807–811. [CrossRef]
37. Moharamzadeh, K.; Brook, I.M.; Van Noort, R.; Scutt, A.M.; Smith, K.G.; Thornhill, M.H. Development, optimization and characterization of a full-thickness tissue engineered human oral mucosal model for biological assessment of dental biomaterials. *J. Mater. Sci. Mater. Med.* **2008**, *19*, 1793–1801. [CrossRef] [PubMed]
38. Colley, H.E.; Hearnden, V.; Jones, A.V.; Weinreb, P.H.; Violette, S.M.; Macneil, S.; Thornhill, M.H.; Murdoch, C. Development of tissue-engineered models of oral dysplasia and early invasive oral squamous cell carcinoma. *Br. J. Cancer* **2011**, *105*, 1582–1592. [CrossRef]
39. Lindberg, K.; Badylak, S.F. Porcine small intestinal submucosa (SIS): A bioscaffold supporting in vitro primary human epidermal cell differentiation and synthesis of basement membrane proteins. *Burns* **2001**, *27*, 254–266. [CrossRef]
40. Kulasekara, K.K.; Lukandu, O.M.; Neppelberg, E.; Vintermyr, O.K.; Johannessen, A.C.; Costea, D.E. Cancer progression is associated with increased expression of basement membrane proteins in three-dimensional in vitro models of human oral cancer. *Arch. Oral Biol.* **2009**, *54*, 924–931. [CrossRef]
41. Sun, L.P.; Xu, K.; Cui, J.; Yuan, D.Y.; Zou, B.; Li, J.; Liu, J.L.; Li, K.Y.; Meng, Z.; Zhang, B. Cancer-associated fibroblast-derived exosomal miR-382-5p promotes the migration and invasion of oral squamous cell carcinoma. *Oncol. Rep.* **2019**, *42*, 1319–1328. [CrossRef] [PubMed]
42. Bienkowska, K.J.; Hanley, C.J.; Thomas, G.J. Cancer-Associated Fibroblasts in Oral Cancer: A Current Perspective on Function and Potential for Therapeutic Targeting. *Front. Oral Health* **2021**, *2*, 686337. [CrossRef] [PubMed]
43. Gaballah, K.; Costea, D.E.; Hills, A.; Gollin, S.M.; Harrison, P.; Partridge, M. Tissue engineering of oral dysplasia. *J. Pathol.* **2008**, *215*, 280–289. [CrossRef] [PubMed]
44. Brauchle, E.; Johannsen, H.; Nolan, S.; Thude, S.; Schenke-Layland, K. Design and analysis of a squamous cell carcinoma in vitro model system. *Biomaterials* **2013**, *34*, 7401–7407. [CrossRef] [PubMed]
45. Ozbek, S.; Balasubramanian, P.G.; Chiquet-Ehrismann, R.; Tucker, R.P.; Adams, J.C. The evolution of extracellular matrix. *Mol. Biol. Cell* **2010**, *21*, 4300–4305. [CrossRef] [PubMed]

46. Steinke, M.; Dally, I.; Friedel, G.; Walles, H.; Walles, T. Host-integration of a tissue-engineered airway patch: Two-year follow-up in a single patient. *Tissue Eng. Part A* **2015**, *21*, 573–579. [CrossRef] [PubMed]
47. Freier, K.; Joos, S.; Flechtenmacher, C.; Devens, F.; Benner, A.; Bosch, F.X.; Lichter, P.; Hofele, C. Tissue microarray analysis reveals site-specific prevalence of oncogene amplifications in head and neck squamous cell carcinoma. *Cancer Res.* **2003**, *63*, 1179–1182.
48. Huang, Q.; Yu, G.P.; McCormick, S.A.; Mo, J.; Datta, B.; Mahimkar, M.; Lazarus, P.; Schaffer, A.A.; Desper, R.; Schantz, S.P. Genetic differences detected by comparative genomic hybridization in head and neck squamous cell carcinomas from different tumor sites: Construction of oncogenetic trees for tumor progression. *Genes Chromosomes Cancer* **2002**, *34*, 224–233. [CrossRef] [PubMed]
49. Kress, S.; Baur, J.; Otto, C.; Burkard, N.; Braspenning, J.; Walles, H.; Nickel, J.; Metzger, M. Evaluation of a Miniaturized Biologically Vascularized Scaffold in vitro and in vivo. *Sci. Rep.* **2018**, *8*, 4719. [CrossRef]
50. Miserocchi, G.; Mercatali, L.; Liverani, C.; De Vita, A.; Spadazzi, C.; Pieri, F.; Bongiovanni, A.; Recine, F.; Amadori, D.; Ibrahim, T. Management and potentialities of primary cancer cultures in preclinical and translational studies. *J. Transl. Med.* **2017**, *15*, 229. [CrossRef]
51. Engelmann, L.; Thierauf, J.; Koerich Laureano, N.; Stark, H.J.; Prigge, E.S.; Horn, D.; Freier, K.; Grabe, N.; Rong, C.; Federspil, P.; et al. Organotypic Co-Cultures as a Novel 3D Model for Head and Neck Squamous Cell Carcinoma. *Cancers* **2020**, *12*, 2330. [CrossRef] [PubMed]

current issues in molecular biology

MDPI

Communication

Sequencing Analysis of *MUC6* and *MUC16* Gene Fragments in Patients with Oropharyngeal Squamous Cell Carcinoma Reveals Novel Mutations: A Preliminary Study

Jadwiga Gaździcka [1,*], Krzysztof Biernacki [1], Silvia Salatino [2], Karolina Gołąbek [1], Dorota Hudy [1], Agata Świętek [1,3], Katarzyna Miśkiewicz-Orczyk [4], Anna Koniewska [4], Maciej Misiołek [4] and Joanna Katarzyna Strzelczyk [1]

1 Department of Medical and Molecular Biology, Faculty of Medical Sciences in Zabrze, Medical University of Silesia in Katowice, Jordana 19, 41-808 Zabrze, Poland
2 Molecular Biology, Core Research Laboratories, Natural History Museum, London SW7 5BD, UK
3 Silesia LabMed Research and Implementation Centre, Medical University of Silesia in Katowice, Jordana 19, 41-808 Zabrze, Poland
4 Department of Otorhinolaryngology and Oncological Laryngology, Faculty of Medical Sciences in Zabrze, Medical University of Silesia in Katowice, C. Skłodowskiej 10, 41-800 Zabrze, Poland
* Correspondence: jgazdzicka@sum.edu.pl

Abstract: The growing incidence of oropharyngeal squamous cell carcinoma (OPSCC) calls for better understanding of the mutational landscape of such cases. Mucins (MUCs) are multifunctional glycoproteins expressed by the epithelial cells and may be associated with the epithelial tumour invasion and progression. The present study aimed at the analysis of the sequence of selected *MUC6* and *MUC16* gene fragments in the tumour, as well as the margin, samples obtained from 18 OPSCC patients. Possible associations between the detected mutations and the clinicopathological and demographic characteristics of the study group were analysed. Sanger sequencing and bioinformatic data analysis of the selected *MUC6* and *MUC16* cDNA fragments were performed. Our study found 13 and 3 mutations in *MUC6* and *MUC16*, respectively. In particular, one novelty variant found that the *MUC6* gene (chr11:1018257 A>T) was the most frequent across our cohort, in both the tumour and the margin samples, and was then classified as a high impact, stop-gain mutation. The current study found novel mutations in *MUC6* and *MUC16* providing new insight into the genetic alternation in mucin genes among the OPSCC patients. Further studies, including larger cohorts, are recommended to recognise the pattern in which the mutations affect oropharyngeal carcinogenesis.

Keywords: oropharyngeal squamous cell carcinoma (OPSCC); mutation; *MUC6*; *MUC16*

Citation: Gaździcka, J.; Biernacki, K.; Salatino, S.; Gołąbek, K.; Hudy, D.; Świętek, A.; Miśkiewicz-Orczyk, K.; Koniewska, A.; Misiołek, M.; Strzelczyk, J.K. Sequencing Analysis of *MUC6* and *MUC16* Gene Fragments in Patients with Oropharyngeal Squamous Cell Carcinoma Reveals Novel Mutations: A Preliminary Study. *Curr. Issues Mol. Biol.* **2023**, *45*, 5645–5661. https://doi.org/10.3390/cimb45070356

Academic Editors: Violeta Popovici and Emma Adriana Ozon

Received: 7 June 2023
Revised: 29 June 2023
Accepted: 2 July 2023
Published: 4 July 2023

1. Introduction

Mucins are multifunctional glycoproteins expressed by epithelial cells in a variety of tissues [1] and classified as membrane-bound or secreted mucins, the latter further divided into gel-forming and non-gel-forming ones [2]. Membrane-bound mucins are important in numerous biological processes, including molecular cell signalling [3]. Acting like receptors, they can conduct signals from the environment to the cell, therefore influencing proliferation, differentiation or apoptosis [4]. Some studies suggest an association between mutations in *MUC16* and the immune response and the cell cycle in cancer patients [5,6]. MUC16 has a role in the maintenance of the mucosa and acts as a barrier against external agents [7], such as bacterial adherence [8,9]. On the other hand, secreted mucins (e.g., MUC6) play a key role in the protection of the epithelium against infection agents, chemical injury or dehydration by forming a mucus layer [10]. MUC6 expresses O-glycans, which may take an important part in bacterial growth control [11]. Importantly, during carcinogenesis mucins may have an important role in cancer cell differentiation and metastasis [12].

In 2020, oropharyngeal cancer was diagnosed in 98,412 patients worldwide and caused the death of 48,143 people [13]. Most oropharyngeal cancers are squamous cell carcinomas (OPSCCs), arising from the mucosa epithelium of the oropharynx. As for all other types of head and neck squamous cell carcinomas (HNSCCs), multiple risk factors are known for OPSCCs: use of tobacco and alcohol, poor oral hygiene, ageing, environmental pollutants and viral infection, such as human papillomavirus (HPV) [14]. Favourably, HPV vaccines are effective against the virus, especially for genotypes HPV-16 and HPV-18, associated with OPSCC [15]. The median age for HPV-positive OPSCC is 53 and 58 years for HPV-negative [16]. OPSCC patients with HPV infection, whose number is increasing worldwide [17], have more favourable prognoses [18,19]. HPV-negative tumours located in the head and neck demonstrated different somatic mutations compared to HPV-positive HNSCC [20]. Sequencing analysis of HNSCC from different anatomic sites, with or without HPV, revealed a markedly diverse landscape of mutations [21]. Comparison of HPV-positive tonsillar OPSCC and HPV-negative oral squamous cell carcinoma (OSCC) revealed a mutation in *MUC12* shared by both types of cancers, and a higher number of mucin genes with mutations were noticed in HPV-positive OPSCC patients [22].

The present study aimed to analyse the sequence of some selected fragments of *MUC6* and *MUC16* genes in the tumour and the margin samples obtained from the OPSCC patients. Some possible associations between the detected mutations and the clinicopathological or demographic characteristics of the study group were also analysed. Moreover, the identified variants in *MUC6* and *MUC16* genes were evaluated in relation to their HPV status. Figure 1 presents the study flow chart.

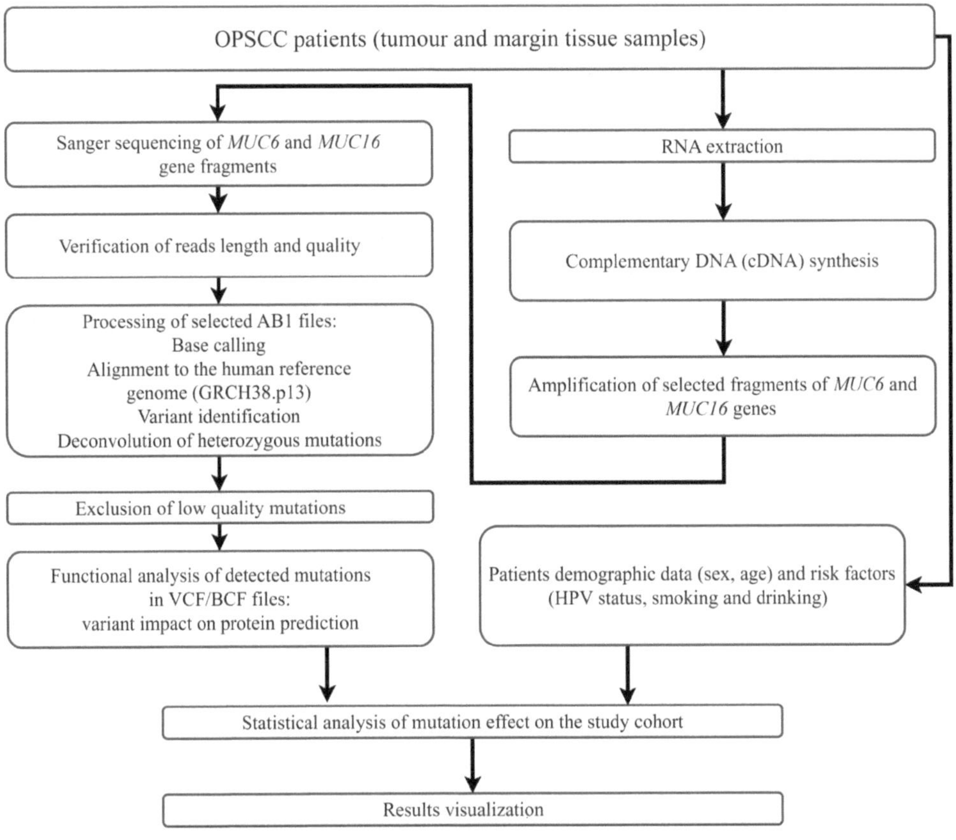

Figure 1. Flow chart of the study.

2. Materials and Methods

2.1. Patients and Samples

The study population comprised 18 OPSCC patients recruited at the Department of Otorhinolaryngology and Oncological Laryngology in Zabrze, Medical University of Silesia in Katowice (Poland). The main inclusion criterion was the diagnosis of primary OPSCC, while the exclusion criteria included preoperative chemotherapy and radiotherapy. All of the cases derived from a Polish, white population. The key data (age, sex, medical history and use of tobacco and alcohol) were collected through an ad hoc questionnaire. The study was approved of by the Ethics Committee of the Medical University of Silesia (no. KNW/0022/KB1/49/16 and KNW/0022/KB1/49/II/16/17). Informed consent was obtained from all the patients enrolled in the study. This work was supported by a grant from the Medical University of Silesia (KNW-2-O07/N/9/N).

Two tissue samples were obtained from each of the OPSCC patients: a tumour sample and a histologically normal surgical margin sample. All samples were collected during the surgical resection. The anatomical location of the OPSCC samples comprised the palatine tonsils only. The tumour samples contained the histopathologically proven primary OPSCC cells. An R0 resection (microscopically margin negative resection) was performed in all the patients and the margin samples were histopathologically confirmed as free of cancerous cells and dysplasia. The tumour stage was categorised according to the International Union Against Cancer (UICC) classification of head and neck tumours (7th Edition) [23]. After resection, all specimens were immediately immersed in RNAlater® (Sigma-Aldrich, Saint Louis, MO, USA) and frozen at −80 °C pending RNA extraction. Figure S1 shows the detailed procedures of our laboratory work.

2.2. MUC 6 and MUC16 Sequencing

2.2.1. RNA Extraction

Homogenisation of each tumour and of the margin sample was the first step preceding RNA extraction and use of ceramic beads Lysing Matrix D (MP Biomedicals, Irvine, CA, USA) in FastPrep®-24 homogeniser (MP Biomedicals, Irvine, CA, USA). RNA extraction was performed with the use of an RNA isolation kit (BioVendor, Brno, Czech Republic) according to the manufacturer's protocol. Estimation of the quality and quantity of the extracted RNA was performed on a NanoPhotometer® Pearl spectrophotometer (IMPLEN, Munich, Germany).

2.2.2. Complementary DNA (cDNA) Synthesis

A total of 5 ng RNA was reverse-transcribed into complementary DNA with the High Capacity cDNA Reverse Transcription Kit with RNase Inhibitor (Applied Biosystems™, Waltham, MA, USA) according to the manufacturer's instructions. The reaction was run in a 20 µL volume. Briefly, 10 µL of the previously extracted RNA was added to 2 µL of 10× Buffer RT, 2 µL of 10× RT Random Primers, 0.8 µL of 25× dNTP mix (100 mM), 1 µL of MultiScribe™ Reverse Transcriptase, 1 µL of RNase inhibitor, and 3.2 µL of nuclease-free H_2O. The thermal parameters of the reaction were as follows: 25 °C for 10 min, 37 °C for 120 min, 85 °C for 5 min and cooldown to 4 °C–∞. The reaction was performed in Mastercycler personal (Eppendorf AG, Hamburg, Germany).

2.2.3. Amplification of Selected Fragments of MUC6 and MUC16 Genes

Both sets of primers were designed using Primer-BLAST 3 (NCBI, Bethesda, MD, USA) to target all mRNA sequences of *MUC6* and *MUC16* genes. The primers were synthesised by Genomed (Genomed Joined-Stock Company, Warsaw, Poland), and their sequences are listed in Table 1. The first criterion for the amplified product selection was the presence of known variants in the central part of the amplified sequence. The secondary condition was the presence of the amplified fragment throughout all the known transcription variants of the gene.

Table 1. Primer sequences for the *MUC6* and *MUC16* transcripts fragments.

Gene	Forward Primer 5′—3′	Reverse Primer 5′—3′
MUC6 fragment (606 bp)	GAAGGATGTTGCCGTCATGG	ACTGAATACACAACGCCCCA
MUC16 fragment (653 bp)	ACAGGCTGGGTCACAAGTTC	GGCGAGGTTGTAGCATGGAT

The PCR reaction was run in a 20 μL volume. The amplification reaction was accomplished with Platinum™ II Hot-Start PCR Master Mix (2X) (Invitrogen, Waltham, MA, USA), according to the manufacturer's protocol. Briefly, 10 μL of Platinum™ II Hot-Start PCR Master Mix was mixed with 0.4 μL 10 μM forward primer, 0.4 μL 10 μM reverse primer, 1 μL cDNA and 8.2 μL nuclease-free water. The PCR conditions were as follows: denaturation at 94 °C for 2 min, and 35 cycles of 94 °C for 15 s, 60 °C for 15 s and 68 °C for 15 s.

2.2.4. Purification of the Selected MUC6 and MUC16 Gene Amplificated Fragments

Purification of the amplification products was performed with ExoSAP-IT™ Express PCR Product Cleanup Reagent (Applied Biosystems™, Waltham, MA, USA), according to the manufacturer's instructions. Briefly, 5 μL of each PCR product was mixed with 2 μL ExoSAP-IT™ Express PCR Product Cleanup Reagent, vortex and incubated at 37 °C for 4 min and 80 °C for 1 min.

2.2.5. Cycle Sequencing Reaction

The cycle sequencing reaction was performed with BigDye™ Terminator v3.1 Cycle Sequencing Kit (Applied Biosystems™, Waltham, MA, USA), according to the manufacturer's protocol. The 3 μL of the purified PCR product was added to the reaction mixture containing 2 μL BigDye™ Terminator v3.1 Ready Reaction Mix, 1 μL of 5 × Sequencing Buffer, 3 μL deionised water and 1 μL forward/reverse primer (3.2 μM). The sequencing reaction was run in the following conditions: 96 °C for 1 min, and 25 cycles of 96 °C for 10 s, 50 °C for 5 s and 60 °C for 4 min. The reaction was run in QuantStudio 5 Real-Time PCR system (Applied Biosystems™, Waltham, MA, USA).

2.2.6. Purification of the Templates and Capillary Electrophoresis

BigDye XTerminator™ Purification Kit (Applied Biosystems™, Waltham, MA, USA) was used to purify the DNA sequencing reaction and prepared according to manufacturer's protocol. Briefly, 55 μL of SAM/BigDye XTerminator™ bead working solution was added to each sample on the sequencing plate and vortexed for 30 min at 2000 rpm (IKA MS3 Digital, IKA Werke GmbH&Co. KG, Staufen, Germany). The last step was to centrifuge the plate at 1000× *g* for 2 min. The supernatant of each sample was analysed in a 3130 Genetic Analyzer (Applied Biosystems™, Waltham, MA, USA) using 3130 POP-7™ Performance-Optimized Polymer (Thermo Fisher Scientific, Waltham, MA, USA). The results were saved in the 3130 Genetic Analyzer software (Applied Biosystems™, Waltham, MA, USA).

2.3. Bioinformatic Analysis

Analysis of the Sanger sequencing data (AB1 files), including base calling, alignment to the human reference sequence (genome assembly GRCh38.p13), variant identification and deconvolution of heterozygous mutations, were performed using Tracy [24]. BCFtools (version 1.8) was used to handle VCF/BCF files, whereas all downstream functional annotations and predictions, including the variant impact on protein, SIFT (version sift5.2.2) [25] and PolyPhen-2 (version 2.2.2) scores [26], were determined using Variant Effect Predictor (VEP v103) [27]. All the acquired traces were manually inspected before the analysis to reject those showing very short sequences (fewer than 200 base pairs) using Chromatogram Explorer Lite (version v5.0.2, Heracle BioSoft, Pitesti, Romania).

2.4. Statistical Analysis and Data Visualisation

A comparison between the tumour and of the corresponding margin samples was performed using the Kruskal–Wallis test for age, TNM staging and histological grading, and *MUC6* and *MUC16* mutation occurrences. Fisher's exact test was used to compare the tumour and the margin samples in regard of sex, smoking, alcohol drinking, *MUC6* and *MUC16* mutation occurrences and HPV status. Lollipop plots of mutation locations in *MUC6* and *MUC16* cDNA fragments were generated using the TrackViewer R library [28]. Hierarchical clustering using Euclidean distance calculation and its visualisation as heatmaps with dendrograms was performed using the gplots R library [29]. Analyses were performed with the use of R 4.2.2 in RStudio version 2022.12.0 build 353 (PBC, Boston, MA, USA) using the stats R library [30].

3. Results

3.1. Study Group

The study group included 18 patients, 4 of whom (22.22%) were women and 14 (77.78%) men, with a mean age of 62.83 ± 8.13. Seven of the patients (38.89%) were smokers. Ten cases (55.56%) admitted consuming alcohol; five (27.78%) consumed alcohol occasionally, and five (27.78%) drank alcohol on a regular basis. Our previous study delivered information about the HPV status of the cohort [31], where there were 12 (66.67%) HPV-positive patients, 5 (27.28%) HPV-negative individuals with only 1 case (5.56%) on which no information was available. Clinical data of the patients are presented in Table 2.

Table 2. Detailed clinical data of the patients.

Parameter	n	%
T1	3	16.67
T2	6	33.33
T3	9	50.00
N0	7	38.89
N1	1	5.56
N2	9	50.00
N3	1	5.56
G1	4	22.22
G2	8	44.44
G3	6	33.33

3.2. Sequencing Analysis

The analysis excluded samples with poor Sanger data quality and the final dataset was composed of 30 samples: 16 (53.33%) with OPSCC tumour and 14 (46.67%) margin samples. We obtained 25 sufficient quality traces of the *MUC6* fragment and 22 of the *MUC16* fragment. The list of used and rejected (based on quality) traces of the sequencing samples is shown in Table S1, along with the tumour and the margin sample ID for each patient.

A total of 30 different mutations were found across the cohort in the *MUC6* cDNA fragment. Similarly, 15 mutations were detected in the *MUC16* cDNA fragment. After evaluation and removal of the low-quality variants, the changes narrowed down to 13 mutations in the *MUC6* fragment and 3 variants the *MUC16* fragment. The example electropherograms of Sanger sequencing traces for *MUC6* and *MUC16* fragments with mutations are presented in Figure S2.

3.2.1. Sequencing Analysis of MUC6

Analysis of the sequencing results for *MUC6* reported a total of 30 variants, 17 of which had low quality (hence, not used for further analyses) and the remainder 13 with good quality. Most single nucleotide variations (SNVs) were missense variants with transversion C>T. All mutations found in the *MUC6* fragment are presented in Table 3, while more detailed information is gathered in Table S2.

Table 3. Characteristics of the mutations identified in the *MUC6* gene.

Type Tissue (ID Sample)	POS	REF	ALT	HGVSc	HGVSp	Consequence
tumour (11, 29, 39, 45, 47, 53, 55, 57, 61, 79); margin (16, 34, 42, 48, 50, 54, 58, 62, 64, 90)	1018257	A	T	ENST00000421673.7:c.4544C>A	ENSP00000406861.2:p.Leu1515Ter	stop gained
margin (11)	1018258	C	T	ENST00000421673.7:c.4543T>A	ENSP00000406861.2:p.Ala1515Thr	missense variant
margin (34)	1018465	T	C	ENST00000421673.7:c.4336C>G	ENSP00000406861.2:p.Asn1446Asp	missense variant
tumour (39)	1018322	A	G	ENST00000421673.7:c.4479A>C	ENSP00000406861.2:p.Thr1493%3D	synonymous variant
tumour (39)	1018506	C	T,A	ENST00000421673.7:c.4295C>A	ENSP00000406861.2:p.Ser1432Asn	missense variant
margin (42)	1018543	A	G	ENST00000421673.7:c.4258G>C	ENSP00000406861.2:p.Cys1420Arg	missense variant
tumour (45)	1018111	C	T,A	ENST00000421673.7:c.4690G>A	ENSP00000406861.2:p.Ala1564Thr	missense variant
tumour (47)	1018243	C	T	ENST00000421673.7:c.4558G>A	ENSP00000406861.2:p.Glu1520Lys	missense variant
tumour (55)	1018180	C	T	ENST00000421673.7:c.4621T>A	ENSP00000406861.2:p.Ala1541Thr	missense variant
margin (56)	1018381	C	T	ENST00000421673.7:c.4420G>A	ENSP00000406861.2:p.Ala1474Thr	missense variant
margin (56)	1018513	C	A,T	ENST00000421673.7:c.4288A>T	ENSP00000406861.2:p.Ala1430Ser	missense variant
margin (58)	1018297	G	A	ENST00000421673.7:c.4504A>T	ENSP00000406861.2:p.Leu1502%3D	synonymous variant
margin (90)	1018276	T	C,A	ENST00000421673.7:c.4525A>G	ENSP00000406861.2:p.Thr1509Ala	missense variant

ID—sample identifier; POS—positions in chromosome; REF—reference nucleotide; ALT—detected nucleotide.

We observed similar occurrences of mutations in the tumour and the margin samples ($p = 0.568$). The highest number of variants in a single tumour sample was three, while in a single margin sample it was two. Moreover, we reported that all of the detected variants in *MUC6* corresponded to amino acidic changes. The mutation occurrences are summarised in Table 4.

Table 4. The occurrences of mutations detected in the fragments of *MUC6* and *MUC16*.

Fragment	Sample	n (%)	Mean	SD	Minimum	Maximum	*p*-Value
MUC6 cDNA	Tumour	10 (62.5%)	1.00	0.97	0	3	0.568
	Margin	11 (78.57%)	1.14	0.77	0	2	0.568
MUC16 cDNA	Tumour	7 (43.75%)	0.75	0.93	0	2	0.946
	Margin	8 (57.14%)	0.64	0.63	0	2	0.946

Interestingly, we found that most of the patients (16 cases) hosted the same stop-gained mutation in the *MUC6* gene (chr11:1018257 A>T; ENST00000421673.7:c.4544C>A), which was present neither in The Genome Aggregation Database (gnomAD) [32] nor in The Single Nucleotide Polymorphism database (dbSNP) [33]. This change was found in ten tumour samples and ten margin samples. Moreover, four patients had this mutation in their tumour and margin samples simultaneously. Furthermore, two of the known variants were found within the *MUC6* fragment: one had a moderate impact (COSV70139063) and the second had a low impact (COSV70138702) (Table S2). Distribution of the detected mutations in the *MUC6* cDNA fragment is presented in Figure 2, while the distribution of the single amino acid variants (SAVs), along with the MUC6 protein sequence, is presented in Figure 3.

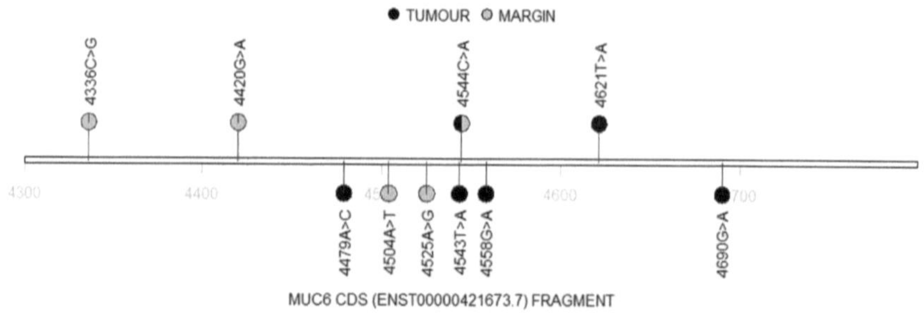

Figure 2. Lollipop diagram demonstrating the nucleotide changes in the mutated *MUC6* cDNA (black: tumour; grey: margin).

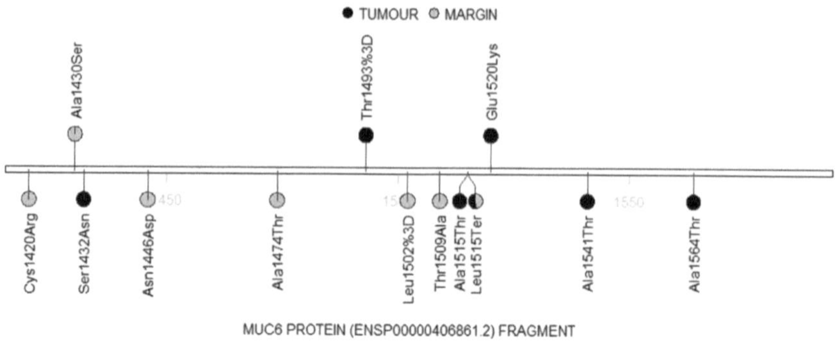

Figure 3. Lollipop diagram demonstrating the amino acid changes in the MUC6 protein fragment (black: tumour; grey: margin).

We used the SIFT and PolyPhen prediction tools to analyse the effect of amino acid changes upon the structure and function of mucin. Using the PolyPhen score, one of the detected variants (c.4258G>C) was flagged as a probably damaging MUC6 protein structure. The values of SIFT and PolyPhen score analysis were 0.06 and 0.921, respectively. Moreover, SIFT predicted five variants as "tolerated" (two tolerated and three tolerated with low confidence) and three with "deleterious effect" (one deleterious and two deleterious with low confidence). PolyPhen predicted eight variants as "benign". Detailed information on the SIFT and PolyPhen analyses are presented in Table S2.

3.2.2. Sequencing Analysis of MUC16

Analysis of the sequencing results for *MUC16* reported a total of 15 variants, 12 of which had low quality (hence, not used for further analyses) and the remaining 3 had good quality. Two of the three SNVs were missense variants and one was a synonymous variant. Found in 12 patients, the most common variant type was a missense one located at position 8,964,498 of chromosome 19 (ENST00000397910.8:c.12272T>A). Moreover, three patients hosted this variant in the tumour and the margin samples simultaneously. Our analysis reported that no known variants occurred in the selected fragment of *MUC16*. All mutations identified in *MUC16* are shown in Table 5, while detailed information is gathered in Table S3.

All the detected mutations corresponded to protein alterations and no significant difference in mutation occurrence was observed between the tumour samples and the margin samples ($p = 0.946$). Distribution of the detected mutations in the *MUC16* cDNA fragment is shown in Figure 4, while the distribution of the mutations found in the MUC16 protein sequence is shown in Figure 5.

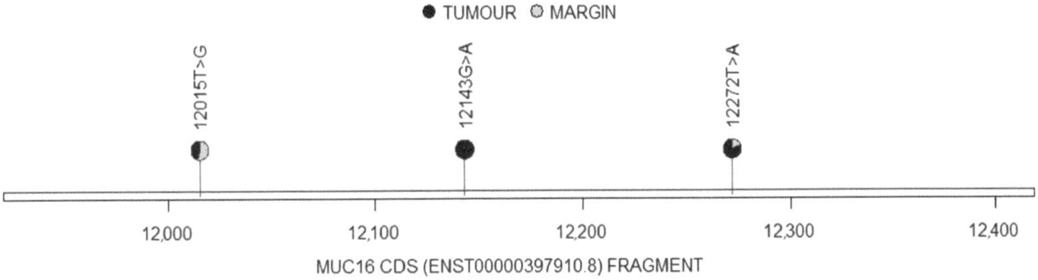

Figure 4. Lollipop diagram demonstrating the nucleotide changes in the mutated *MUC16* cDNA (black: tumour; grey: margin).

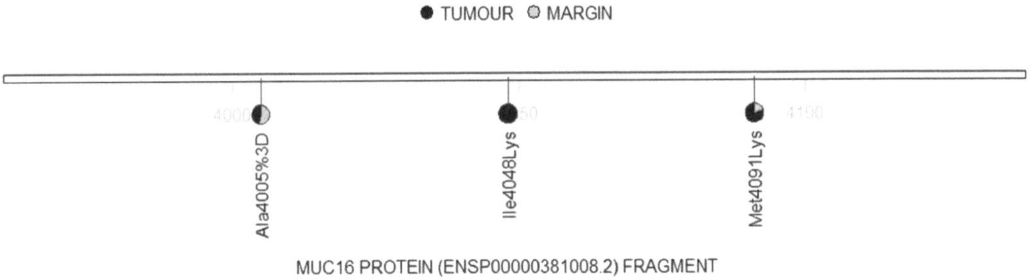

Figure 5. Lollipop diagram demonstrating the amino acid changes in the MUC16 protein fragment (black: tumour; grey: margin).

Table 5. Characteristics of the mutations identified in the *MUC16* gene.

Type Tissue (ID Sample)	POS	REF	ALT	HGVSc	HGVSp	Consequence
tumour (11, 15, 33, 37, 47, 57, 63); margin (16, 30, 48, 56, 62, 64, 80, 90)	8964498	A	T	ENST00000397910.8:c.12272T>A	ENSP00000381008.2:p.Met4091Lys	missense variant
tumour (11); margin (16)	8964755	A	C/G,T	ENST00000397910.8:c.12015T>G	ENSP00000381008.2:p.Ala4005%3D	synonymous variant
tumour (15, 33, 37, 63); margin (64)	8964627	A	T	ENST00000397910.8:c.12143G>A	ENSP00000381008.2:p.Ile4048Lys	missense variant

ID—sample identifier; POS—positions in chromosome; REF—reference nucleotide; ALT—detected nucleotide.

Additionally, the first detected variant (ENST00000397910.8:c.12272T>A) was predicted with SIFT tool as "tolerated" (tolerated with low confidence, value 0.09) being a replacement of isoleucine (Ile) with lysine (Lys). The second detected variant (ENST00000397910.8:c.12143G>A) was predicted as "deleterious effect" (deleterious low confidence, the value of the SIFT score analysis was 0.0), as a consequence of the change from methionine (Met) to lysine (Lys). Analysis with the PolyPhen tool pointed to both variants as "benign" (0.013 and 0.028, respectively).

3.3. HPV Presence and MUC6 and MUC16 Mutations

We analysed the potential association between the HPV status and mutations of the mucin genes. No significant differences were found between the occurrences of mutations in *MUC6* or *MUC16* in HPV-positive tumour, as compared to HPV-negative tumour samples. Similarly, no significant results were observed in the margin samples.

3.4. Impact of the Common Occurrences of MUC6 and MUC16 Mutations on Clinicopathological and Demographic Characteristics of the Study Groups

There were no significant differences in mutation occurrences of *MUC6* between the tumour and the margin samples. No significant correlation was found between the parameters (age, TNM staging and histological grading) and the *MUC6* mutation frequency observed throughout the study group. In addition, no significant differences were observed between the mutation occurrence depending on demographic parameters (such as sex, smoking status, alcohol drinking status: casual, regular or general) in tumour or margin samples. Similarly, no significant differences were observed between mutation occurrences of *MUC16* and the clinicopathological or demographic characteristics for tumour or margin samples.

3.5. Mutation Clustering

The heatmaps and dendrograms presenting the mutations found in *MUC6* and *MUC16* are shown in Figures 6 and 7, respectively. Similarly, analyses for both genes, *MUC6* and *MUC16*, are presented in Figure 8. The hierarchical cluster analysis performed for the identified variants revealed no significant similarities nor mutation patterns in the *MUC6* coding sequence (Figure 6) or both *MUC6* and *MUC16* (Figure 8) corresponding to the tumour or margin groups. The mutational patterns did not correspond to any of the clustering data available in this study (TNM staging and histological grading, sex, alcohol drinking or smoking).

MUC16 cDNA (Figure 7) mutations patterns showed that the presence of ENST00000397910.8:c.12272T>A and ENST00000397910.8:c.12143G>A together persisted frequently in the tumour samples (four out of seven samples) but not in the margin tissue samples (one out of eight samples). In one patient, both mutations were found in the tumour and the margin samples (sample IDs 63 and 64). Another patient had both mutations in the tumour sample but not in the corresponding margin sample (samples IDs 15 and 16).

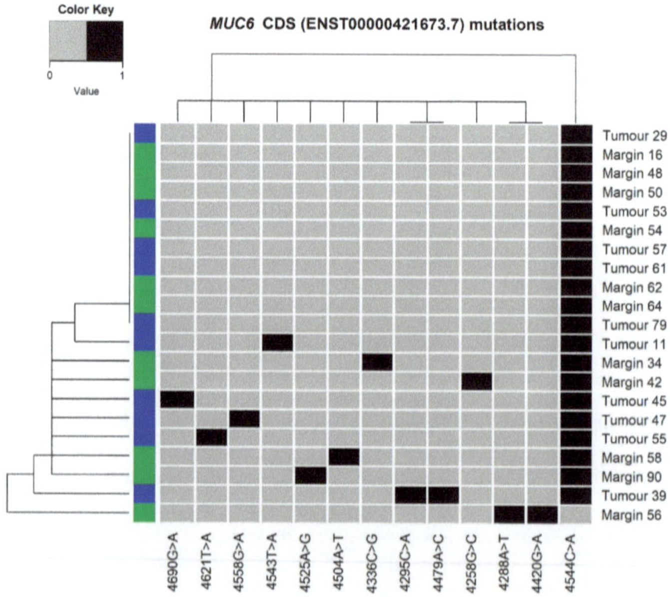

Figure 6. Heatmap and dendrograms presenting the mutations found in the selected fragment of *MUC6* coding sequence (CDS ENST00000421673.7). Tissue type shown as blue (tumour) or green (margin).

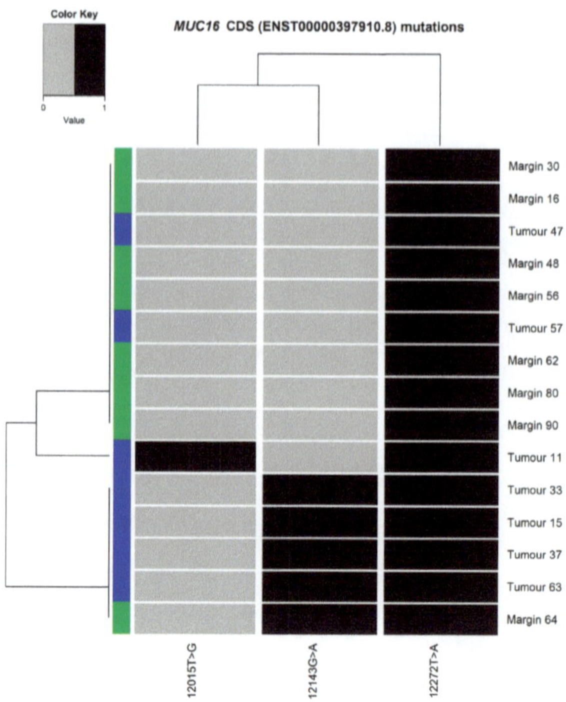

Figure 7. Heatmap and dendrograms presenting the mutations found in the selected fragment of *MUC16* coding sequence (CDS ENST00000397910.8). The tissue type illustrated as blue (tumour) or green (margin).

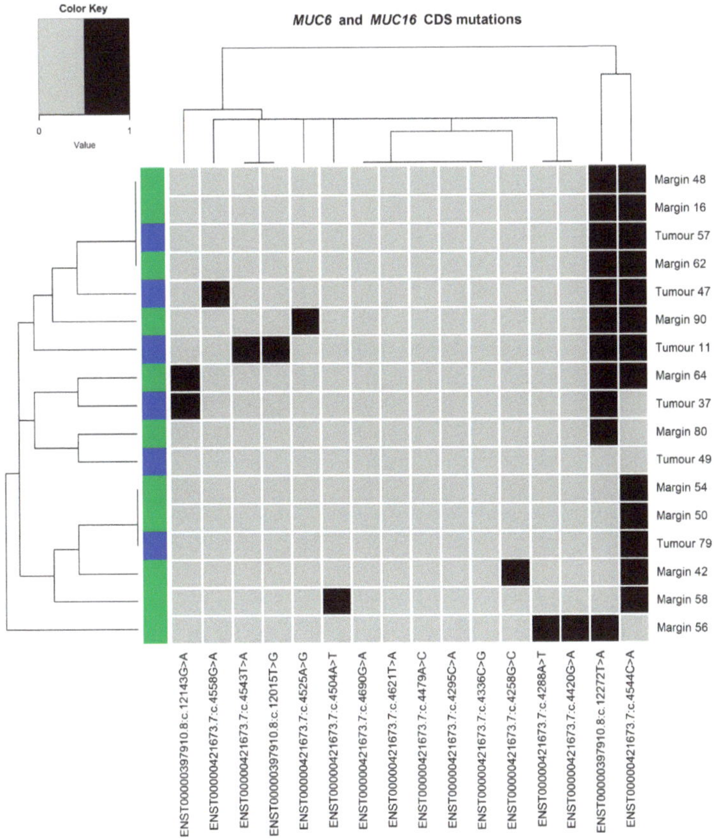

Figure 8. Heatmap and dendrograms presenting the *MUC6* coding sequence (CDS ENST000004 21673.7) and *MUC16* coding sequence (CDS ENST00000397910.8) mutations. The tissue type shown as blue (tumour) or green (margin).

4. Discussion

The growing number of OPSCC cases worldwide requires continuous research to better understand the molecular pathogenesis and to find potential targets for the effective cancer therapy. Mucins are glycosylated proteins known to facilitate tumour invasion and metastasis [4]. In the present study, we conducted a sequencing analysis of the selected fragments of *MUC6* and *MUC16* in 30 samples collected from the OPSCC patients. We detected 13 and 3 somatic mutations in *MUC6* and *MUC16*, respectively. The changes were mostly missense mutations. The selection of genes for sequencing analysis based on literature, which showed that mucin genes often mutated in HNSCC patients. However, the potential impact of somatic mutations in these genes on OPSCC is still poorly understood. Kannan et al. [34] detected numerous missense mutations in mucin genes, including *MUC6* and *MUC16*, in tonsil samples obtained from HPV-16 positive patients with OPSCC. In addition, Ährlund-Richter et al. [35] showed that *MUC6* and *MUC16* were mutated in over 30% of primary HPV-positive tonsillar squamous cell carcinoma (TSCC) and base of tongue squamous cell carcinoma (BOTSCC) with and without recurrence. In addition, Haft et al. [36] found mutations in other mucin genes (*MUC4* and *MUC5B*) in HPV-positive OPSCC patients from The Cancer Genome Atlas database (TCGA). It has been found that mutated *MUC6* and *MUC16* were involved in pathways associated with the extracellular matrix and carbohydrates in patients with TSCC and BOTSCC [35]. It was shown that nine out of nineteen mucin genes (including *MUC6* and *MUC16*) were frequently mutated in various cancer types, including HNSCC [37]. Overexpression of mucins is associated with proliferation, migration and invasion in various epithelial cancers [38]. Moreover, the frequency of mutations in mucins has been suggested to have an impact on cancer patients' survival [39], indicating that mucins may be an important target of further analyses.

Our sequencing analysis of the *MUC6* fragment revealed some known mutations, confirmed also in COSMIC (Catalogue of Somatic Mutations in Cancer, version 96) [40] and detected in different cancers. The somatic variant COSV70128702 has been previously reported in colorectal cancer [41], while COSV70139063 was found in patients with basal cell carcinoma [42]. Interestingly, we identified a novel stop-gained mutation in the *MUC6* gene (ENST00000421673.7:c.4544C>A), which is not present in the gnomAD database [32] or in dbSNP [33] and classified as one with a potentially high impact on the downstream protein product. Detected in 16 patients, this somatic variant was the most common in our study. Moreover, it was found in both: the tumour and the margin samples. It suggests that it could be a population-specific polymorphism. However, further studies are needed to confirm or reject this hypothesis. As predicted by PolyPhen, another variant (ENST00000421673.7:c.4258G>C) may impact MUC6 protein structure in probably the damaging ways. Further analyses are required to investigate the potential influence of such changes on protein function or gene expression, especially in the case of the identified stop-gain mutation in *MUC6*. Mutations in *MUC6* were analysed in different cancers. Rokutan et al. [43] observed *MUC6* mutating in 20% of gastric dysplasia/intraepithelial neoplasia samples. Moreover, *MUC6* mutations were found in Epstein–Barr virus (EBV)-associated lymphoepithelioma-like cholangiocarcinoma [44], and directly associated with thyroid cancer [45]. On the other hand, the mutation status of *MUC6* and three other genes (*ATR*, *ERBB3* and *KDR*) was obtained as a marker for the recurrence of non-small-cell lung cancer (NSCLC) patients [46]. In addition, Shi et al. [47] identified that mutation in *MUC6* in Chinese patients with hepatocellular carcinoma, was associated with early recurrence. Interestingly, patients with stomach adenocarcinoma and mutated *MUC6* had better overall survival prognose, as compared to patients with wild-type *MUC6* [37]. Our study reveals no associations between *MUC6* mutations and the clinicopathological or demographic parameters of the OPSCC patients. However, the study group size was limited, therefore we plan to conduct further studies and statistical analysis with a larger study group.

Our study identified a few mutations in the fragment of *MUC16* cDNA. We found that ENST00000397910.8:c.12272T>A was the most common missense variant detected in *MUC16* in the tumour and the margin samples of OPSCC patients, which suggests that it could also be a population-specific polymorphism. Another variant ENST00000397910.8:c. 12143G>A was found mostly in the tumour samples. Interestingly, we observed that these two mutations also occurred in one margin sample. The margin samples were verified as histopathologically free of cancerous cells. However, this tissue might still contain a minimal residual number of cancer cells, which are not detected by routine diagnostic methods [48]. Further investigations are required to verify this hypothesis. However, Kloss-Brandstätter et al. [49] detected mutation in mtDNA in OSCC patients and showed that some mutations were observed in tumour and resection margin samples. Our analysis revealed no association between mutation frequency in *MUC16* and the clinicopathological parameters, such as T or histological grading. Similarly, no significant associations were observed between the detected SNVs of *MUC16* and the demographic parameters. Interestingly, some studies observed that mutated *MUC16* influence on survival or prognosis in various cancers; Liu et al. [5] found that *MUC16* mutations were associated with better overall survival in patients with hepatocellular carcinoma, whereas Wang et al. [50] found that *MUC16* mutations were associated with overall survival in patients with melanoma. In addition, mutated *MUC16* was shown to improve the survival prognoses in patients with skin cutaneous melanoma [37], as well as in patients with gastric cancer [6,51]. Therefore, we design further studies assuming the use of an expanded cohort to determine whether such mutations could affect the OPSCC patient survival.

It has been observed that HPV-positive HNSCC had a mutational landscape different than HPV-negative tumours [22,52]. In our study, the frequencies of mutations in *MUC6* or *MUC16* did not correlate with HPV presence, although that might probably be due to the small size of our cohort. However, this could also be due to the location of the samples, all collected from the tonsils. Similarly to our results, Plath et al. [53] observed no significant differences in the total mutation counts and HPV-16 status in HNSCC patients, including the oropharynx site. Nichols et al. [22] found more mutations in an HPV-positive OPSCC patient than in an HPV-negative oral squamous cell carcinoma patient. In another study, HPV-positive patients with TSCC and BOTSCC hosted significantly fewer mutations than patients without HPV and the distribution of mutations was also different in the various genes analysed [51]. On the other hand, Gillison et al. [54] found no different significant mutation rates between HPV-positive HNSCC (mostly located in oropharyngeal) and HPV-negative HNSCC (mostly oral cavity). However, they reported that the most significantly mutated genes were different among patients with and without HPV, and suggested that the viral-host interaction may have an influence on genome characteristic of HPV-positive patients with oral and oropharyngeal squamous cell carcinoma [54].

The main limitations of the present study were small sample size and only one cDNA fragment of the selected mucins investigated. Further studies on a larger group are necessary to confirm the obtained results and to analyse the impact of the detected mutations on 5-year survival and recurrence.

5. Conclusions

The present study investigated multiple mutations identified from the selected fragments of *MUC6* and *MUC16* in the tumour and the surgical margin samples collected from the tonsillar tissue of OPSCC patients. We found that the most common stop-gain variant across the cohort was a novel mutation, located in the exonic region of the *MUC6* gene.

Our findings indicate that some of the investigated variants have a potentially deleterious impact on the structure of their protein products and will therefore require further investigations. The current study provides a new insight into the genetic alternation in mucin genes among the OPSCC patients. However, more extensive studies are needed to understand how such mutations may affect oropharyngeal carcinogenesis and survival and to verify whether they are polymorphisms specific to the Polish population.

Supplementary Materials: The following supporting information can be downloaded at: https://www.mdpi.com/article/10.3390/cimb45070356/s1, Figure S1: The workflow diagram of wet-laboratory steps (samples preparation and sequencing reaction); Figure S2: Electropherograms of Sanger sequencing traces for *MUC6* ENST00000421673.7 and *MUC16* ENST00000397910.8; Table S1: Accepted and rejected sequencing results; Table S2: Detailed information of investigated *MUC6* (ENSG00000184956) mutations; Table S3: Detailed information of investigated *MUC16* (ENSG00000181143) mutations.

Author Contributions: Conceptualisation, J.G. and J.K.S.; Methodology, J.G., J.K.S. and K.B.; Formal Analysis, S.S. and K.B.; Investigation, K.G., D.H., A.Ś., J.G. and K.B.; Resources, K.M.-O. and A.K.; Writing—Original Draft Preparation, J.G.; Writing—Review and Editing, K.G. and S.S.; Visualisation, K.B.; Supervision, M.M. and J.K.S.; Project Administration, J.G. All authors have read and agreed to the published version of the manuscript.

Funding: This research was funded by the Medical University of Silesia (KNW-2-O07/N/9/N).

Institutional Review Board Statement: The study was conducted in accordance with the Declaration of Helsinki and approved by the Ethics Committee of the Medical University of Silesia (approval codes: KNW/0022/KB1/49/16 and KNW/0022/KB1/49/II/16/17).

Informed Consent Statement: Informed consent was obtained from all the subjects involved in the study.

Data Availability Statement: The data used to support the findings of this study are available from the corresponding author upon request.

Acknowledgments: We would like to thank Raju Misra, Head of the Molecular Biology Laboratories at the Natural History Museum in London, and his team for cooperation in the analysis of the DNA sequencing results.

Conflicts of Interest: The authors declare no conflict of interest. The funders had no role in the design of the study; in the collection, analyses, or interpretation of data; in the writing of the manuscript; or in the decision to publish the results.

References

1. Hansson, G.C. Mucins and the Microbiome. *Annu. Rev. Biochem.* **2020**, *89*, 769–793. [CrossRef] [PubMed]
2. Linden, S.K.; Sutton, P.; Karlsson, N.G.; Korolik, V.; McGuckin, M.A. Mucins in the mucosal barrier to infection. *Mucosal. Immunol.* **2008**, *1*, 183–197. [CrossRef] [PubMed]
3. Rao, C.V.; Janakiram, N.B.; Mohammed, A. Molecular Pathways: Mucins and Drug Delivery in Cancer. *Clin. Cancer Res.* **2017**, *23*, 1373–1378. [CrossRef] [PubMed]
4. Hollingsworth, M.A.; Swanson, B.J. Mucins in cancer: Protection and control of the cell surface. *Nat. Rev. Cancer* **2004**, *4*, 45–60. [CrossRef] [PubMed]
5. Liu, B.; Dong, Z.; Lu, Y.; Ma, J.; Ma, Z.; Wang, H. Prognostic Value of MUC16 Mutation and Its Correlation with Immunity in Hepatocellular Carcinoma Patients. *Evid. Based Complement. Altern. Med.* **2022**, *2022*, 3478861. [CrossRef] [PubMed]
6. Li, X.; Pasche, B.; Zhang, W.; Chen, K. Association of MUC16 Mutation With Tumor Mutation Load and Outcomes in Patients With Gastric Cancer. *JAMA Oncol.* **2018**, *4*, 1691–1698. [CrossRef]
7. Aithal, A.; Rauth, S.; Kshirsagar, P.; Shah, A.; Lakshmanan, I.; Junker, W.M.; Jain, M.; Ponnusamy, M.P.; Batra, S.K. MUC16 as a novel target for cancer therapy. *Expert Opin. Ther. Targets* **2018**, *22*, 675–686. [CrossRef]
8. Gipson, I.K.; Spurr-Michaud, S.; Tisdale, A.; Menon, B.B. Comparison of the transmembrane mucins MUC1 and MUC16 in epithelial barrier function. *PLoS ONE* **2014**, *9*, e100393. [CrossRef]
9. Blalock, T.D.; Spurr-Michaud, S.J.; Tisdale, A.S.; Heimer, S.R.; Gilmore, M.S.; Ramesh, V.; Gipson, I.K. Functions of MUC16 in corneal epithelial cells. *Investig. Ophthalmol. Vis. Sci.* **2007**, *48*, 4509–4518. [CrossRef]
10. Syed, Z.A.; Zhang, L.; Ten Hagen, K.G. In vivo models of mucin biosynthesis and function. *Adv. Drug Deliv. Rev.* **2022**, *184*, 114182. [CrossRef]
11. Morozov, V.; Borkowski, J.; Hanisch, F.G. The Double Face of Mucin-Type O-Glycans in Lectin-Mediated Infection and Immunity. *Molecules* **2018**, *23*, 1151. [CrossRef] [PubMed]
12. Bhatia, R.; Gautam, S.K.; Cannon, A.; Thompson, C.; Hall, B.R.; Aithal, A.; Banerjee, K.; Jain, M.; Solheim, J.C.; Kumar, S.; et al. Cancer-associated mucins: Role in immune modulation and metastasis. *Cancer Metastasis Rev.* **2019**, *38*, 223–236. [CrossRef] [PubMed]
13. Sung, H.; Ferlay, J.; Siegel, R.L.; Laversanne, M.; Soerjomataram, I.; Jemal, A.; Bray, F. Global Cancer Statistics 2020: GLOBOCAN Estimates of Incidence and Mortality Worldwide for 36 Cancers in 185 Countries. *CA Cancer J. Clin.* **2021**, *71*, 209–249. [CrossRef] [PubMed]
14. Johnson, D.E.; Burtness, B.; Leemans, C.R.; Lui, V.W.Y.; Bauman, J.E.; Grandis, J.R. Head and neck squamous cell carcinoma. *Nat. Rev. Dis. Primers.* **2020**, *6*, 92. [CrossRef] [PubMed]
15. Zhou, J.Z.; Jou, J.; Cohen, E. Vaccine Strategies for Human Papillomavirus-Associated Head and Neck Cancers. *Cancers* **2021**, *14*, 33. [CrossRef]
16. Windon, M.J.; D'Souza, G.; Rettig, E.M.; Westra, W.H.; van Zante, A.; Wang, S.J.; Ryan, W.R.; Mydlarz, W.K.; Ha, P.K.; Miles, B.A.; et al. Increasing prevalence of human papillomavirus-positive oropharyngeal cancers among older adults. *Cancer* **2018**, *124*, 2993–2999. [CrossRef]
17. Carlander, A.F.; Jakobsen, K.K.; Bendtsen, S.K.; Garset-Zamani, M.; Lynggaard, C.D.; Jensen, J.S.; Grønhøj, C.; Buchwald, C.V. A Contemporary Systematic Review on Repartition of HPV-Positivity in Oropharyngeal Cancer Worldwide. *Viruses* **2021**, *13*, 1326. [CrossRef]
18. Mahal, B.A.; Catalano, P.J.; Haddad, R.I.; Hanna, G.J.; Kass, J.I.; Schoenfeld, J.D.; Tishler, R.B.; Margalit, D.N. Incidence and Demographic Burden of HPV-Associated Oropharyngeal Head and Neck Cancers in the United States. *Cancer Epidemiol. Biomark. Prev.* **2019**, *28*, 1660–1667. [CrossRef]
19. Lechner, M.; Liu, J.; Masterson, L.; Fenton, T.R. HPV-associated oropharyngeal cancer: Epidemiology, molecular biology and clinical management. *Nat. Rev. Clin. Oncol.* **2022**, *19*, 306–327. [CrossRef]
20. Farah, C.S. Molecular landscape of head and neck cancer and implications for therapy. *Ann. Transl. Med.* **2021**, *9*, 915. [CrossRef]
21. Stransky, N.; Egloff, A.M.; Tward, A.D.; Kostic, A.D.; Cibulskis, K.; Sivachenko, A.; Kryukov, G.V.; Lawrence, M.S.; Sougnez, C.; McKenna, A.; et al. The mutational landscape of head and neck squamous cell carcinoma. *Science* **2011**, *333*, 1157–1160. [CrossRef] [PubMed]
22. Nichols, A.C.; Chan-Seng-Yue, M.; Yoo, J.; Xu, W.; Dhaliwal, S.; Basmaji, J.; Szeto, C.C.; Dowthwaite, S.; Todorovic, B.; Starmans, M.H.; et al. A Pilot Study Comparing HPV-Positive and HPV-Negative Head and Neck Squamous Cell Carcinomas by Whole Exome Sequencing. *ISRN Oncol.* **2012**, *2012*, 809370. [CrossRef] [PubMed]

23. Sobin, L.H.; Gospodarowicz, M.K.; Wittekind, C.; International Union Against Cancer. *TNM Classification of Malignant Tumours*, 7th ed.; Wiley-Blackwell: Chichester, UK; Hoboken, NJ, USA, 2010.
24. Rausch, T.; Fritz, M.H.-Y.; Untergasser, A.; Benes, V. Tracy: Basecalling, alignment, assembly and deconvolution of sanger chromatogram trace files. *BMC Genom.* **2020**, *21*, 230. [CrossRef] [PubMed]
25. Ng, P.C.; Henikoff, S. SIFT: Predicting amino acid changes that affect protein function. *Nucleic Acids Res.* **2003**, *31*, 3812–3814. [CrossRef] [PubMed]
26. Adzhubei, I.A.; Schmidt, S.; Peshkin, L.; Ramensky, V.E.; Gerasimova, A.; Bork, P.; Kondrashov, A.S.; Sunyaev, S.R. A method and server for predicting damaging missense mutations. *Nat. Methods* **2010**, *7*, 248–249. [CrossRef]
27. McLaren, W.; Gil, L.; Hunt, S.E.; Riat, H.S.; Ritchie, G.R.; Thormann, A.; Flicek, P.; Cunningham, F. The Ensembl Variant Effect Predictor. *Genome Biol.* **2016**, *17*, 122. [CrossRef]
28. Ou, J.; Zhu, L.J. trackViewer: A Bioconductor package for interactive and integrative visualization of multi-omics data. *Nat. Methods* **2019**, *16*, 453–454. [CrossRef] [PubMed]
29. Warnes, G.; Bolker, B.; Bonebakker, L.; Gentleman, R.; Huber, W.; Liaw, A.; Lumley, T.; Maechler, M.; Magnusson, A.; Moeller, S.; et al. Gplots: Various R Programming Tools for Plotting Data. R Package Version 3.1.3. 2022. Available online: https://CRAN.R-project.org/package=gplots (accessed on 7 February 2023).
30. R Core Team. *R: A Language and Environment for Statistical Computing*; R Foundation for Statistical Computing: Vienna, Austria, 2013; ISBN 3-900051-07-0. Available online: http://www.R-project.org/ (accessed on 7 February 2023).
31. Strzelczyk, J.K.; Biernacki, K.; Gaździcka, J.; Chełmecka, E.; Miśkiewicz-Orczyk, K.; Zięba, N.; Strzelczyk, J.; Misiołek, M. The Prevalence of High- and Low-Risk Types of HPV in Patients with Squamous Cell Carcinoma of the Head and Neck, Patients with Chronic Tonsillitis, and Healthy Individuals Living in Poland. *Diagnostics* **2021**, *11*, 2180. [CrossRef] [PubMed]
32. Karczewski, K.J.; Francioli, L.C.; Tiao, G.; Cummings, B.B.; Alföldi, J.; Wang, Q.; Collins, R.L.; Laricchia, K.M.; Ganna, A.; Birnbaum, D.P.; et al. The mutational constraint spectrum quantified from variation in 141,456 humans. *Nature* **2020**, *581*, 434–443. [CrossRef]
33. The National Center for Biotechnology Information. Available online: https://www.ncbi.nlm.nih.gov/pmc/articles/PMC29783/ (accessed on 17 March 2022).
34. Kannan, A.; Hertweck, K.L.; Philley, J.V.; Wells, R.B.; Dasgupta, S. Genetic Mutation and Exosome Signature of Human Papilloma Virus Associated Oropharyngeal Cancer. *Sci. Rep.* **2017**, *7*, 46102. [CrossRef]
35. Ährlund-Richter, A.; Holzhauser, S.; Dalianis, T.; Näsman, A.; Mints, M. Whole-Exome Sequencing of HPV Positive Tonsillar and Base of Tongue Squamous Cell Carcinomas Reveals a Global Mutational Pattern along with Relapse-Specific Somatic Variants. *Cancers* **2021**, *14*, 77. [CrossRef] [PubMed]
36. Haft, S.; Ren, S.; Xu, G.; Mark, A.; Fisch, K.; Guo, T.W.; Khan, Z.; Pang, J.; Ando, M.; Liu, C.; et al. Mutation of chromatin regulators and focal hotspot alterations characterize human papillomavirus-positive oropharyngeal squamous cell carcinoma. *Cancer* **2019**, *125*, 2423–2434. [CrossRef] [PubMed]
37. Liu, B.; Hu, F.F.; Zhang, Q.; Hu, H.; Ye, Z.; Tang, Q.; Guo, A.Y. Genomic landscape and mutational impacts of recurrently mutated genes in cancers. *Mol. Genet. Genom. Med.* **2018**, *6*, 910–923. [CrossRef] [PubMed]
38. Reynolds, I.S.; Fichtner, M.; McNamara, D.A.; Kay, E.W.; Prehn, J.H.M.; Burke, J.P. Mucin glycoproteins block apoptosis; promote invasion, proliferation, and migration; and cause chemoresistance through diverse pathways in epithelial cancers. *Cancer Metastasis Rev.* **2019**, *38*, 237–257. [CrossRef]
39. King, R.J.; Yu, F.; Singh, P.K. Genomic alterations in mucins across cancers. *Oncotarget* **2017**, *8*, 67152–67168. [CrossRef]
40. Catalogue of Somatic Mutations in Cancer. Available online: https://cancer.sanger.ac.uk/cosmic (accessed on 14 October 2022).
41. Giannakis, M.; Mu, X.J.; Shukla, S.A.; Qian, Z.R.; Cohen, O.; Nishihara, R.; Bahl, S.; Cao, Y.; Amin-Mansour, A.; Yamauchi, M.; et al. Genomic Correlates of Immune-Cell Infiltrates in Colorectal Carcinoma. *Cell Rep.* **2016**, *15*, 857–865. [CrossRef]
42. Sharpe, H.J.; Pau, G.; Dijkgraaf, G.J.; Basset-Seguin, N.; Modrusan, Z.; Januario, T.; Tsui, V.; Durham, A.B.; Dlugosz, A.A.; Haverty, P.M.; et al. Genomic analysis of smoothened inhibitor resistance in basal cell carcinoma. *Cancer Cell* **2015**, *27*, 327–341. [CrossRef]
43. Rokutan, H.; Abe, H.; Nakamura, H.; Ushiku, T.; Arakawa, E.; Hosoda, F.; Yachida, S.; Tsuji, Y.; Fujishiro, M.; Koike, K.; et al. Initial and crucial genetic events in intestinal-type gastric intramucosal neoplasia. *J. Pathol.* **2019**, *247*, 494–504. [CrossRef]
44. Chiang, N.J.; Hou, Y.C.; Tan, K.T.; Tsai, H.W.; Lin, Y.J.; Yeh, Y.C.; Chen, L.T.; Hou, Y.F.; Chen, M.H.; Shan, Y.S. The immune microenvironment features and response to immunotherapy in EBV-associated lymphoepithelioma-like cholangiocarcinoma. *Hepatol. Int.* **2022**, *16*, 1137–1149. [CrossRef]
45. Hou, S.; Xie, X.; Zhao, J.; Wu, C.; Li, N.; Meng, Z.; Cai, C.; Tan, J. Downregulation of miR-146b-3p Inhibits Proliferation and Migration and Modulates the Expression and Location of Sodium/Iodide Symporter in Dedifferentiated Thyroid Cancer by Potentially Targeting MUC20. *Front. Oncol.* **2021**, *10*, 566365. [CrossRef]
46. Cho, S.H.; Yoon, S.; Lee, D.H.; Kim, S.W.; Kim, K. Recurrence-associated gene signature in patients with stage I non-small-cell lung cancer. *Sci. Rep.* **2021**, *11*, 19596. [CrossRef] [PubMed]
47. Shi, H.; Zhang, W.; Hu, B.; Wang, Y.; Zhang, Z.; Sun, Y.; Mao, G.; Li, C.; Lu, S. Whole-exome sequencing identifies a set of genes as markers of hepatocellular carcinoma early recurrence. *Hepatol. Int.* **2022**, *17*, 393–405. [CrossRef] [PubMed]
48. Pierik, A.S.; Leemans, C.R.; Brakenhoff, R.H. Resection margins in head and neck cancer surgery: An update of residual disease and field cancerization. *Cancers* **2021**, *13*, 2635. [CrossRef] [PubMed]

49. Kloss-Brandstätter, A.; Weissensteiner, H.; Erhart, G.; Schäfer, G.; Forer, L.; Schönherr, S.; Pacher, D.; Seifarth, C.; Stöckl, A.; Fendt, L.; et al. Validation of Next-Generation Sequencing of Entire Mitochondrial Genomes and the Diversity of Mitochondrial DNA Mutations in Oral Squamous Cell Carcinoma. *PLoS ONE* **2015**, *10*, e0135643. [CrossRef]
50. Wang, X.; Yu, X.; Krauthammer, M.; Hugo, W.; Duan, C.; Kanetsky, P.A.; Teer, J.K.; Thompson, Z.J.; Kalos, D.; Tsai, K.Y.; et al. The Association of MUC16 Mutation with Tumor Mutation Burden and Its Prognostic Implications in Cutaneous Melanoma. *Cancer Epidemiol. Biomark. Prev.* **2020**, *29*, 1792–1799. [CrossRef]
51. Yang, Y.; Zhang, J.; Chen, Y.; Xu, R.; Zhao, Q.; Guo, W. MUC4, MUC16, and TTN genes mutation correlated with prognosis, and predicted tumor mutation burden and immunotherapy efficacy in gastric cancer and pan-cancer. *Clin. Transl. Med.* **2020**, *10*, e155. [CrossRef]
52. Bersani, C.; Sivars, L.; Haeggblom, L.; DiLorenzo, S.; Mints, M.; Ährlund-Richter, A.; Tertipis, N.; Munck-Wikland, E.; Näsman, A.; Ramqvist, T.; et al. Targeted sequencing of tonsillar and base of tongue cancer and human papillomavirus positive unknown primary of the head and neck reveals prognostic effects of mutated FGFR3. *Oncotarget* **2017**, *8*, 35339–35350. [CrossRef]
53. Plath, M.; Gass, J.; Hlevnjak, M.; Li, Q.; Feng, B.; Hostench, X.P.; Bieg, M.; Schroeder, L.; Holzinger, D.; Zapatka, M.; et al. Unraveling most abundant mutational signatures in head and neck cancer. *Int. J. Cancer* **2021**, *148*, 115–127. [CrossRef]
54. Gillison, M.L.; Akagi, K.; Xiao, W.; Jiang, B.; Pickard, R.K.L.; Li, J.; Swanson, B.J.; Agrawal, A.D.; Zucker, M.; Stache-Crain, B.; et al. Human papillomavirus and the landscape of secondary genetic alterations in oral cancers. *Genome Res.* **2019**, *29*, 1–17. [CrossRef]

current issues in molecular biology

MDPI

Article

Investigating Cox-2 and EGFR as Biomarkers in Canine Oral Squamous Cell Carcinoma: Implications for Diagnosis and Therapy

Rita Files [1,†], Catarina Santos [1,†], Felisbina L. Queiroga [1,2,3], Filipe Silva [1,2], Leonor Delgado [4,5], Isabel Pires [1,2,*] and Justina Prada [1,2]

1 Department of Veterinary Sciences, University of Trás-os-Montes and Alto Douro, 5000-801 Vila Real, Portugal; ritafiles2000@gmail.com (R.F.); catarinasls13@gmail.com (C.S.); fqueirog@utad.pt (F.L.Q.); fsilva@utad.pt (F.S.); jprada@utad.pt (J.P.)
2 Animal and Veterinary Research Centre (CECAV), Associate Laboratory for Animal and Veterinary Sciences (AL4AnimalS), University of Trás-os-Montes and Alto Douro, 5000-801 Vila Real, Portugal
3 Centre for the Study of Animal Science, CECA-ICETA, University of Porto, 4200-427 Porto, Portugal
4 UNIPRO—Oral Pathology and Rehabilitation Research Unit, University Institute of Health Sciences—CESPU (IUCS-CESPU), 4585-116 Gandra, Portugal; mleonor.delgado@iucs.cespu.pt
5 Pathology Department, INNO Serviços Especializados em Veterinária, 4710-503 Braga, Portugal
* Correspondence: ipires@utad.pt
† These authors contributed equally to this work.

Abstract: Oral squamous cell carcinoma (OSCC) is a common and highly aggressive dog tumor known for its local invasiveness and metastatic potential. Understanding the molecular mechanisms driving the development and progression of OSCC is crucial for improving diagnostic and therapeutic strategies. Additionally, spontaneous oral squamous cell carcinomas in dogs are an excellent model for studying human counterparts. In this study, we aimed to investigate the significance of two key molecular components, Cox-2 and EGFR, in canine OSCC. We examined 34 tumor sections from various dog breeds to assess the immunoexpression of Cox-2 and EGFR. Our findings revealed that Cox-2 was highly expressed in 70.6% of cases, while EGFR overexpression was observed in 44.1%. Cox-2 overexpression showed association with histological grade of malignancy (HGM) ($p = 0.006$) and EGFR with vascular invasion ($p = 0.006$). COX-2 and EGFR concurrent expression was associated with HGM ($p = 0.002$), as well as with the presence of vascular invasion ($p = 0.002$). These data suggest that Cox-2 and EGFR could be promising biomarkers and potential therapeutic targets, opening avenues for developing novel treatment strategies for dogs affected by OSCC. Further studies are warranted to delve deeper into these findings and translate them into clinical practice.

Keywords: oral squamous cell carcinoma; EGFR; COX-2; histological grade; canine

Citation: Files, R.; Santos, C.; Queiroga, F.L.; Silva, F.; Delgado, L.; Pires, I.; Prada, J. Investigating Cox-2 and EGFR as Biomarkers in Canine Oral Squamous Cell Carcinoma: Implications for Diagnosis and Therapy. *Curr. Issues Mol. Biol.* **2024**, *46*, 485–497. https://doi.org/10.3390/cimb46010031

Academic Editors: Violeta Popovici and Emma Adriana Ozon

Received: 6 December 2023
Revised: 29 December 2023
Accepted: 2 January 2024
Published: 4 January 2024

1. Introduction

With the rapid developments in the field of veterinary oncology, there is a great need for a better understanding of the molecular alterations behind the development of animal cancer [1]. Thus, the oral cavity is one of the most frequent sites of canine neoplastic proliferation, accounting for around 5 to 7% of tumors in dogs [1].

In dogs, oral squamous cell carcinoma (OSCC) is the second most prevalent malignant oral epithelial neoplasm (17% to 25%) [1]. Oral squamous cell carcinomas in dogs predominantly appear on the gingiva, affecting both the upper and lower areas, the tongue, and the tonsils. These carcinomas are also found in the lips, hard or soft palate, and the pharynx. Biological behavior depends on their location. Generally, oral tumors in dogs have a 15% rate of metastasizing, while those on the tongue can exhibit a higher rate, up to 40%, of metastasizing to nearby lymph nodes and the ones of tonsils 77–96% [2,3]. In humans, the location of OSCC also plays a significant role in determining the prognosis. SCCs of the tongue tend to metastasize more rapidly than other parts of the oral cavity. This could be attributed to the dense network of lymphatics in the tongue and the movement of tongue muscles, which may facilitate the spread of cancerous cells [4].

In canine populations, oral squamous cell carcinoma (OSCC) is more commonly observed in larger dog breeds, especially those older than seven years. Breeds like English springer Spaniels, Shetland sheepdog, and German shepherds show a higher incidence of OSCC [3]. OSCCs affecting the tongue seem more prevalent in breeds such as Poodles, Labrador Retrievers, and Samoyeds [5] and those affecting the tonsils in German Shepherds [6].

Dogs and cats, unlike laboratory rodents, manifest spontaneous cancers that closely mimic the heterogeneity observed in human tumors. Notably, as household pets, dogs cohabitate in shared environments with humans, displaying clinical manifestations, traits, and biological patterns akin to human cancer. This hints at the potential existence of common risk factors between humans and dogs. However, this relationship remains not fully elucidated due to the limited number of specific studies within Comparative and Evolutionary Oncology (CEO) [7,8]. Pets live integrated lives with their owners, thereby encountering shared environmental and socio-economic elements that could predispose them to cancer. Both pets and humans face similar environmental hazards, including toxins and carcinogens like air pollutants or pesticides in food and water [7,9].

Human OSCC is primarily associated with risk factors such as alcohol consumption, tobacco use, UV radiation exposure, and viral infections like HPV and EBV. Alcohol and tobacco are the more geographically prevalent risk factors [10–12]. Additionally, individuals with Fanconi anemia, a rare hereditary disease, display increased susceptibility to OSCC [10–12]. OSCC is closely linked to the oral microenvironment, stemming from contact with saliva, an acidic biological fluid derived from salivary gland secretions widely employed in the diagnosis of oral tumors [10–12].

Studies in cats have shown that exposure to household tobacco smoke potentially doubles the risk of them developing oral squamous cell carcinoma. Although a direct statistical significance of this correlation has not been conclusively established, there is a noticeable link between tobacco smoke exposure and high expression of p53 protein in feline OSCCs [13]. In dogs, there is no concrete evidence linking oral SCC in dogs with tobacco smoke exposure. Consequently, dogs are being considered as a comparative model for researching OSCCs not linked to alcohol and tobacco, which are about 10–15% of the total OSCC cases in humans [14].

The COX enzyme plays a crucial role in converting arachidonic acid into prostaglandins (PGs) in the body. This enzyme has two forms: COX-1, constitutively expressed in most cells, and COX-2, an inducible variant expressed at high levels in inflamed tissues [15–17].

The epidermal growth factor receptor (EGFR), also known as HER1 or erbB1, belongs to the family of tyrosine kinase receptors. It can form heterodimers with other members of the ErbB family, such as ErbB2, ErbB3, and ErbB4 [18,19]. These receptors play crucial roles in fundamental cellular activities like cell proliferation, division, and differentiation. [20].

COX-2 and EGFR are frequently overexpressed in several malignant tumors associated with various diseases [10,15–17,21]. When these molecules are overexpressed in tumors, they share functions in several crucial steps, including angiogenesis, apoptosis inhibition, immune response suppression, increased cell proliferation, invasive potential, cell differentiation, and migration [10,15–17,21]. Its importance in oncology is remarkable, with reports of its overexpression in various types of human cancer, including colon [22], stomach, breast, lung, esophagus, pancreas, bladder, prostate, and OSCC [22–25]. In addition, its expression has also been identified in some canine epithelial tumors, such as adenocarcinomas, mammary gland carcinomas, prostate and ovarian tumors, transitional cell carcinoma, and squamous cell carcinoma [19,26]. Recent studies have highlighted an interconnection between the EGFR and COX-2 pathways, with EGFR signaling inducing COX-2 expression and increasing prostaglandin production [27]. Similarly, COX-2-derived prostaglandin E2 (PGE2) can amplify EGFR signaling [26,28,29]. In addition, it has been observed that EGFR inhibition in canine squamous cell carcinomas reduces COX-2 expression, demonstrating the interdependence of these pathways [28,30]. COX-2 and EGFR are promising pharmacological and chemopreventive targets for treating various pathological conditions, including cancer [30].

Thus, they are promising future biomarkers in veterinary oncology due to the importance of both molecules in progression and malignancy, decreased survival and poor tumor prognosis [13,17]. Few studies have been carried out on the expression of these two molecules in canine OSCC [19,28]. Therefore, in this study, we aimed to investigate the importance of these key molecular components, Cox-2 and EGFR, in canine OSCC.

2. Materials and Methods

2.1. Animals and Tissue Specimens

We included 34 samples of canine tumors, histologically classified as Oral Squamous Cell Carcinoma (OSCC), from the archives of the Histopathology Laboratory of the University of Trás-os-Montes and Alto Douro (UTAD). Portuguese veterinary clinics and hospitals provided these samples. They were excised from 34 dogs and had been previously fixed in 10% formalin and embedded in paraffin.

Clinical data such as age, gender, and breed were recorded for each animal. Obtaining clinical staging or follow-up information for the animals included in the study was not possible.

For microscopic examination, 4 μm-thick tissue sections were stained with hematoxylin and eosin. Each specimen was reviewed by two independent pathologists (IP and JP). Our analysis included all slides and meticulously evaluated all the tumor sections.

The histopathologic diagnosis criteria were based on the internationally recognized classification system for animal tumors established by the World Health [31].

Additionally, ten samples of normal canine oral mucosa were included, and collected in the post-mortem routine examination.

2.2. Histopathological Evaluation

Histological grading was determined using a modified version of the multifactorial system developed by Anneroth [32]. The assessed parameters included keratinization/differentiation, nuclear pleomorphism, mitotic count, and the tumor–host relationship, including the invasion pattern, invasion stage, and lymphoplasmacytic infiltration.

The levels of keratinization were stratified based on the proportion of tumor cells exhibiting keratinization, yielding the following grades: I (>50% keratinized cells), II (20–50% keratinized), and III (0–20% keratinized) [citations needed]. Nuclear pleomorphism was classified as I (minimal, >75% mature cells), II (moderate, 50–75% mature cells), or III (marked nuclear pleomorphism, <50% mature cells). Mitotic count was measured across ten high-power fields (HPF) and classified as I (0 to 1 mitosis/HPF), II (2 to 3 mitoses/HPF), or III (≥four mitoses/HPF) [32,33].

The pattern of invasion was classified as I (well-defined with pushing borders), II (infiltration by solid cords, bands, and strands), and III (infiltration by small groups, strands, or individual cells). The stage of invasion was categorized as: I (corresponding to carcinoma in situ or questionable invasion), II (apparent invasion limited to the lamina propria), and III (invasion beyond the lamina propria, involving muscle. Lymphoplasmacytic infiltration was evaluated and categorized as I (marked), II (moderate), or III (mild to absent) [32,33].

The sum of these parameters was then used to classify the tumors into three grades: Grade I (scores 5–10) for well-differentiated tumors, Grade II (scores 11–15) for moderately differentiated tumors, and Grade III (scores > 16) for poorly differentiated tumors [32,33]. The presence or absence of emboli was also recorded.

2.3. Immunohistochemistry

For immunohistochemistry, sections 3 μm in thickness were used. The primary antibodies included COX-2 (Clone SP21, Transduction Laboratories®, Lexington, Kentucky,

USA; dilution 1:40; 24 h at 4 °C) and EGFR (clone 31G7, Invitrogen®, Paisley, Scotland, UK; dilution 1:100; 45 min at room temperature). These antibodies have been validated in canine tissues [34,35].

Visualization of the primary antibodies was achieved using the NovolinkTM Polymer Detection System (Leica Biosystems®, Newcastle, UK), with 3,3′-diaminobenzidine tetrachloride (DAB) as the chromogen, following manufacturer instructions. Subsequently, tissue sections were counterstained with Gill's hematoxylin and cover-slipped.

The specificity of the staining was confirmed using negative controls (omitting the primary antibody) and positive controls (kidney samples for COX-2 and normal skin and mammary tumor samples for EGFR).

2.4. Quantification of Immunolabeling

For Cox-2, immunolabeling was quantified using a semi-quantitative method adapted from [36], based on the percentage of positive tumor cells (extension) and the intensity of staining. The percentage of the positive cells was given scores ranging from 1 to 3 (1 for ≤10%, 2 for 11–50%, 3 for >51%), while the intensity of staining was also scored from 1 to 3 (weak, moderate, and strong). These scores were combined to produce a final score, calculated as the product of extension and intensity, categorizing the samples as Low (score < 6), and high expression (score ≥ 6).

The immunoreactivity of the EGFR antibody was considered positive when membranous staining above the background level in greater than 1% of tumor cells was detected. The intensity of the staining was evaluated as previously described. High expression was considered in cases where the staining of the membrane was of strong intensity [37].

All samples were independently evaluated by two observers (IP and JP), who were blinded to clinical and pathological characteristics, using a Nikon Eclipse E600 microscope coupled with a Nikon DXM1200 digital camera, provided by Nikon Instruments Inc., Melville, NY, USA. A third reviewer (LD) was consulted in cases of inconsistent findings. A consensus discussion was then held to determine the final score.

2.5. Statistical Analysis

Statistical analysis was conducted using SPSS software (Statistical Package for the Social Sciences), version 19.0 (IBM SPSS Statistics for Windows, IBM Corp®, Armonk, NY, USA). Categorical variables were analyzed using the chi-square test and Fisher's exact test. A significance level of $p < 0.05$ was considered statistically significant for all associations.

3. Results

3.1. Clinical Information

Of the animals with tumors, 47.1% (16 cases) were female, while 52.9% (18 cases) were male, with data missing for four animals. The age range of the animals was 1 to 17 years, with a mean age of 10.600 and a standard deviation of 3.2660. The breeds were as follows: 18 cases (52.9%) were of non-specified breed, 2 cases were Poodles, 2 cases were Labrador Retrievers, and there was 1 case each of the following breeds: Beagle, Boxer, Border Collie, Siberian Husky, Pekingese, Pinscher, and Yorkshire Terrier.

3.2. Histopathological Classification of the Tumors

The classification, based on the criteria mentioned earlier, resulted in the following distribution: 9 cases (26.5%) were categorized as well-differentiated tumors (grade I), 9 cases (26.5%) as moderately differentiated tumors (grade II), and 16 cases (47.1%) as poorly differentiated tumors (grade III). Vascular emboli were present in 6 tumors.

3.3. COX-2 Immunoreactivity

Normal oral mucosa was negative for Cox-2 in all cases. In oral squamous cell carcinomas, immunoreactivity for COX-2 was diffusely and uniformly present in the cytoplasm of tumor cells, with some variability observed across the histological samples, being more intense in invasive areas. Three cases were negative for COX-2 expression. Among the positive cases, 3 showed focal labeling, 12 had multifocal labeling, and 14 (48.3%) displayed diffuse labeling. Regarding labeling intensity, 4 (13.8%) cases showed weak staining, 6 (20.7%) moderate staining, and 19 (65.5%) strong staining.

In well-differentiated tumors (grade I), most cases were either negative ($n = 5$) or showed weak COX-2 intensity ($n = 1$). In contrast, moderately differentiated squamous cell carcinomas ($n = 9$) predominantly had multifocal (22.2%) or diffuse labeling (55.6%), with strong COX-2 expression observed in 6 cases (66.7%) and moderate reactivity in 2 cases (22.2%). Higher-grade tumors, exhibited multifocal labeling (43.8%) or diffuse labeling (50%), with strong intensity in 10 cases (62.5%). The differences in labeling extent ($p = 0.003$) and intensity ($p = 0.009$) between histological grades were statistically significant. No differences were observed in tumors with vascular invasion.

For data analysis, COX-2 expression was categorized as low in 10 (29.4%) cases and high in 24 (70.6%) cases (Figure 1). When analyzing the association between COX-2 immunoreactivity and the histological grading of the tumors, a significant correlation was found with higher tumor grading (poorly differentiated tumors) and COX-2 immunoreactivity ($p = 0.006$). Figure 2 shows Cox-2 immunoexpression in tumors with different histological grades of malignancy.

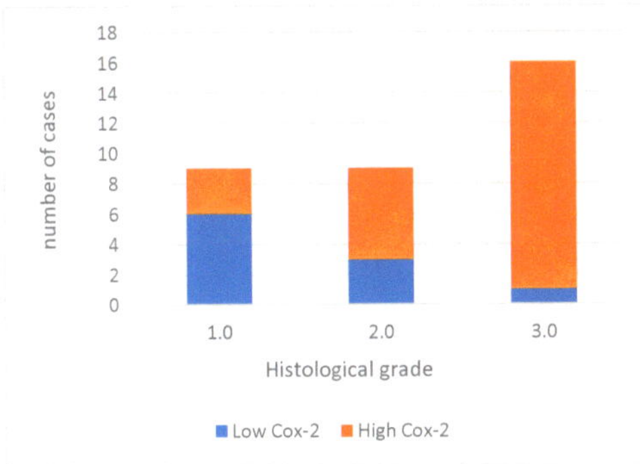

Figure 1. Cox-2 immunoexpression in tumors with different histological grades of malignancy.

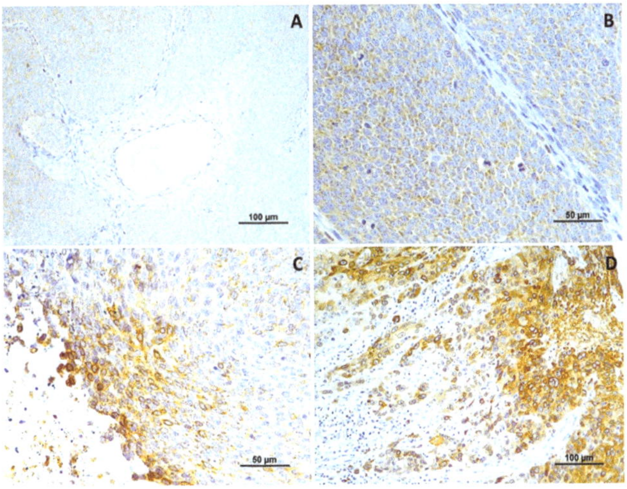

Figure 2. Cox-2 immunoexpression in tumors with different histological grades of malignancy. (**A**) low score in well-differentiated tumors (grade I); (**B**) high score in moderately differentiated tumors (grade II); (**C**,**D**) high score in poorly differentiated tumors (grade III).

3.4. EGFR Immunoreactivity

In normal oral mucosa, EGFR was present in all cases with a moderate membranous reaction. In OSSC, immunoreactivity for EGFR was observed in all cases, with cytoplasmic patterns in 15 tumors (44.1%) and, more frequently, membranous patterns in 19 (55.9%). The intensity of labeling varied, being weak in 8 cases, moderate in 11 cases (32.4%), and strong in 15 cases (44.1%).

Most well-differentiated tumors exhibited a cytoplasmic pattern with weak intensity (71.5%). In contrast, moderately and poorly differentiated squamous cell carcinomas predominantly showed a membranous reaction with strong intensity. The differences in labeling intensity among the histological grade groups were statistically significant ($p = 0.008$). However, no significant differences were observed concerning the location of immunoreactivity.

For data analysis, EGFR expression was categorized into low expression in 19 cases (55.9%) and high expression in 15 cases (44.1%). Well-differentiated tumors predominantly exhibited low labeling (85.7%). Moderately differentiated tumors had high labeling in 5 out of 9 cases, and most poorly differentiated tumors displayed high labeling (56.3%) (Figure 3). High EGFR expression tends to be more common in tumors with a higher histological grade of malignancy, especially in grade 3.0. However, the differences between tumor grade classes are not statistically significant. ($p = 0.067$). A significant association was noted with vascular invasion ($p = 0.006$). Figure 4 shows EGFR immunoexpression in tumors with different histological grades of malignancy.

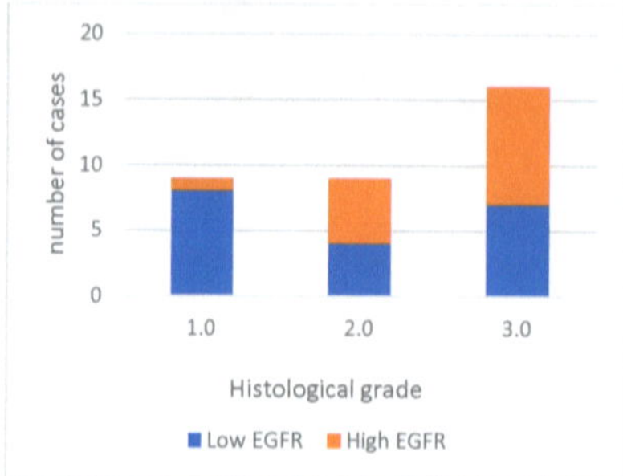

Figure 3. EGFR immunoexpression in tumors with different histological grades of malignancy.

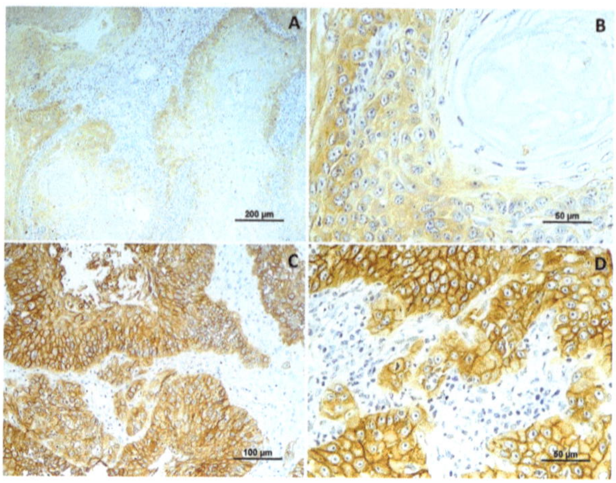

Figure 4. EGFR immunoexpression in tumors with different histological grades of malignancy. (**A**) low score in well-differentiated tumors (grade I); (**B**) low score in a in moderately differentiated tumors (grade II); (**C**,**D**) high score in poorly differentiated tumors (grade III).

3.5. Concurrent Cyclooxygenase-2 (COX-2)/Epidermal Growth Factor Receptor (EGFR) Expression

Of the 34 tumors analyzed, 12 exhibited high immunoreactivity for both COX-2 and EGFR markers. Fifteen cases demonstrated a discordant expression pattern, with either high EGFR and low COX-2 expression, or high COX-2 and low EGFR expression (Figure 5). Furthermore, seven cases showed low immunoreactivity for both COX-2 and EGFR.

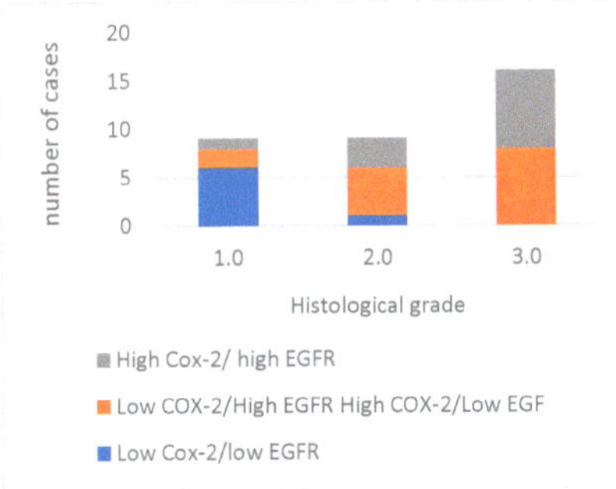

Figure 5. Concurrent expression COX-2 and EGFR in tumors with different histological grades of malignancy.

Comparing the concurrent expression of COX-2 and EGFR in the histological groups, as the histological grade increases, the number of cases with high expression of both COX-2 and EGFR also increases, with the most noticeable difference occurring in grade 3.0. The absence of cases with low COX-2/low EGFR expression in grades 2.0 and 3.0 is also noted. This association between the concurrent expression of COX-2 and EGFR and the histological grade ($p = 0.002$) and with the presence of vascular invasion ($p = 0.002$) of the tumors is statistically significant.

4. Discussion

Oral squamous cell carcinoma (OSCC) stands out as one of the most prevalent malignant tumors of man, making up 1–2% of all malignant tumors worldwide [37,38]. Squamous cell carcinomas (SCCs) are the second most common cancer of the canine oral cavity, resulting in significant morbidity and mortality [39].

In this study, involving 34 cases of -OSCC in dogs, it was observed that the average age of the affected dogs was 10.6 years. No specific gender was more prone to this condition, which is consistent with similar studies [3]. Our study found a higher occurrence of OSCC in mixed breed dogs, although it was also present in various pure breeds, including Poodles, Labrador Retrievers, Beagles, Boxers, Border Collies, Siberian Huskies, Pekingese, Pinschers, and Yorkshire Terriers. Despite the limited sample size, the prevalence of oral cancers in mixed breeds has also been reported in other studies of dogs [14,40]. Contrary to what is generally reported in the literature [3], our study noted that OSCC affected purebred dogs of all sizes, from minor to medium and large. Nevertheless, it is challenging to draw definitive conclusions due to the small sample size and insufficient data on tumor locations and the possibility of mixed breeds being offspring of predisposed breeds.

The main aim of this study was to explore the immunohistochemical expression of COX-2 and EGFR in canine OSCC and assess their correlation with the histological grade of malignancy, as well as to investigate any potential association between these two molecules.

COX-2, an inducible isozyme that plays a crucial role in inflammatory processes, has been associated with malignant diseases [41]. Its overexpression is associated with greater cancer cell growth, increased cell invasion, and an unfavorable prognosis, particularly in canine mammary carcinoma [42]. Our results showed high expression in 70.6% of cases and an association between the histological grade of malignancy and the intensity of Cox-2. These results align with other studies on canine [43,44] and feline [45,46] squamous cell carcinomas. Similar results have also been observed in several other canine tumors, such as mammary tumors [47], melanocytic tumors [48], and rectal and bladder tumors [49], including transitional cell carcinoma [50]. Further research is essential to deepen the relationship between COX-2 expression and indicators of SCC aggressiveness.

Furthermore, in humans, COX-2 has also been implicated in malignancy in several types of cancer, including urothelial tumors [51], laryngeal carcinoma [52], esophageal carcinoma [53] and OSCC [54,55]. Previous studies have highlighted the association COX-2 with angiogenesis and blood vessel formation, and overexpression of COX-2 has been associated with inhibition of apoptosis in tumor cells [56]. The generation of PGE2 by COX-2 has immunosuppressive properties, facilitating the evasion of surveillance mechanisms [57]. The potential use of inhibitors is promising for attenuating resistance to chemotherapy [57].

Thus, our results could suggest that COX-2 inhibitors can treat and increase survival in animals with OSCC. The inhibition of COX-2 activity presents itself as a promising strategy in treating malignant diseases, given the availability of selective and non-selective inhibitors that have exhibited positive effects in high COX-2-expressing cancers. These inhibitors can shift the immune response from supporting tumor growth to destroying it, thereby transforming the tumor microenvironment [58,59]. Piroxicam, a nonsteroidal anti-inflammatory drug (NSAID), has proven beneficial in treating OSCC in dogs. Other potential options, some already licensed for managing pain and inflammation in canines, include mavacoxib [60,61], celecoxib, firocoxib, and enflicoxib [62], among others. However, it is crucial to carefully consider various factors, like the specific subtype of carcinoma, the exact nature and dose of the COX-2 inhibitor, the stage of the tumor, and the effectiveness and practicality of combining these inhibitors with other treatments [63]

EGFR is a cell surface tyrosine kinase fundamental in cell proliferation, angiogenesis, and metastasis—a factor in tumor growth [26,64]. In our study, we observed EGFR overexpression in 56.3% of cases, and we also showed that high EGFR expression tends to be associated with tumors with grade III of malignancy. This aligns with research in human oral [65–67] and cutaneous squamous cell carcinomas [68]. The association between EGFR immunoexpression and the degree of malignancy is also consistent with previous studies on several canine cancer [69,70].

In human cancers, EGFR has been associated with a poor prognosis in gastric carcinoma and head and neck SCC [71,72]. Studies in canine cutaneous squamous cell carcinomas have highlighted the role of EGFR in promoting the growth and survival of tumor cells [73]. It is suggested that EGFR may influence prognosis through its direct expression and the modulation of other regulatory molecules that drive tumor growth [74].

In the field of human medicine, anti-EGFR therapies are promising in the treatment of squamous cell carcinoma [71,75,76]. The application of anti-EGFR therapies, including monoclonal antibodies target at the receptor's surface or tyrosine kinase inhibitors targeting its intracellular domain, has shown encouraging results in canine tumors [77–80] and oral squamous cell carcinoma in cats, particularly with Cetuximab [77]. However, conclusive evidence regarding their absolute efficacy is still lacking. Our findings underscore the importance of focusing scientific research on developing targeted molecular therapies in veterinary medicine, specifically utilizing tyrosine kinase inhibitors (TKIs) and anti-EGFR monoclonal antibodies. Recent strides in comparative oncology have been promoting the transfer of these small molecule inhibitors and monoclonal antibodies from human to veterinary applications [78]. Despite the potential risk of triggering an immune response against these antibodies, leading to adverse effects and reduced treatment efficacy, research in canine mammary cancer suggests that adaptation is viable. The antibodies retain their affinity for EGFR and their anticancer properties in tumor cell lines, thus addressing this challenge. This assertion is supported by our protein alignment data and the significant similarity between canine and human EGFR genes [77,79–81].

In our study, in addition to examining the expression of these two molecules in relation to histopathological characteristics, we investigated the correlation between COX-2 and EGFR in canine OSCC. Our results showed a statistically significant association ($p = 0.002$), and we also showed that the expression of these molecules increased along with the histological grade, in line with previous research in human SCC [82] and canine mammary cancer [35]. Furthermore, we have established a link between these two molecules and vascular invasion, underlining their role in tumor malignancy [83].

A definitive therapy for oral squamous cell carcinoma (OSCC) in dogs remains elusive; directing efforts toward targeting EGFR and COX-2 in veterinary clinical oncology holds promise in unveiling fresh perspectives on cancer biology and the efficacy of advanced targeted therapies.

In summary, the emerging roles of COX-2 and EGFR as promising biomarkers in predicting tumor progression suggest potential avenues for future therapies. Utilizing canine models provides a fresh strategy for advancing cancer treatment. Since dogs encounter cancer at rates similar to humans, share mutual risk factors with their human companions, and exhibit remarkably comparable immune systems, they serve as valuable models for investigating human malignancies and novel biomarkers. This underscores their promising role in cancer research.

The signaling pathways involving COX-2 and EGFR remain inadequately understood in canine OSCC, underscoring the necessity for further research. Future studies should explore additional biomarkers and incorporate methodologies like quantitative PCR (qPCR) to analyze COX-2 and EGFR gene expression and Western blot analysis for protein expression. Additionally, incorporating prognostic studies will provide deeper insights into the clinical relevance of our findings.

5. Conclusions

These findings revealed the association of COX-2 and EGFR with malignancy and established a correlation between these two molecules. This study proves pivotal in advancing our understanding of new biomarkers in OSCC. In the future, exploring the expression of these two molecules in relation to dog survival holds promise for further insights. Studying the relationship between risk factors impacting humans and those impacting dogs is crucial. This investigation seeks to uncover potential treatments that could simultaneously address issues in both species. Furthermore, understanding the pathways involved in both can aid in utilizing suitable therapies for these shared conditions.

Author Contributions: Conceptualization, F.L.Q. and I.P.; data curation, I.P.; formal analysis, F.S., I.P. and J.P.; funding acquisition, F.L.Q., F.S. and I.P.; investigation, J.P.; methodology, C.S., I.P. and J.P.; project administration, I.P.; resources, L.D. and I.P.; software, F.S.; supervision, I.P. and J.P.; validation, R.F., C.S., F.L.Q., F.S., L.D., I.P. and J.P.; visualization, L.D. and I.P.; writing—original draft, R.F. and I.P.; writing—review and editing, F.L.Q., F.S., I.P. and J.P. All authors have read and agreed to the published version of the manuscript.

Funding: This work was supported by the projects UIDB/CVT/00772/2020 and LA/P/0059/2020 funded by the Portuguese Foundation for Science and Technology (FCT).

Institutional Review Board Statement: Not applicable.

Informed Consent Statement: Not applicable.

Data Availability Statement: The data information can be asked for from the corresponding author.

Conflicts of Interest: The authors declare no conflicts of interest.

References

1. Rao, C.V.; Janakiram, N.B.; Madka, V.; Devarkonda, V.; Brewer, M.; Biddick, L.; Lightfoot, S.; Steele, V.E.; Mohammed, A. Simultaneous Targeting of 5-LOX-COX and EGFR Blocks Progression of Pancreatic Ductal Adenocarcinoma. *Oncotarget* **2015**, *6*, 33290–33305. [CrossRef] [PubMed]
2. Dobson, J.M.; Lascelles, B.D.X. *BSAVA Manual of Canine and Feline Oncology*; British Small Animal Veterinary Association: Quedgeley, UK, 2011.
3. Munday, J.S.; Löhr, C.V.; Kiupel, M. Tumors of the Alimentary Tract. In *Tumors in Domestic Animals*; Wiley: Hoboken, NJ, USA, 2016; pp. 499–601. ISBN 978-1-119-18120-0.

4. Genden, E.M.; Ferlito, A.; Bradley, P.J.; Rinaldo, A.; Scully, C. Neck Disease and Distant Metastases. *Oral Oncol.* **2003**, *39*, 207–212. [CrossRef] [PubMed]
5. Squamous Cell Carcinomas in Dogs and Cats. Available online: https://www.embracepetinsurance.com/health/squamous-cell-carcinoma (accessed on 28 December 2023).
6. Vos, J.H.; Van Der Gaag, I. Canine and Feline Oral-Pharyngeal Tumours. *J. Vet. Med. Ser. A* **1987**, *34*, 420–427. [CrossRef] [PubMed]
7. Grigolato, R.; Accorona, R.; Lombardo, G.; Corrocher, G.; Garagiola, U.; Massari, F.; Nicoli, S.; Rossi, S.; Calabrese, L. Oral Cancer in Non-Smoker Non-Drinker Patients. Could Comparative Pet Oncology Help to Understand Risk Factors and Pathogenesis? *Crit. Rev. Oncol./Hematol.* **2021**, *166*, 103458. [CrossRef]
8. Giuliano, A. Companion Animal Model in Translational Oncology; Feline Oral Squamous Cell Carcinoma and Canine Oral Melanoma. *Biology* **2021**, *11*, 54. [CrossRef] [PubMed]
9. Bonfanti, U.; Bertazzolo, W.; Gracis, M.; Roccabianca, P.; Romanelli, G.; Palermo, G.; Zini, E. Diagnostic Value of Cytological Analysis of Tumours and Tumour-like Lesions of the Oral Cavity in Dogs and Cats: A Prospective Study on 114 Cases. *Vet. J.* **2015**, *205*, 322–327. [CrossRef]
10. Al-Maghrabi, B.; Gomaa, W.; Abdelwahed, M.; Al-Maghrabi, J. Increased COX-2 Immunostaining in Urothelial Carcinoma of the Urinary Bladder Is Associated with Invasiveness and Poor Prognosis. *Anal. Cell. Pathol.* **2019**, *2019*, 5026939. [CrossRef]
11. Chen, S.H.; Hsiao, S.Y.; Chang, K.Y.; Chang, J.Y. New Insights into Oral Squamous Cell Carcinoma: From Clinical Aspects to Molecular Tumorigenesis. *Int. J. Mol. Sci.* **2021**, *22*, 2252. [CrossRef]
12. Johnson, D.E.; Burtness, B.; Leemans, C.R.; Lui, V.W.Y.; Bauman, J.E.; Grandis, J.R. Head and Neck Squamous Cell Carcinoma. *Nat. Rev. Dis. Primers* **2020**, *6*, 92. [CrossRef]
13. Bertone, E.R.; Snyder, L.A.; Moore, A.S. Environmental and Lifestyle Risk Factors for Oral Squamous Cell Carcinoma in Domestic Cats. *J. Vet. Intern. Med.* **2003**, *17*, 557–562. [CrossRef]
14. Boss, M.-K.; Harrison, L.G.; Gold, A.; Karam, S.D.; Regan, D.P. Canine Oral Squamous Cell Carcinoma as a Spontaneous, Translational Model for Radiation and Immunology Research. *Front. Oncol.* **2023**, *12*, 1033704. [CrossRef] [PubMed]
15. Gadgeel, S.M.; Ali, S.; Philip, P.A.; Ahmed, F.; Wozniak, A.; Sarkar, F.H. Response to Dual Blockade of Epidermal Growth Factor Receptor (EGFR) and Cycloxygenase-2 in Nonsmall Cell Lung Cancer May Be Dependent on the EGFR Mutational Status of the Tumor. *Cancer* **2007**, *110*, 2775–2784. [CrossRef] [PubMed]
16. Gao, L.; Wang, T.H.; Chen, C.P.; Xiang, J.J.; Zhao, X.B.; Gui, R.Y.; Liao, X.H. Targeting COX-2 Potently Inhibits Proliferation of Cancer Cells in Vivo but Not in Vitro in Cutaneous Squamous Cell Carcinoma. *Transl. Cancer Res.* **2021**, *10*, 2219–2228. [CrossRef] [PubMed]
17. Patrignani, P.; Tacconelli, S.; Sciulli, M.G.; Capone, M.L. New Insights into COX-2 Biology and Inhibition. *Brain Res. Rev.* **2005**, *48*, 352–359. [CrossRef]
18. Chaudhary, S.; Pothuraju, R.; Rachagani, S.; Siddiqui, J.A.; Atri, P.; Mallya, K.; Nasser, M.W.; Sayed, Z.; Lyden, E.R.; Smith, L.; et al. Dual Blockade of EGFR and CDK4/6 Delays Head and Neck Squamous Cell Carcinoma Progression by Inducing Metabolic Rewiring. *Cancer Lett.* **2021**, *510*, 79–92. [CrossRef] [PubMed]
19. Yan, W.; Wistuba, I.I.; Emmert-buck, M.R.; Erickson, H.S. SCC Highlights and Insights. *Am. J. Cancer Res.* **2011**, *1*, 275–300. [PubMed]
20. Souza, J.L.; Martins-Cardoso, K.; Guimarães, I.S.; De Melo, A.C.; Lopes, A.H.; Monteiro, R.Q.; Almeida, V.H. Interplay Between EGFR and the Platelet-Activating Factor/PAF Receptor Signaling Axis Mediates Aggressive Behavior of Cervical Cancer. *Front. Oncol.* **2020**, *10*, 557280. [CrossRef]
21. Szweda, M.; Rychlik, A.; Babińska, I.; Pomianowski, A. Cyclooxygenase-2 as a Biomarker with Diagnostic, Therapeutic, Prognostic, and Predictive Relevance in Small Animal Oncology. *J. Vet. Res.* **2020**, *64*, 151–160. [CrossRef]
22. Sheng, J.; Sun, H.; Yu, F.-B.; Li, B.; Zhang, Y.; Zhu, Y.-T. The Role of Cyclooxygenase-2 in Colorectal Cancer. *Int. J. Med. Sci.* **2020**, *17*, 1095–1101. [CrossRef]
23. Oh, J.H.; Cho, J.-Y. Comparative Oncology: Overcoming Human Cancer through Companion Animal Studies. *Exp. Mol. Med.* **2023**, *55*, 725–734. [CrossRef]
24. Hu, Z.; Hu, Y.; Jiang, H. Overexpression of COX-2 and Clinicopathological Features of Gastric Cancer: A Meta-Analysis. *Transl. Cancer Res.* **2020**, *9*, 2200–2209. [CrossRef] [PubMed]
25. Ali, S.M.A.; Mirza, Y. Overexpression of EGFR, COX2 and P53 in Oral Squamous Cell Carcinoma Patients of Pakistan and Correlation with Prognosis. *Ann. Oncol.* **2019**, *30*, vii21–vii22. [CrossRef]
26. Huang, Z.; Rui, X.; Yi, C.; Chen, Y.; Chen, R.; Liang, Y.; Wang, Y.; Yao, W.; Xu, X.; Huang, Z. Silencing LCN2 Suppresses Oral Squamous Cell Carcinoma Progression by Reducing EGFR Signal Activation and Recycling. *J. Exp. Clin. Cancer Res.* **2023**, *42*, 60, Correction in *J. Exp. Clin. Cancer Res.* **2023**, *42*, 104. [CrossRef] [PubMed]
27. Li, N.; Li, H.; Su, F.; Li, J.; Ma, X.; Gong, P. Relationship between Epidermal Growth Factor Receptor (EGFR) Mutation and Serum Cyclooxygenase-2 Level, and the Synergistic Effect of Celecoxib and Gefitinib on EGFR Expression in Non-Small Cell Lung Cancer Cells. *Int. J. Clin. Exp. Pathol.* **2015**, *8*, 9010–9020. [PubMed]
28. Fukuda, K.; Otani, S.; Takeuchi, S.; Arai, S.; Nanjo, S.; Tanimoto, A.; Nishiyama, A.; Naoki, K.; Yano, S. Trametinib Overcomes KRAS-G12V–Induced Osimertinib Resistance in a Leptomeningeal Carcinomatosis Model of EGFR-Mutant Lung Cancer. *Cancer Sci.* **2021**, *112*, 3784–3795. [CrossRef] [PubMed]

29. Normanno, N.; Maiello, M.R.; De Luca, A. Epidermal Growth Factor Receptor Tyrosine Kinase Inhibitors (EGFR-TKIs): Simple Drugs with a Complex Mechanism of Action? *J. Cell. Physiol.* **2003**, *194*, 13–19. [CrossRef] [PubMed]
30. Sigismund, S.; Avanzato, D.; Lanzetti, L. Emerging Functions of the EGFR in Cancer. *Mol. Oncol.* **2018**, *12*, 3–20. [CrossRef]
31. Head, K.; Cullen, J.M.; Dubielzig, R.; Else, R.; Misdorp, W.; Patnaik, A.K.; Tateyama, S.; Gaag, I. *Histological Classification of Tumors of the Alimentary System of Domestic Animals*; WHO Histological Classification of Tumors of Domestic Animal; WHO: Geneva, Switzerland, 2003.
32. Anneroth, G.; Batsakis, J.; Luna, M. Review of the Literature and a Recommended System of Malignancy Grading in Oral Squamous Cell Carcinomas. *Eur. J. Oral Sci.* **1987**, *95*, 229–249. [CrossRef]
33. Monteiro, A.; Delgado, L.; Monteiro, L.; Pires, I.; Prada, J.; Raposo, T. Immunohistochemical Expression of Tensin-4/CTEN in Squamous Cell Carcinoma in Dogs. *Vet. Sci.* **2023**, *10*, 86. [CrossRef]
34. Carvalho, M.I.; Pires, I.; Prada, J.; Ferreira, A.F.; Queiroga, F.L. Positive Interplay Between CD3+ T-Lymphocytes and Concurrent COX-2/EGFR Expression in Canine Malignant Mammary Tumors. *Anticancer Res.* **2015**, *35*, 2915–2920.
35. Guimarães, M.J.; Carvalho, M.I.; Pires, I.; Prada, J.; Gil, A.G.; Lopes, C.; Queiroga, F.L. Concurrent Expression of Cyclo-Oxygenase-2 and Epidermal Growth Factor Receptor in Canine Malignant Mammary Tumours. *J. Comp. Pathol.* **2014**, *150*, 27–34. [CrossRef] [PubMed]
36. Prada, J.; Queiroga, F.L.; Gregório, H.; Pires, I. Evaluation of Cyclooxygenase-2 Expression in Canine Mast Cell Tumours. *J. Comp. Pathol.* **2012**, *147*, 31–36. [CrossRef] [PubMed]
37. Ch'ng, S.; Low, I.; Ng, D.; Brasch, H.; Sullivan, M.; Davis, P.; Tan, S.T. Epidermal Growth Factor Receptor: A Novel Biomarker for Aggressive Head and Neck Cutaneous Squamous Cell Carcinoma. *Hum. Pathol.* **2008**, *39*, 344–349. [CrossRef] [PubMed]
38. Tan, Y.; Wang, Z.; Xu, M.; Li, B.; Huang, Z.; Qin, S.; Nice, E.C.; Tang, J.; Huang, C. Oral Squamous Cell Carcinomas: State of the Field and Emerging Directions. *Int. J. Oral Sci.* **2023**, *15*, 44. [CrossRef] [PubMed]
39. Munday, J.S.; Dunowska, M.; Laurie, R.E.; Hills, S. Genomic Characterisation of Canine Papillomavirus Type 17, a Possible Rare Cause of Canine Oral Squamous Cell Carcinoma. *Vet. Microbiol.* **2016**, *182*, 135–140. [CrossRef] [PubMed]
40. Blume, G.R.; Eloi, R.S.A.; Oliveira, L.B.; Sonne, L.; Rezende, L.P.O.; Sant'Ana, F.J.F. Lesions of the Oral Cavity of Dogs: 720 Cases. *Pesqui. Vet. Bras.* **2023**, *43*, e07073. [CrossRef]
41. Ceccarelli, C.; Piazzi, G.; Paterini, P.; Pantaleo, M.A.; Taffurelli, M.; Santini, D.; Martinelli, G.N.; Biasco, G. Concurrent EGFr and Cox-2 Expression in Colorectal Cancer: Proliferation Impact and Tumour Spreading. *Ann. Oncol.* **2005**, *16*, iv74–iv79. [CrossRef]
42. Heller, D.A.; Clifford, C.A.; Goldschmidt, M.H.; Holt, D.E.; Shofer, F.S.; Smith, A.; Sorenmo, K.U. Cyclooxygenase-2 Expression Is Associated with Histologic Tumor Type in Canine Mammary Carcinoma. *Vet. Pathol.* **2005**, *42*, 776–780. [CrossRef]
43. Mohammed, S.I.; Khan, K.N.M.; Sellers, R.S.; Hayek, M.G.; DeNicola, D.B.; Wu, L.; Bonney, P.L.; Knapp, D.W. Expression of Cyclooxygenase-1 and 2 in Naturally-Occurring Canine Cancer. *Prostaglandins Leukot. Essent. Fat. Acids* **2004**, *70*, 479–483. [CrossRef]
44. De Almeida, E.M.P.; Piché, C.; Sirois, J.; Doré, M. Expression of Cyclo-Oxygenase-2 in Naturally Occurring Squamous Cell Carcinomas in Dogs. *J. Histochem. Cytochem.* **2001**, *49*, 867–875. [CrossRef]
45. Hayes, A.; Scase, T.; Miller, J.; Murphy, S.; Sparkes, A.; Adams, V. COX-1 and COX-2 Expression in Feline Oral Squamous Cell Carcinoma. *J. Comp. Pathol.* **2006**, *135*, 93–99. [CrossRef] [PubMed]
46. Nasry, W.; Wang, H.; Jones, K.; Dirksen, W.; Rosol, T.; Rodriguez-Lecompte, J.; Martin, C. CD147 and Cyclooxygenase Expression in Feline Oral Squamous Cell Carcinoma. *Vet. Sci.* **2018**, *5*, 72. [CrossRef] [PubMed]
47. Souza, C.H.D.M.; Toledo-piza, E.; Amorin, R.; Barboza, A.; Tobias, K.M. Inflammatory Mammary Carcinoma in 12 Dogs: Clinical Features, Cyclooxygenase-2 Expression, and Response to Piroxicam Treatment. *Can. Vet. J.* **2009**, *50*, 506–510.
48. Martínez, C.M.; Peñafiel-Verdú, C.; Vilafranca, M.; Ramírez, G.; Méndez-Gallego, M.; Buendía, A.J.; Sánchez, J. Cyclooxygenase-2 Expression Is Related with Localization, Proliferation, and Overall Survival in Canine Melanocytic Neoplasms. *Vet. Pathol.* **2011**, *48*, 1204–1211. [CrossRef] [PubMed]
49. Knottenbelt, C.; Mellor, D.; Nixon, C.; Thompson, H.; Argyle, D.J. Cohort Study of COX-1 and COX-2 Expression in Canine Rectal and Bladder Tumours. *J. Small Anim. Pract.* **2006**, *47*, 196–200. [CrossRef] [PubMed]
50. Cekanova, M.; Uddin, J.; Bartges, J.W.; Callens, A.; Legendre, A.M.; Rathore, K.; Wright, L.; Carter, A.; Marnett, L.J. Molecular Imaging of Cyclooxygenase-2 in Canine Transitional Cell Carcinomas In Vitro and In Vivo. *Cancer Prev. Res.* **2013**, *6*, 466–476. [CrossRef] [PubMed]
51. Mohammed, S.I.; Craig, B.A.; Mutsaers, A.J.; Glickman, N.W.; Snyder, P.W.; DeGortari, A.E.; Schlittler, D.L.; Coffman, K.T.; Bonney, P.L.; Knapp, D.W. Effects of the Cyclooxygenase Inhibitor, Piroxicam, in Combination with Chemotherapy on Tumor Response, Apoptosis, and Angiogenesis in a Canine Model of Human Invasive Urinary Bladder Cancer. *Mol. Cancer Ther.* **2003**, *2*, 183–188.
52. Ranelletti, F.O.; Almadori, G.; Rocca, B.; Ferrandina, G.; Ciabattoni, G.; Habib, A.; Galli, J.; Maggiano, N.; Gessi, M.; Lauriola, L. Prognostic Significance of Cyclooxygenase-2 in Laryngeal Squamous Cell Carcinoma. *Int. J. Cancer* **2001**, *95*, 343–349. [CrossRef]
53. Takatori, H.; Natsugoe, S.; Okumura, H.; Matsumoto, M.; Uchikado, Y.; Setoyama, T.; Sasaki, K.; Tamotsu, K.; Owaki, T.; Ishigami, S.; et al. Cyclooxygenase-2 Expression Is Related to Prognosis in Patients with Esophageal Squamous Cell Carcinoma. *Eur. J. Surg. Oncol.* **2008**, *34*, 397–402. [CrossRef]

54. Sappayatosok, K.; Maneerat, Y.; Swasdison, S.; Viriyavejakul, P.; Dhanuthai, K.; Zwang, J.; Chaisri, U. Expression of Pro-Inflammatory Protein, iNOS, VEGF and COX-2 in Oral Squamous Cell Carcinoma (OSCC), Relationship with Angiogenesis and Their Clinico-Pathological Correlation. *Med. Oral Patol. Oral Y Cir. Bucal* **2009**, *14*, E319–E324.
55. Seyedmajidi, M.; Shafaee, S.; Siadati, S.; Khorasani, M.; Bijani, A.; Ghasemi, N. Cyclo-Oxygenase-2 Expression in Oral Squamous Cell Carcinoma. *J. Cancer Res. Ther.* **2014**, *10*, 1024–1029. [CrossRef] [PubMed]
56. Lorch, J.H.; Klessner, J.; Park, J.K.; Getsios, S.; Wu, Y.L.; Stack, M.S.; Green, K.J. Epidermal Growth Factor Receptor Inhibition Promotes Desmosome Assembly and Strengthens Intercellular Adhesion in Squamous Cell Carcinoma Cells. *J. Biol. Chem.* **2004**, *279*, 37191–37200. [CrossRef] [PubMed]
57. Kyzas, P.A.; Stefanou, D.; Agnantis, N.J. COX-2 Expression Correlates with VEGF-C and Lymph Node Metastases in Patients with Head and Neck Squamous Cell Carcinoma. *Mod. Pathol.* **2005**, *18*, 153–160. [CrossRef] [PubMed]
58. Silveira, T.L.; Pang, L.Y.; Di Domenico, A.; Veloso, E.S.; Silva, I.L.D.; Puerto, H.L.D.; Ferreria, E.; Argyle, D.J. COX-2 Silencing in Canine Malignant Melanoma Inhibits Malignant Behaviour. *Front. Vet. Sci.* **2021**, *8*, 633170. [CrossRef] [PubMed]
59. Hussain, M.; Javeed, A.; Ashraf, M.; Al-Zaubai, N.; Stewart, A.; Mukhtar, M.M. Non-Steroidal Anti-Inflammatory Drugs, Tumour Immunity and Immunotherapy. *Pharmacol. Res.* **2012**, *66*, 7–18. [CrossRef] [PubMed]
60. Hurst, E.A.; Pang, L.Y.; Argyle, D.J. The Selective Cyclooxygenase-2 Inhibitor Mavacoxib (Trocoxil) Exerts Anti-tumour Effects In Vitro Independent of Cyclooxygenase-2 Expression Levels. *Vet. Comp. Oncol.* **2019**, *17*, 194–207. [CrossRef] [PubMed]
61. Pang, L.Y.; Argyle, S.A.; Kamida, A.; Morrison, K.O.; Argyle, D.J. The Long-Acting COX-2 Inhibitor Mavacoxib (Trocoxil^{TM}) Has Anti-Proliferative and pro-Apoptotic Effects on Canine Cancer Cell Lines and Cancer Stem Cells In Vitro. *BMC Vet. Res.* **2014**, *10*, 184. [CrossRef]
62. Solà, J.; Menargues, À.; Homedes, J.; Salichs, M.; Álvarez, I.; Romero, L.; Vela, J.M. Selective Inhibition of Cyclooxygenase-2 by Enflicoxib, Its Enantiomers and Its Main Metabolites in Vitro in Canine Blood. *Vet. Pharm. Ther.* **2022**, *45*, 235–244. [CrossRef]
63. Lyndin, M.; Kravtsova, O.; Sikora, K.; Lyndina, Y.; Kuzenko, Y.; Awuah, W.A.; Abdul-Rahman, T.; Hyriavenko, N.; Sikora, V.; Romaniuk, A. COX2 Effects on Endometrial Carcinomas Progression. *Pathol.—Res. Pract.* **2022**, *238*, 154082. [CrossRef]
64. Bazzani, L.; Donnini, S.; Finetti, F.; Christofori, G.; Ziche, M. PGE2/EP3/SRC Signaling Induces EGFR Nuclear Translocation and Growth through EGFR Ligands Release in Lung Adenocarcinoma Cells. *Oncotarget* **2017**, *8*, 31270–31287. [CrossRef]
65. Costa, V.; Kowalski, L.P.; Coutinho-Camillo, C.M.; Begnami, M.D.; Calsavara, V.F.; Neves, J.I.; Kaminagakura, E. EGFR Amplification and Expression in Oral Squamous Cell Carcinoma in Young Adults. *Int. J. Oral Maxillofac. Surg.* **2018**, *47*, 817–823. [CrossRef] [PubMed]
66. Adorno-Farias, D.; Badilla, S.M.; Vidal, G.P.; Fernandez-Ramires, R. Epidermal Growth Factor Receptor (EGFR) Is Overexpressed in Both Dysplasia and Neoplastic Oral Diseases. *Oral Maxillofac. Pathol.* **2022**, *13*, 73–78.
67. Hashmi, A.A.; Hussain, Z.F.; Aijaz, S.; Irfan, M.; Khan, E.Y.; Naz, S.; Faridi, N.; Khan, A.; Edhi, M.M. Immunohistochemical Expression of Epidermal Growth Factor Receptor (EGFR) in South Asian Head and Neck Squamous Cell Carcinoma: Association with Various Risk Factors and Clinico-Pathologic and Prognostic Parameters. *World J. Surg. Oncol.* **2018**, *16*, 118. [CrossRef] [PubMed]
68. Cañueto, J.; Cardeñoso, E.; García, J.L.; Santos-Briz, Á.; Castellanos-Martín, A.; Fernández-López, E.; Blanco Gómez, A.; Pérez-Losada, J.; Román-Curto, C. Epidermal Growth Factor Receptor Expression Is Associated with Poor Outcome in Cutaneous Squamous Cell Carcinoma. *Br. J. Dermatol.* **2017**, *176*, 1279–1287. [CrossRef] [PubMed]
69. Queiroga, F.L.; Perez-Alenza, M.D.; González-Gil, A.; Silván, G.; Peña, L.; Illera, J.C. Quantification of Epidermal Growth Factor Receptor (EGFR) in Canine Mammary Tumours by ELISA Assay: Clinical and Prognostic Implications. *Vet. Comp. Oncol.* **2017**, *15*, 383–390. [CrossRef] [PubMed]
70. Sabattini, S.; Mancini, F.R.; Marconato, L.; Bacci, B.; Rossi, F.; Vignoli, M.; Bettini, G. EGFR Overexpression in Canine Primary Lung Cancer: Pathogenetic Implications and Impact on Survival. *Vet. Comp. Oncol.* **2014**, *12*, 237–248. [CrossRef] [PubMed]
71. Abu-Humaidan, A.H.A.; Ekblad, L.; Wennerberg, J.; Sørensen, O.E. EGFR Modulates Complement Activation in Head and Neck Squamous Cell Carcinoma. *BMC Cancer* **2020**, *20*, 121. [CrossRef]
72. Lu, X.; Huang, L.; Zhang, W.; Ning, X. Tepoxalin a Dual 5-LOX-COX Inhibitor and Erlotinib an EGFR Inhibitor Halts Progression of Gastric Cancer in Tumor Xenograft Mice. *Am. J. Transl. Res.* **2018**, *10*, 3847–3856.
73. Sanz Ressel, B.L.; Massone, A.R.; Barbeito, C.G. Dysregulated Expression of Phosphorylated Epidermal Growth Factor Receptor and Phosphatase and Tensin Homologue in Canine Cutaneous Papillomas and Squamous Cell Carcinomas. *J. Comp. Pathol.* **2020**, *174*, 26–33. [CrossRef]
74. Wang, W.M.; Yang, S.S.; Shao, S.H.; Nie, H.Q.; Zhang, J.; Su, T. Metformin Downregulates the Expression of Epidermal Growth Factor Receptor Independent of Lowering Blood Glucose in Oral Squamous Cell Carcinoma. *Front. Endocrinol.* **2022**, *13*, 828608. [CrossRef]
75. Takei, J.; Kaneko, M.K.; Ohishi, T.; Kawada, M.; Harada, H.; Kato, Y. A Novel Anti-EGFR Monoclonal Antibody (EMab-17) Exerts Antitumor Activity against Oral Squamous Cell Carcinomas via Antibody-Dependent Cellular Cytotoxicity and Complement-Dependent Cytotoxicity. *Oncol. Lett.* **2020**, *19*, 2809–2816. [CrossRef] [PubMed]
76. Connell, C.; Smyth, E.C. Anti-EGFR plus Chemotherapy in Unselected Advanced Oesophageal Squamous Cell Carcinoma: Less POWERful than Expected. *Ann. Oncol.* **2020**, *31*, 161–162. [CrossRef] [PubMed]
77. Altamura, G.; Borzacchiello, G. Anti-EGFR Monoclonal Antibody Cetuximab Displays Potential Anti-Cancer Activities in Feline Oral Squamous Cell Carcinoma Cell Lines. *Front. Vet. Sci.* **2022**, *9*, 1040552. [CrossRef] [PubMed]

78. Londhe, P.; Gutwillig, M.; London, C. Targeted Therapies in Veterinary Oncology. *Vet. Clin. N. Am. Small Anim. Pract.* **2019**, *49*, 917–931. [CrossRef] [PubMed]
79. Singer, J.; Fazekas, J.; Wang, W.; Weichselbaumer, M.; Matz, M.; Mader, A.; Steinfellner, W.; Meitz, S.; Mechtcheriakova, D.; Sobanov, Y.; et al. Generation of a Canine Anti-EGFR (ErbB-1) Antibody for Passive Immunotherapy in Dog Cancer Patients. *Mol. Cancer Ther.* **2014**, *13*, 1777–1790. [CrossRef]
80. Singer, J.; Weichselbaumer, M.; Stockner, T.; Mechtcheriakova, D.; Sobanov, Y.; Bajna, E.; Wrba, F.; Horvat, R.; Thalhammer, J.G.; Willmann, M.; et al. Comparative Oncology: ErbB-1 and ErbB-2 Homologues in Canine Cancer Are Susceptible to Cetuximab and Trastuzumab Targeting. *Mol. Immunol.* **2012**, *50*, 200–209. [CrossRef]
81. Beirão, B.C.B.; Raposo, T.; Jain, S.; Hupp, T.; Argyle, D.J. Challenges and Opportunities for Monoclonal Antibody Therapy in Veterinary Oncology. *Vet. J.* **2016**, *218*, 40–50. [CrossRef]
82. Chiang, K.H.; Shieh, J.M.; Shen, C.J.; Chang, T.W.; Wu, P.T.; Hsu, J.Y.; Tsai, J.P.; Chang, W.C.; Chen, B.K. Epidermal Growth Factor-Induced COX-2 Regulates Metastasis of Head and Neck Squamous Cell Carcinoma through Upregulation of Angiopoietin-like 4. *Cancer Sci.* **2020**, *111*, 2004–2015. [CrossRef]
83. Cai, S.; Zhang, Y.X.; Han, K.; Ding, Y.Q. Expressions and Clinical Significance of COX-2, VEGF-C, and EFGR in Endometrial Carcinoma. *Arch. Gynecol. Obstet.* **2017**, *296*, 93–98. [CrossRef]

current issues in molecular biology

Article

The Potential Association of CDKN2A and Ki-67 Proteins in View of the Selected Characteristics of Patients with Head and Neck Squamous Cell Carcinoma

Dariusz Nałęcz [1,*], Agata Świętek [2,3], Dorota Hudy [2], Zofia Złotopolska [1], Michał Dawidek [4], Karol Wiczkowski [2,5] and Joanna Katarzyna Strzelczyk [2]

1 Department of Otolaryngology and Maxillofacial Surgery, St. Vincent De Paul Hospital, 1 Wójta Radtkego St., 81-348 Gdynia, Poland; zzlotopolska@gmail.com
2 Department of Medical and Molecular Biology, Faculty of Medical Sciences in Zabrze, Medical University of Silesia in Katowice, 19 Jordana St., 41-808 Zabrze, Poland; agata.swietek@sum.edu.pl (A.Ś.); dorota.hudy@sum.edu.pl (D.H.); karolwicz@gmail.com (K.W.); jstrzelczyk@sum.edu.pl (J.K.S.)
3 Silesia LabMed Research and Implementation Centre, Medical University of Silesia in Katowice, 19 Jordana St., 41-808 Zabrze, Poland
4 Department of Head and Neck Reconstructive Surgery and Robotic Surgery, 1 Powstania Styczniowego St., 81-519 Gdynia, Poland; mdawidek@szpitalepomorskie.eu
5 Students' Scientific Association at the Department of Medical and Molecular Biology, Medical University of Silesia in Katowice, 19 Jordana St., 41-808 Zabrze, Poland
* Correspondence: dnalecz@szpitalepomorskie.eu

Abstract: Head and neck squamous cell carcinoma (HNSCC) is the sixth most prevalent type of cancer worldwide. Not all mechanisms associated with cell cycle disturbances have been recognized in HNSCC. The aim of this study was to examine the concentration of CDKN2A and Ki-67 proteins in 54 tumor and margin samples of HNSCC and to evaluate their association with the clinical and demographic variables. The ELISA method was used to measure concentrations of CDKN2A and Ki-67 in the tissue homogenates. A significantly higher CDKN2A concentration was found in OSCC tumor samples as compared with OPSCC+HPSCC+LSCC. An inverse correlation was observed for Ki-67. We showed an association between the CDKN2A level and the clinical parameters N in tumors. The patients with concomitant diseases had significantly higher levels of Ki-67 as compared with patients with no concomitant diseases. An analysis of the effect of drinking habits on Ki-67 level demonstrated a statistical difference between regular or occasional users of stimulants and patients who do not use any stimulants in the tumor and margin samples. Moreover, we found an association between CDKN2A and Ki-67 concentrations and the HPV status in tumor and margin samples. The levels of the proteins tested may be dependent on environmental factors. Our results showed that changes in protein levels in HNSCC subtypes may reflect different molecular pathways of tumor development or may also be responsible for the involvement of CDKN2A and Ki-67 in the carcinogenesis process.

Keywords: HNSCC; carcinogenesis; protein level; tumor; surgical margin; CDKN2A; Ki-67

Citation: Nałęcz, D.; Świętek, A.; Hudy, D.; Złotopolska, Z.; Dawidek, M.; Wiczkowski, K.; Strzelczyk, J.K. The Potential Association of CDKN2A and Ki-67 Proteins in View of the Selected Characteristics of Patients with Head and Neck Squamous Cell Carcinoma. *Curr. Issues Mol. Biol.* **2024**, *46*, 13267–13280. https://doi.org/10.3390/cimb46110791

Academic Editors: Madhav Bhatia, Arumugam R. Jayakumar and Violeta Popovici

Received: 27 September 2024
Revised: 7 November 2024
Accepted: 18 November 2024
Published: 20 November 2024

1. Introduction

Head and neck cancer is the sixth most prevalent type of cancer worldwide where more than 90% of cases represent squamous cell carcinoma [1]. The average age upon diagnosis is 60 years, yet the incidence of such cancers in adults younger than 45 years has increased over the last years due to the higher numbers of oropharyngeal cancers associated with oncogenic human papillomavirus [1,2]. The prognosis for HNSCC prepared by the Polish Ministry of Health projects the growth in incidence by another 10% before 2025 [3]. The main risk factors are HPV infection (especially type 16 and 18), use of tobacco and alcohol abuse, leading particularly to the development of squamous cell carcinoma [1,4]. EBV infection has a role in the development of nasopharyngeal cancers with the greatest carcinogenesis potential upon concomitant infection with HPV [5]. Due to non-specific symptoms, such as xerostomia, dysphagia, gingival bleeding, inflammation of the oral cavity mucous tissue as well as poor social awareness of head and neck cancers, the vast majority of patients are diagnosed at stages III and IV while HNSCC overall survival (OS) is around 1.5 years [6]. A total of 58% of patients with HNSCC show a 5-year survival rate [7]. Better clinical outcomes strongly demand the personalization of treatment as well as risk stratification among patients [8].

Ki-67 is a nuclear antigen showing a role in cell proliferation. The protein is present during the active phases of the cell cycle (G1, S, G2, M) and reaches its peak value during mitosis. During the cell resting phase (G0), no activity of Ki-67 is observed [9]. This protein is considered an important indicator of the cell division rate and used as a diagnostic and prognostic marker in some cancers. Moreover, it is particularly helpful in determination of the tumor's mitotic activity [10–13].

The *CDKN2A* (cyclin-dependent kinase inhibitor 2A) gene encodes $p16^{INK4A}$ (p16, CDKN2A) and $p14^{ARF}$ (ARF) proteins, expressed in multiple cell types. The $p14^{ARF}$ protein plays the role of a tumor suppressor, regulating the cell cycle during G1 and G2 phase transition through HDM2 inhibition [14]. The $p16^{INK4A}$ protein takes part in multiple cellular processes, including promotion of proliferation, inhibition of apoptosis and inducing angiogenesis in the cancer cells [15]. Its role is to inhibit the cyclin-dependent kinases (CDK4 and CDK6), thereby activating the retinoblastoma (Rb) family of proteins. This results in suppression of the cell transition from the G1 to the S phase of the cell cycle [14]. Changes in expression of *CDKN2A* gene are associated with the disruption of the normal cell cycle functions, leading to uncontrolled cell proliferation and therefore the implication of neoplastic transformation, so far described in a wide range of cancer types, including melanoma, lymphoma, head and neck squamous cell carcinoma, esophageal squamous cell carcinoma, oral cavity cancer, epithelial ovarian cancer, pancreatic adenocarcinoma, gastric cancer, colorectal cancer, non-small cell lung cancer, prostate cancer and many others [16–22].

Given the role of CDKN2A and Ki-67 proteins in important cellular processes, including the proliferation and regulation of the cell cycle, it was hypothesized that the levels of those proteins would be changed in the tumor samples as compared with the margin samples depending on selected demographic and clinical–pathological characteristics in our study group with head and neck squamous cell carcinoma (HNSCC). This study aimed at the evaluation of CDKN2A and Ki-67 concentrations and the possible association of the clinical and demographic variables in tumors and the resected surgical margin samples from patients with primary HNSCC.

Based on our knowledge (PubMed, Medline databases), this has been the first study to analyze the concentrations of CDKN2A and Ki-67 proteins in tumor and surgical margin homogenates obtained from patients with HNSCC by immunoassay (ELISA).

This study aimed at the evaluation of CDKN2A and Ki-67 concentrations and the possible association of the clinical and demographic variables in tumors and the resected surgical margin samples from patients with primary HNSCC.

2. Materials and Methods

2.1. Study Population

This study comprised 108 samples (54 tumor samples and 54 samples of the corresponding margin) from patients diagnosed with HNSCC. The HNSCC group included 30 cases of oral squamous cell carcinoma (OSCC), 2 cases of oropharyngeal squamous cell carcinoma (OPSCC), 17 cases of laryngeal squamous cell carcinoma (LSCC), 2 cases of hypopharyngeal squamous cell carcinoma (HPSCC), 2 cases of nasal squamous cell carcinoma (NCSCC) and 1 case of skin squamous cell carcinoma (SSCC). The cancer samples and the corresponding margins were collected after surgical resection and were histologically diagnosed by a pathologist. The samples collected were histologically assessed and classified as primary HNSCC. The histologically confirmed cancer-free specimens were taken from the surgical margin at a distance of at least 10 mm from the tumor margin. All the patients were diagnosed, and samples were collected at the Department of Otolaryngology and Maxillary Surgery, St. Vincent De Paul Hospital, Gdynia. The tumor and margin samples were divided into two parts. DNA was first isolated from the samples intended for HPV analysis and then frozen at −80 °C. Samples intended for protein analyses were immediately frozen at −80 °C until further analyses. The study was approved by the Bioethical Committee, Regional Medical Chamber in Gdansk (no. KB-42/21). The laboratory analyses were all carried out in the Department of Medical and Molecular Biology, Faculty of Medical Sciences in Zabrze, Medical University of Silesia. The main inclusion criteria for the HNSCC group included diagnosis of primary squamous cell carcinoma and no preoperative radio- or chemotherapy. All the participants delivered their written informed consent to take part in the study which was also the inclusion criteria. Demographic data of the study group are illustrated in Table 1.

Table 1. Demographic data of the study group.

	N	%
Average age (range)	64 (53–72)	
Female	18	33.33
Male	36	66.67
Smokers	45	83.33
Non-smokers	9	16.67
Alcohol users (occasional)	36	66.67
Alcohol users (regular)	15	27.78
Alcohol non-users	3	5.56

Tumor samples were assessed in accordance with the TNM classification. The tumor stage was categorized according to the 8th edition of the AJCC Cancer Staging Manual [23]. Three patients were staged T1 (5.56%), 10 samples with T2 (18.52%), 19 cases with T3 (35.19%) and 20 patients with T4 (37.04%). Lymph node metastasis was found in 20 patients, i.e., N1 in 6 (11.11%), N2 in 12 subjects (22.22%) and N3 in 2 samples (3.70%). In 15 patients (27.78%) the cancer's grade was G1, almost 60% of patients (32; 59.26%) were graded G2, 6 subjects (11.11%) were G3 and 1 sample (1.85%) was G4.

2.2. *Tissue Homogenisation*

The tumor and the surgical margin samples were homogenized in cooled PBS buffer (EURx, Gdansk, Poland) at the ratio of 9:1 (PBS volume/tissue weight). Homogenization was conducted using a Bio-Gen PRO200 homogenizer (PRO Scientific Inc., Oxford, CT, USA) at the rate of 10,000 rpm. Subsequently, the homogenates were sonicated using a UP100H ultrasonifier (Hielscher, Teltow, Germany).

2.3. *Determination of CDKN2A and Ki-67 Protein Concentrations*

The enzyme-linked immunosorbent assay (ELISA) was used to determine the concentrations of selected proteins in the homogenates. Commercially available ELISA kits were used for CDKN2A and Ki-67 proteins (SEA794Hu and SEC047Hu, respectively, by Cloud-Clone Corp., Houston, TX, USA). The analyses were performed according to the manufacturer's instructions. The sensitivity of the detectable CDKN2A dose was 0.262 ng/mL and 0.32 ng/mL for Ki-67. The intra-assay and inter-assay precisions for all kits were <10% and <12%, respectively. All the standards and samples were evaluated in duplicate. The absorbance readings of the samples were recorded at 450 nm with the use of a Synergy H1 microplate reader (Bio-Tek, Winooski, VT, USA). The results were calculated with Gen5 2.06 software (BioTek, Winooski, VT, USA).

2.4. Total Protein Concentration Determinations

The quantification of total protein in the homogenates made use of an AccuOrange™ Protein Quantitation Kit (Biotium, Fremont, CA, USA) according to the manufacturer's protocol. The assay detection range was 0.1–15 μg/mL. Determinations were carried out in duplicate with the previously prepared tissue homogenates, according to the manufacturer's instructions, without any dilutions. Fluorescence was evaluated with the use of a Synergy H1 microplate reader (BioTek, Winooski, VT, USA) with an excitation wavelength of 480 nm and an emission wavelength of 598 nm. Concentrations of the analyzed proteins were normalized for each sample with reference to the total protein in the tissue lysates, and the values were presented as ng/μg (Ki-67) or pg/μg (CDKN2A).

2.5. DNA Isolation and HPV Confirmation

The samples were first homogenized with the use of a Lysing Matrix A (MP Biomedicals, Irvine, CA, USA) and DNA was then isolated using the commercial GeneMATRIX Tissue DNA Purification Kit (Eurx, Gdansk, Poland) according to the manufacturer's protocol. The quantity and quality of the isolated DNA was assessed using a spectrophotometer NanoPhotometer Pearl (Implen, Munich, Germany). The genetic material was frozen and stored at −20 °C until further analysis.

The presence of HPV was confirmed with a GeneFlowTM HPV Array Test Kit (DiagCor Bioscience Ltd., Kowloon Bay, Hong Kong, China) incorporating the flow-through system FT-PRO (DiagCor Bioscience Ltd., Kowloon Bay, Hong Kong, China) according to the manufacturer's instructions. The DNA was first used in a PCR reaction on a Mastercycler Personal Thermal Cycler (Eppendorf, Hamburg, Germany). The PCR products were then denatured and hybridized. Following enzyme conjugation and color development, the results were screened using CapturePro Image, CaptureREAD 3.1. (DiagCor Bioscience Ltd., Kowloon Bay, Hong Kong, China). Positive and negative controls were included in all runs.

2.6. Statistical Analyses

ELISA results were evaluated with the Shapiro–Wilk test to determine the distribution of the variables. A Student's *t*-test, a Mann–Whitney *U* test or a Kruskal–Wallis with Dunn–Sidak post hoc were used to verify the significance of differences in the means or medians between the groups. Correlation was established with Spearman's rank correlation coefficient. The assumed level of significance was $p < 0.05$. The results are presented as mean ± SD or median with quartile range in text. The STATISTICA version 13 software (TIBCO Software Inc., Palo Alto, CA, USA) was used to perform all the analyses. Significant data are shown as box plots with the median in the middle and the 1st and 3rd quartile as a box with minimum and maximum values as whiskers. Data referred to in the text are presented as median with the 1st and 3rd quartile as follows, median (quartile 1st–quartile 3rd).

3. Results

3.1. Levels of Ki-67 Protein in Tumor and Margin Samples

No significant differences were found in Ki-67 levels in HNSCC tumor samples as compared to the margin samples. A significantly lower concentration of Ki-67 was observed in the OSCC cancer subtype compared with the joint group of OPSCC with HPSCC and LSCC cancer subtypes (0.0005 (0.0002–0.0017) vs. 0.0032 (0.0014–0.0047); $p = 0.0128$) in the tumor samples. The results for the tumor samples are presented in Figure 1.

We found that Ki-67 concentration was significantly higher in patients with concomitant diseases as compared with patients without concomitant diseases, as presented in Figure 2 (0.0021 (0.0013–0.0041) vs. 0.0002 (0.0002–0.0004); $p = 0.0011$).

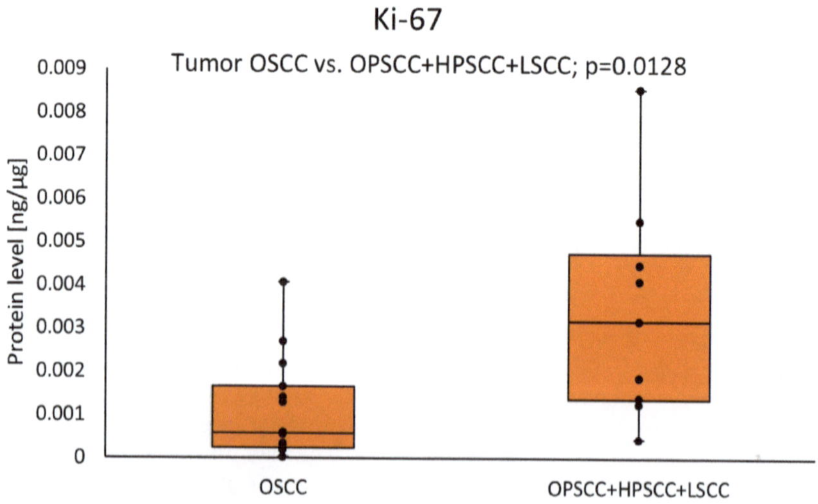

Figure 1. The Ki-67 protein level in the tumor samples according to the tumor subtypes. Statistical analysis with a Mann–Whitney *U* test and differences with $p < 0.05$ are considered statistical.

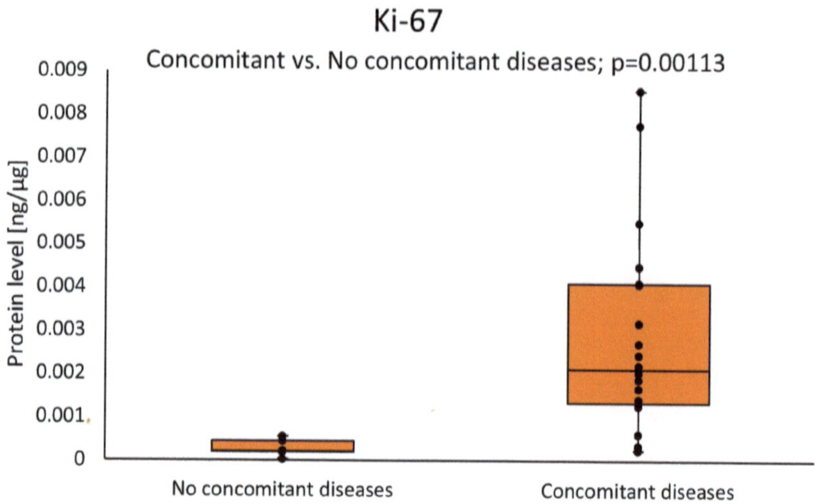

Figure 2. The Ki-67 protein level in the tumor samples according to concomitant diseases status. Statistical analysis was undertaken with a Mann–Whitney *U* test and differences with $p < 0.05$ are considered statistical.

The median concentration of Ki-67 was significantly higher in the abstinent individuals as compared with occasional drinkers (0.0169 (0.0106–0.0350) vs. 0.0014 (0.0003–0.0023); $p = 0.0081$), and in abstinents compared with regular drinkers (0.0169 (0.0106–0.0350) vs. 0.0012 (0.0004–0.0034); $p = 0.0121$) in the tumor samples. Moreover, the median concentration of Ki-67 was higher in occasional drinkers compared with regular drinkers (0.0024 (0.0014–0.0039) vs. 0.0009 (0.0004–0.0018); $p = 0.0462$) in the margin. The results are presented in Figure 3.

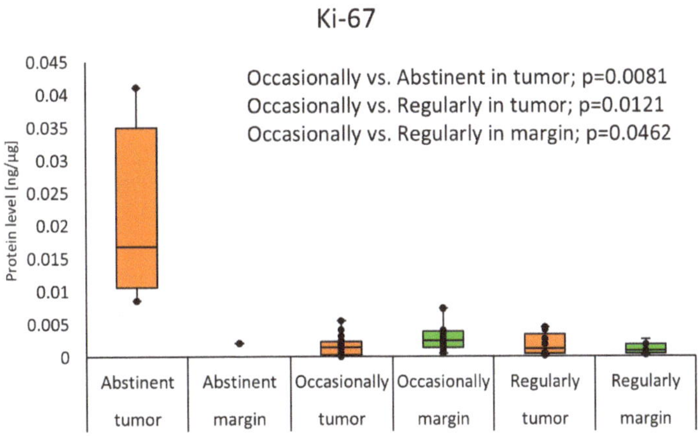

Figure 3. The Ki-67 protein level in tumor and margin samples according to drinking status. Statistical analysis was undertaken with Kruskal–Wallis and Dunn–Sidak post hoc, differences with $p < 0.05$ are considered statistical. The orange color indicates the tumor; green indicates the margin.

No other significant differences were found between Ki-67 concentrations and the demographic such as smoking, HPV-DNA and p16 status or clinical–pathological parameters, such as T, N and G classification, except for a positive correlation between Ki-67 protein level in the tumor and the patient's age (0.59; $p = 0.0012$).

3.2. Levels of CDKN2A Protein in Tumor and Margin Samples

No significant differences were found in the CDKN2A level in HNSCC tumors as compared to the margin samples. We observed a significant difference between OSCC and OPSCC with HPSCC and LSCC tumor groups in CDKN2A level (3.8029 (2.8172–7.9817) vs. 2.9439 (1.8819–3.7261); $p = 0.0437$). The CDKN2A level was higher in the OSCC group compared with the joint group of OPSCC, HPSCC and LSCC. The results are given in Figure 4.

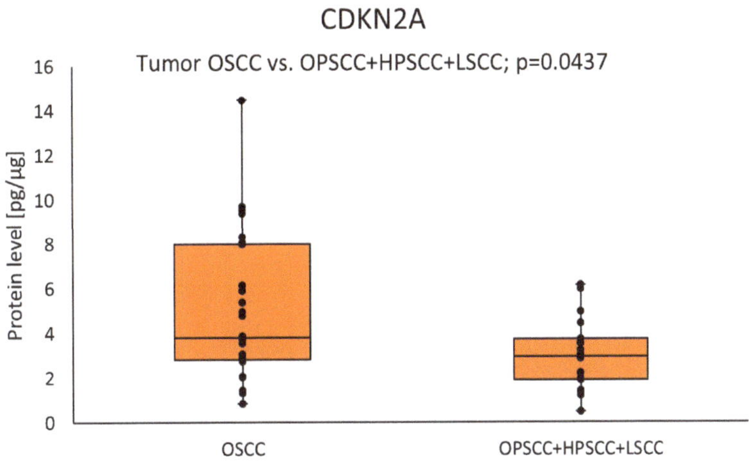

Figure 4. The CDKN2A protein level in the tumor samples according to tumor subtypes. Statistical analysis was undertaken with a Mann–Whitney U test and differences of $p < 0.05$ are considered statistical.

We observed higher levels of CDKN2A protein in the tumor samples of patients with lower nodal status, N0 vs. N2+N3 (3.8763 (2.8281–8.1203) vs. 2.3377 (1.3179–3.7278); p = 0.0362). The results are presented in Figure 5.

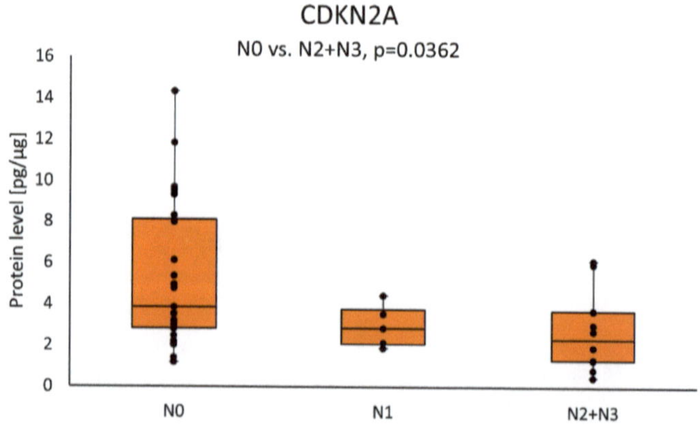

Figure 5. The CDKN2A protein level in the tumor samples according to the patient's nodal status. Statistical analysis was undertaken with Kruskal–Wallis and Dunn–Sidak post hoc, and differences of $p < 0.05$ are considered statistical.

The tumor samples from HPV-positive patients showed a higher level of CDKN2A as compared with those with negative HPV status (7.7355 (6.1427–10.592) vs. 3.0240 (1.7708–4.0156); p = 0.0034) (Figure 6).

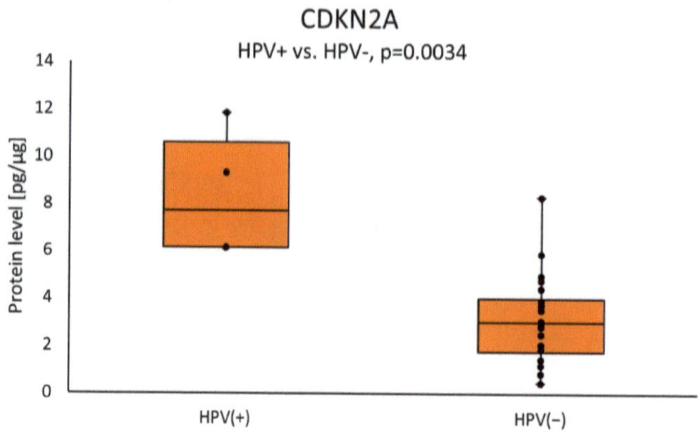

Figure 6. The CDKN2A protein level in the tumor samples according to HPV status. Statistical analysis was undertaken with a Mann–Whitney U test and differences of $p < 0.05$ are considered statistical.

No significant differences in CDKN2A concentration were observed in the margin samples in each of the combinations mentioned above and no other significant differences were found for the demographic or clinical–pathological parameters, such as T and G classification. However, a positive correlation was observed between the CDKN2A protein level in the tumor samples and the smoking years (0.29; p = 0.0376), while the margin samples showed a correlation with the number of cigarettes smoked per day (0.30; p = 0.0363) and a medium positive correlation with pack-years (0.31; p = 0.0367).

4. Discussion

CDKN2A and Ki-67 are some of the key proteins associated with regulation of the cell cycle [24–26]. However, the exact role of CDKN2A and Ki-67 in prognosis and biological function has not been yet well established in the HNSCC subtypes.

Our analysis showed no statistical differences in the expression levels of CDKN2A and Ki-67 proteins in the tumor samples as compared to the surgical margin samples. However, significantly higher levels of CDKN2A protein were observed in patients with OSCC compared with OPSCC+HPSCC+LSCC. Previous studies have shown that alterations in CDKN2A expression are more frequent in OSCC than in other solid tumors such as pancreatic tumors, bladder cancer, renal cell carcinoma, non-small cell lung cancer, melanoma, glioma, gastroesophageal junction and gastric adenocarcinomas [27–29]. This is mainly due to mutations, loss of heterozygosity and DNA hypermethylation of the CDKN2A gene. It is assumed that epigenetic changes in CDKN2A have been linked to genetic or epigenetic changes in other cancer-related genes [14,30]. Some studies have reported increased CDKN2A expression in HNSCC [31,32], while others report decreased CDKN2A expression [33–35]. These differences may be due to the type of the material used, including HNSCC tumor tissues, tissues from animal models and the detection method applied, including immunohistochemistry, reverse phase protein array (RPPA) and fluorescence in situ hybridization. Moreover, we can speculate that discrepancies in CDKN2A protein expression may result from the changes in transcription and translation processes and may also be due to tumor heterogeneity [36]. On the other hand, our data show a statistically significant difference in higher Ki-67 levels in the OPSCC+HPSCC+LSCC group as compared with OSCC. This variation in Ki-67 expression reflects the underlying molecular pathways, clinical courses, treatment modalities and outcome differences among the HNSCC subtypes. The study by Ahmed et al. revealed reduced Ki-67 expression in laryngeal tumors compared with oral and pharynx squamous cell carcinomas. Presumably, their results differed from ours because their team studied the genes and used qPCR techniques [37]. Other studies showed abnormal DNA methylation, with hypermethylation occurring predominantly in OSCC cases, while hypomethylation was observed more frequently in LSCC and OPSCC, which may partly explain the differential behavior of HNSCC subtypes, including the altered cell cycle and proliferation [38–40].

Moreover, we reported the increased concentration of Ki-67 in patients with concomitant diseases (including cardiovascular diseases, kidney diseases, gastrointestinal diseases, endocrine diseases and others), compared to patients without concomitant diseases. Increased Ki-67 levels have been found in some diseases. such as diabetes, atherosclerosis, rheumatoid arthritis and pancreatitis [41]. A study of oral cancer in rats showed higher expression of Ki-67 in diabetic animals as compared with healthy rats [33]. Other studies have shown that inflammation and/or cancer can affect changes in Ki-67 expression and the cell cycle [41]. We could speculate that higher Ki-67 levels in the group with concomitant diseases may have resulted from a local tissue response to the accompanying inflammation or compensation for the cell damage or death.

Our study demonstrated that the median concentration of Ki-67 in the tumor samples was significantly higher in abstinent individuals as compared with occasional drinkers as well as regular drinkers. Furthermore, the median level of Ki-67 in the margin samples was higher in occasional drinkers compared with regular drinkers. Alcohol consumption is a well-known risk factor for head and neck squamous cell carcinoma, however the underlying molecular mechanisms remain unclear. It has been suggested that ethanol may induce basal cell proliferation, which can consequently cause DNA damage and lead to cancer [42]. Studies of tongue carcinoma, esophageal squamous cell carcinoma and head and neck squamous cell carcinoma cell lines have shown that alcohol inhibits cell proliferation and could have an effect on cell cycle inhibition in the G2/M phase [43,44]. On the other hand, Liu et al. observed increased proliferation in oroesophageal squamous cell carcinoma cell lines exposed to ethanol [45]. The disparate results are possibly dependent on the alcohol dose used and the type of cells. Moreover, it seems that acetaldehyde, which is the first metabolite of ethanol, is able to modify the processes of methylation, DNA synthesis and repair, and may interact with proteins and thus influence the proliferation processes [46,47].

In our study, we observed a positive correlation between Ki-67 levels and age. To date, few studies have described a correlation between Ki-67 expression and age in patients with HNSCC. It has been shown that older patients may have different patterns of inflammation, immune surveillance and cellular aging, which, among other factors, may influence the tumor growth [48]. A study by Jing et al. demonstrated that Ki-67 expression was significantly connected with patient age, presenting that people under 60 years exhibited lower Ki-67 expression [49]. In another recent study Wang et al. showed no statistically significant variance in the mean Ki-67 expression between patients with HNSCC aged 60 or younger and those older than 60 [50]. Similar results were obtained in another histological and immunohistochemical study in OSCC, where no differences in proliferative activity were noted between younger and older patients [51].

Our study reported that CDKN2A protein concentration in the tumor samples in the group of patients with the nodal status N2 or N3 was significantly lower as compared with samples in the group of patients with the nodal status N0. Interestingly, CDKN2A is also occasionally used preoperatively during diagnostic biopsy to predict cancer aggressiveness [52]. The literature available reports that decreased CDKN2A expression is associated with poor prognosis, higher stage and progression of HNSCC [53–56]. Another study has reported that decreased CDKN2A/p16 expression was associated with OSCC progression and lymph node metastasis [57]. In a Chinese study, the authors reported that methylation, and therefore decreased CDKN2A expression, were associated with the lymph node metastasis in squamous cell carcinomas of the buccal mucosa [58]. In contrast, another team found no significance between CDKN2A methylation and clinicopathological data in OSCC [59]. Interestingly, other studies have confirmed that CDKN2A methylation might be associated with metastasis in other cancer types, including gastric cancer, breast cancer and endometrial cancer [60–62].

In our study, CDKN2A protein levels in the study group were higher in HPV-positive DNA tumor tissue. This correlation is attributed to the influence of the E7 oncogene of HPV, which can bind to the tumor suppressor protein pRb and form a complex, which may result in increased levels of CDKN2A. Previous reviews estimated that 22–26% of HNSCCs were HPV-positive [63]. Some studies have demonstrated the utility of p16 as a surrogate marker of HPV infection in HNSCC while other studies have found no such association [64,65]. This suggests the existence of other HPV-independent mechanisms that may affect CDKN2A overexpression, which has been confirmed in cervical cancer and cancers of the oral cavity and pharynx. HNSCC is a heterogeneous disease entity due to its HPV infection status. Therefore, we suggest that there are differences in the oncogenic pathways, probably in addition to viral oncoproteins, which are associated with HPV genes involved in different processes, able to affect multiple molecular pathways.

Furthermore, we reported a positive correlation between the CDKN2A protein level in the tumor samples and years of smoking, while in the marginal samples we observed a positive correlation with the number of cigarettes smoked per day and an average positive correlation with pack-years. As is well known, smoking is one of the best-studied non-genetic risk factors for cancer, which can cause changes in the gene expression profile leading to malignant transformation [66]. The results of most studies have shown that CDKN2A levels can be decreased under the influence of smoking due to changes in the form of deletion or methylation of the gene promoter [18,67–71]. Discrepancies compared with the results of our study may be due to factors such as differences in sampling techniques, preparation and detection methods. Based on our results, the effect of cigarette smoke components may be through modulation of CDKN2A expression in response to epithelial damage. We suspect that alternative genetic and epigenetic mechanisms may exist in smokers that may be involved in altered gene and protein expression, including CDKN2A.

To summarize, the small sample size was the main limitation of this study. Larger and more diverse cohorts should be used to validate our results for the use of these parameters in the monitoring of the clinical course of the disease. Furthermore, such an analysis should include the cell lines to better understand the role these proteins play in head and neck squamous cell carcinoma.

Along with the continuous development of medicine, molecular diagnostics is beginning to play an increasingly important role in oncology. Importantly, the latest ESMO treatment guidelines recognize the role of molecular diagnostics [72]. This underscores the need for further research to help develop available diagnostic and prognostic markers. In the future, a better understanding of the molecular biology of HNSCC should lead to personalized treatment based on genomic alterations or gene expression profiles. This will further enable acceleration and simplification of the diagnostic pathway, reducing the overall time to treatment from the time of cancer diagnosis. In addition, molecular biology analyses can also be used as predictors of possible response to chemotherapy [73].

To the best of our knowledge based on PubMed and Medline databases, this is the first study to have analyzed the concentrations of CDKN2A and Ki-67 proteins in homogenates of tumors and matched surgical margins from patients with HNSCC using an immunoassay (ELISA). In earlier studies, the authors focused on examining the methylation pattern of the CDKN2A gene using methylation-specific PCR and analyzing CDKN2A mutations using polysomic and genomic data, as well as p16 immunoexpression analysis [74–79].

5. Conclusions

Changes in the protein levels in HNSCC subtypes may reflect the diversity of molecular pathways of tumor development depending on the subtype or may also be responsible for the involvement of CDKN2A and Ki-67 in the carcinogenesis process. The levels of the proteins tested may be dependent also on environmental factors such as alcohol consumption, smoking and HPV status. The results underscore the necessity for a more personalized approach in the treatment of HNSCC, considering the specific molecular and environmental influences on the tumor biology. Future research should comprise larger cohorts and assume long-term follow-up to validate Ki-67 and CDKN2A as reliable prognostic biomarkers and to explore targeted interventions based on these molecular insights.

Author Contributions: Conceptualization, D.N. and J.K.S.; methodology, D.N. and J.K.S.; formal analysis, D.N.; investigation, D.N. and A.Ś.; resources, D.N., M.D. and Z.Z.; writing—original draft preparation D.N., A.Ś., D.H. and K.W.; writing—review and editing, Z.Z., M.D. and J.K.S.; visualization, D.N. and D.H.; supervision, J.K.S.; project administration, D.N. All authors have read and agreed to the published version of the manuscript.

Funding: This research received no external funding.

Institutional Review Board Statement: The study was conducted according to the guidelines of the Declaration of Helsinki, and approved by the Bioethical Committee, Regional Medical Chamber in Gdansk (no. KB-42/21).

Informed Consent Statement: Informed consent was obtained from all subjects involved in the study.

Data Availability Statement: The data used to support the findings of this research are available upon request.

Acknowledgments: We thank all our patients for their voluntary participation in this study.

Conflicts of Interest: The authors declare no conflicts of interest. The funders had no role in the design of the study; in the collection, analyses or interpretation of data; in the writing of the manuscript; or in the decision to publish the results.

Abbreviations

HNSCC	Head and neck squamous cell carcinoma
HPV	Human papillomavirus
CDKN2A/p16	Cyclin-dependent kinase inhibitor 2A
Rb/pRb	Retinoblastoma protein
CDK4	Cyclin-dependent kinase
CDK6	Cyclin-dependent kinase
OSCC	Oral squamous cell carcinoma
LSCC	Laryngeal squamous cell carcinoma
OPSCC	Oropharyngeal squamous cell carcinoma
NCSCC	Nasal squamous cell carcinoma
SSCC	Squamous cell carcinoma
HPSCC	Hypopharyngeal squamous cell carcinoma
AJCC	American Joint Committee on Cancer
PCR	Polymerase chain reaction
ELISA	Enzyme-linked immunosorbent assay
RPPA	Reverse phase protein array
qPCR	Quantitative polymerase chain reaction

References

1. Sabatini, M.E.; Chiocca, S. Human papillomavirus as a driver of head and neck cancers. *Br. J. Cancer* **2020**, *122*, 306–314. [CrossRef] [PubMed]
2. Hashim, D.; Genden, E.; Posner, M.; Hashibe, M.; Boffetta, P. Head and neck cancer prevention: From primary prevention to impact of clinicians on reducing burden. *Ann. Oncol.* **2019**, *30*, 744–756. [CrossRef] [PubMed]
3. Ustawa z Dnia 27 Sierpnia 2004 r. o Świadczeniach Opieki Zdrowotnej Finansowanych ze Środków Publicznych. The Law of 27 August 2004 on Health Care Services Financed from Public Resources. *Prime Minist. Repub. Pol.* **2021**, *1285*, 1292–1559.
4. Ortiz-Cuaran, S.; Bouaoud, J.; Karabajakian, A.; Fayette, J.; Saintigny, P. Precision Medicine Approaches to Overcome Resistance to Therapy in Head and Neck Cancers. *Front. Oncol.* **2021**, *11*, 614332. [CrossRef]
5. Blanco, R.; Carrillo-Beltrán, D.; Corvalán, A.H.; Aguayo, F. High-Risk Human Papillomavirus and Epstein-Barr Virus Coinfection: A Potential Role in Head and Neck Carcinogenesis. *Biology* **2021**, *10*, 1232. [CrossRef]
6. Żurek, M.; Jasak, K.; Jaros, K.; Daniel, P.; Niemczyk, K.; Rzepakowska, A. Clinico-Epidemiological Analysis of Most Prevalent Parotid Gland Carcinomas in Poland over a 20-Year Period. *Int. J. Environ. Res. Public Health* **2022**, *19*, 10247. [CrossRef]
7. Du, E.; Mazul, A.L.; Farquhar, D.; Brennan, P.; Anantharaman, D.; Abedi-Ardekani, B.; Weissler, M.C.; Hayes, D.N.; Olshan, A.F.; Zevallos, J.P. Long-term Survival in Head and Neck Cancer: Impact of Site, Stage, Smoking, and Human Papillomavirus Status. *Laryngoscope* **2019**, *129*, 2506–2513. [CrossRef]
8. Rosenberg, A.J.; Vokes, E.E. Optimizing Treatment De-Escalation in Head and Neck Cancer: Current and Future Perspectives. *Oncologist* **2021**, *26*, 40–48. [CrossRef]
9. Menon, S.S.; Guruvayoorappan, C.; Sakthivel, K.M.; Rasmi, R.R. Ki-67 protein as a tumour proliferation marker. *Clin. Chim. Acta* **2019**, *491*, 39–45. [CrossRef]
10. Li, L.T.; Jiang, G.; Chen, Q.; Zheng, J.N. Ki67 is a promising molecular target in the diagnosis of cancer (review). *Mol. Med. Rep.* **2015**, *11*, 1566–1572. [CrossRef]
11. Luo, Z.W.; Zhu, M.G.; Zhang, Z.Q.; Ye, F.J.; Huang, W.H.; Luo, X.Z. Increased expression of Ki-67 is a poor prognostic marker for colorectal cancer patients: A meta analysis. *BMC Cancer* **2019**, *19*, 123. [CrossRef] [PubMed]
12. Andrés-Sánchez, N.; Fisher, D.; Krasinska, L. Physiological functions and roles in cancer of the proliferation marker Ki-67. *J. Cell Sci.* **2022**, *135*, jcs258932. [CrossRef] [PubMed]
13. Gown, A.M. The Biomarker Ki-67: Promise, Potential, and Problems in Breast Cancer. *Appl. Immunohistochem. Mol. Morphol. AIMM* **2023**, *31*, 478–484. [CrossRef] [PubMed]
14. Zhao, R.; Choi, B.Y.; Lee, M.H.; Bode, A.M.; Dong, Z. Implications of Genetic and Epigenetic Alterations of CDKN2A (p16(INK4a)) in Cancer. *EBioMedicine* **2016**, *8*, 30–39. [CrossRef] [PubMed]
15. Hanahan, D.; Weinberg, R.A. Hallmarks of cancer: The next generation. *Cell* **2011**, *144*, 646–674. [CrossRef]
16. Yap, K.L.; Li, S.; Munoz-Cabello, A.M.; Raguz, S.; Zeng, L.; Mujtaba, S.; Gil, J.; Walsh, M.J.; Zhou, M.M. Molecular interplay of the noncoding RNA ANRIL and methylated histone H3 lysine 27 by polycomb CBX7 in transcriptional silencing of INK4a. *Mol. Cell* **2010**, *38*, 662–674. [CrossRef]
17. Kuwabara, T.; Hiyama, T.; Tanaka, S.; Yoshihara, M.; Arihiro, K.; Chayama, K. Genetic pathways of multiple esophageal squamous cell carcinomas. *Oncol. Rep.* **2011**, *25*, 453–459.
18. Tam, K.W.; Zhang, W.; Soh, J.; Stastny, V.; Chen, M.; Sun, H.; Thu, K.; Rios, J.J.; Yang, C.; Marconett, C.N.; et al. CDKN2A/p16 inactivation mechanisms and their relationship to smoke exposure and molecular features in non-small-cell lung cancer. *J. Thorac. Oncol.* **2013**, *8*, 1378–1388. [CrossRef]

19. Bhagat, R.; Kumar, S.S.; Vaderhobli, S.; Premalata, C.S.; Pallavi, V.R.; Ramesh, G.; Krishnamoorthy, L. Epigenetic alteration of p16 and retinoic acid receptor beta genes in the development of epithelial ovarian carcinoma. *Tumour. Biol.* **2014**, *35*, 9069–9078. [CrossRef]
20. Kostaki, M.; Manona, A.D.; Stavraka, I.; Korkolopoulou, P.; Levidou, G.; Trigka, E.A.; Christofidou, E.; Champsas, G.; Stratigos, A.J.; Katsambas, A.; et al. High-frequency p16(INK) (4A) promoter methylation is associated with histone methyltransferase SETDB1 expression in sporadic cutaneous melanoma. *Exp. Dermatol.* **2014**, *23*, 332–338. [CrossRef]
21. Robaina, M.C.; Faccion, R.S.; Arruda, V.O.; De Rezende, L.M.; Vasconcelos, G.M.; Apa, A.G.; Bacchi, C.E.; Klumb, C.E. Quantitative analysis of CDKN2A methylation, mRNA, and p16(INK4a) protein expression in children and adolescents with Burkitt lymphoma: Biological and clinical implications. *Leuk. Res.* **2015**, *39*, 248–256. [CrossRef] [PubMed]
22. Chung, C.H.; Guthrie, V.B.; Masica, D.L.; Tokheim, C.; Kang, H.; Richmon, J.; Agrawal, N.; Fakhry, C.; Quon, H.; Subramaniam, R.M.; et al. Genomic alterations in head and neck squamous cell carcinoma determined by cancer gene-targeted sequencing. *Ann. Oncol.* **2015**, *26*, 1216–1223. [CrossRef]
23. Amin, M.B.; Greene, F.L.; Edge, S.B.; Compton, C.C.; Gershenwald, J.E.; Brookland, R.K.; Meyer, L.; Gress, D.M.; Byrd, D.R.; Winchester, D.P. The Eighth Edition AJCC Cancer Staging Manual: Continuing to build a bridge from a population-based to a more "personalized" approach to cancer staging. *CA Cancer J. Clin.* **2017**, *67*, 93–99. [CrossRef] [PubMed]
24. Knudsen, E.S.; Pruitt, S.C.; Hershberger, P.A.; Witkiewicz, A.K.; Goodrich, D.W. Cell cycle and beyond: Exploiting new RB1 controlled mechanisms for cancer therapy. *Trends Cancer* **2019**, *5*, 308–324. [CrossRef] [PubMed]
25. Knudsen, E.S.; Kumarasamy, V.; Nambiar, R.; Pearson, J.D.; Vail, P.; Rosenheck, H.; Wang, J.; Eng, K.; Bremner, R.; Schramek, D.; et al. CDK/cyclin dependencies define extreme cancer cell-cycle heterogeneity and collateral vulnerabilities. *Cell Rep.* **2022**, *38*, 110448. [CrossRef]
26. Uxa, S.; Castillo-Binder, P.; Kohler, R.; Stangner, K.; Müller, G.A.; Engeland, K. Ki-67 gene expression. *Cell Death Differ.* **2021**, *28*, 3357–3370. [CrossRef]
27. Stransky, N.; Egloff, A.M.; Tward, A.D.; Kostic, A.D.; Cibulskis, K.; Sivachenko, A.; Kryukov, G.V.; Lawrence, M.S.; Sougnez, C.; McKenna, A.; et al. The mutational landscape of head and neck squamous cell carcinoma. *Science* **2011**, *333*, 1157–1160. [CrossRef]
28. Chai, A.W.Y.; Lim, K.P.; Cheong, S.C. Translational genomics and recent advances in oral squamous cell carcinoma. *Semin. Cancer Biol.* **2020**, *61*, 71–83. [CrossRef]
29. Adib, E.; Nassar, A.H.; Akl, E.W.; Abou Alaiwi, S.; Nuzzo, P.V.; Mouhieddine, T.H.; Sonpavde, G.; Haddad, R.I.; Mouw, K.W.; Giannakis, M.; et al. CDKN2A Alterations and Response to Immunotherapy in Solid Tumors. *Clin. Cancer Res. Off. J. Am. Assoc. Cancer Res.* **2021**, *27*, 4025–4035. [CrossRef]
30. Padhi, S.S.; Roy, S.; Kar, M.; Saha, A.; Roy, S.; Adhya, A.; Baisakh, M.; Banerjee, B. Role of CDKN2A/p16 expression in the prognostication of oral squamous cell carcinoma. *Oral Oncol.* **2017**, *73*, 27–35. [CrossRef]
31. Lechner, M.; Chakravarthy, A.R.; Walter, V.; Masterson, L.; Feber, A.; Jay, A.; Weinberger, P.M.; McIndoe, R.A.; Forde, C.T.; Chester, K.; et al. Frequent HPV-independent p16/INK4A overexpression in head and neck cancer. *Oral Oncol.* **2018**, *83*, 32–37. [CrossRef] [PubMed]
32. Plath, M.; Broglie, M.A.; Forbs, D.; Stoeckli, S.J.; Jochum, W. Prognostic significance of cell cycle-associated proteins p16, pRB, cyclin D1 and p53 in resected oropharyngeal carcinoma. *J. Otolaryngol. Head Neck Surg.* **2018**, *47*, 53. [CrossRef] [PubMed]
33. Vairaktaris, E.; Yapijakis, C.; Psyrri, A.; Spyridonidou, S.; Yannopoulos, A.; Lazaris, A.; Vassiliou, S.; Ferekidis, E.; Vylliotis, A.; Nkenke, E.; et al. Loss of tumour suppressor p16 expression in initial stages of oral oncogenesis. *Anticancer. Res.* **2007**, *27*, 979–984. [PubMed]
34. Schwarz, S.; Bier, J.; Driemel, O.; Reichert, T.E.; Hauke, S.; Hartmann, A.; Brockhoff, G. Losses of 3p14 and 9p21 as shown by fluorescence in situ hybridization are early events in tumorigenesis of oral squamous cell carcinoma and already occur in simple keratosis. *Cytom. Part A J. Int. Soc. Anal. Cytol.* **2008**, *73*, 305–311. [CrossRef] [PubMed]
35. Pérez-Sayáns, M.; Suárez-Peñaranda, J.M.; Padín-Iruegas, M.E.; Gayoso-Diz, P.; Reis-De Almeida, M.; Barros-Angueira, F.; Gándara-Vila, P.; Blanco-Carrión, A.; García-García, A. The Loss of p16 Expression Worsens the Prognosis of OSCC. *Appl. Immunohistochem. Mol. Morphol. AIMM* **2015**, *23*, 724–732. [CrossRef]
36. Mroz, E.A.; Rocco, J.W. Intra-tumor heterogeneity in head and neck cancer and its clinical implications. *World J. Otorhinolaryngol. Head Neck Surg.* **2016**, *2*, 60–67. [CrossRef]
37. Ahmed, M.W.; Kayani, M.A.; Shabbir, G.; Ali, S.M.; Shinwari, W.U.; Mahjabeen, I. Expression of PTEN and its correlation with proliferation marker Ki-67 in head and neck cancer. *Int. J. Biol. Markers* **2016**, *31*, 193–203. [CrossRef]
38. Poage, G.M.; Houseman, E.A.; Christensen, B.C.; Butler, R.A.; Avissar-Whiting, M.; McClean, M.D.; Waterboer, T.; Pawlita, M.; Marsit, C.J.; Kelsey, K.T. Global hypomethylation identifies Loci targeted for hypermethylation in head and neck cancer. *Clin. Cancer Res. Off. J. Am. Assoc. Cancer Res.* **2011**, *17*, 3579–3589. [CrossRef]
39. Campbell, J.D.; Yau, C.; Bowlby, R.; Liu, Y.; Brennan, K.; Fan, H.; Taylor, A.M.; Wang, C.; Walter, V.; Akbani, R.; et al. Genomic, Pathway Network, and Immunologic Features Distinguishing Squamous Carcinomas. *Cell Rep.* **2018**, *23*, 194–212.e6. [CrossRef]
40. Soares-Lima, S.C.; Mehanna, H.; Camuzi, D.; de Souza-Santos, P.T.; Simão, T.A.; Nicolau-Neto, P.; Almeida Lopes, M.S.; Cuenin, C.; Talukdar, F.R.; Batis, N.; et al. Upper Aerodigestive Tract Squamous Cell Carcinomas Show Distinct Overall DNA Methylation Profiles and Different Molecular Mechanisms behind WNT Signaling Disruption. *Cancers* **2021**, *13*, 3014. [CrossRef]
41. Preethi, P.L.; Rao, S.R.; Madapusi, B.T.; Narasimhan, M. Immunolocalization of Ki-67 in different periodontal conditions. *J. Indian Soc. Periodontol.* **2014**, *18*, 161–165. [CrossRef] [PubMed]

42. Feller, L.; Chandran, R.; Khammissa, R.A.; Meyerov, R.; Lemmer, J. Alcohol and oral squamous cell carcinoma. *SADJ J. S. Afr Dent. Assoc. Tydskr. Van Die Suid-Afr. Tandheelkd. Ver.* **2013**, *68*, 176–180.
43. Ferraguti, G.; Terracina, S.; Petrella, C.; Greco, A.; Minni, A.; Lucarelli, M.; Agostinelli, E.; Ralli, M.; de Vincentiis, M.; Raponi, G.; et al. Alcohol and Head and Neck Cancer: Updates on the Role of Oxidative Stress, Genetic, Epigenetics, Oral Microbiota, Antioxidants, and Alkylating Agents. *Antioxidants* **2022**, *11*, 145. [CrossRef] [PubMed]
44. Le, T.D.; Do, T.A.; Yu, R.; Yoo, H. Ethanol elicits inhibitory effect on the growth and proliferation of tongue carcinoma cells by inducing cell cycle arrest. *Korean J. Physiol. Pharmacol. Off. J. Korean Physiol. Soc. Korean Soc. Pharmacol.* **2012**, *16*, 153–158. [CrossRef]
45. Liu, Y.; Chen, H.; Sun, Z.; Chen, X. Molecular mechanisms of ethanol-associated oro-esophageal squamous cell carcinoma. *Cancer Lett.* **2015**, *361*, 164–173. [CrossRef]
46. Reidy, J.; McHugh, E.; Stassen, L.F. A review of the relationship between alcohol and oral cancer. *Surgeon* **2011**, *9*, 278–283. [CrossRef]
47. Waszkiewicz, N.; Zalewska, A.; Szulc, A.; Kepka, A.; Konarzewska, B.; Zalewska-Szajda, B.; Chojnowska, S.; Waszkiel, D.; Zwierz, K. Wpływ alkoholu na jame ustna, ślinianki oraz śline [The influence of alcohol on the oral cavity, salivary glands and saliva]. *Pol. Merkur. Lek.* **2011**, *30*, 69–74.
48. Ahmad, A.; Banerjee, S.; Wang, Z.; Kong, D.; Majumdar, A.P.; Sarkar, F.H. Aging and inflammation: Etiological culprits of cancer. *Curr. Aging Sci.* **2009**, *2*, 174–186. [CrossRef]
49. Jing, Y.; Zhou, Q.; Zhu, H.; Zhang, Y.; Song, Y.; Zhang, X.; Huang, X.; Yang, Y.; Ni, Y.; Hu, Q. Ki-67 is an independent prognostic marker for the recurrence and relapse of oral squamous cell carcinoma. *Oncol. Lett.* **2019**, *17*, 974–980. [CrossRef]
50. Wang, T.; Xue, L.; Li, Z.; Hong, Z.; Hu, N.; Li, Y.; Yan, B. A novel nomogram model based on Ki-67 characteristic expression to predict prognosis in head and neck squamous cell carcinoma. *Front. Oncol.* **2024**, *14*, 1376498. [CrossRef]
51. Deyhimi, P.; Torabinia, N.; Torabinia, A. A comparative study of histological grade and expression of Ki67 protein in oral squamous cell carcinoma in young and old patients. *Dent. Res. J.* **2013**, *10*, 514–517.
52. Loeschke, S.; Ohlmann, A.K.; Bräsen, J.H.; Holst, R.; Warnke, P.H. Prognostic value of HMGA2, P16, and HPV in oral squamous cell carcinomas. *J. Cranio-Maxillo-Facial Surg. Off. Publ. Eur. Assoc. Cranio-Maxillo-Facial Surg.* **2016**, *44*, 1422–1429. [CrossRef] [PubMed]
53. Wang, Y.; Zhou, C.; Li, T.; Luo, J. Prognostic value of CDKN2A in head and neck squamous cell carcinoma via pathomics and machine learning. *J. Cell. Mol. Med.* **2024**, *28*, e18394. [CrossRef] [PubMed]
54. Smith, E.M.; Wang, D.; Kim, Y.; Rubenstein, L.M.; Lee, J.H.; Haugen, T.H.; Turek, L.P. P16INK4a expression, human papillomavirus, and survival in head and neck cancer. *Oral Oncol.* **2008**, *44*, 133–142. [CrossRef]
55. Shah, N.G.; Trivedi, T.I.; Tankshali, R.A.; Goswami, J.V.; Jetly, D.H.; Shukla, S.N.; Shah, P.M.; Verma, R.J. Prognostic significance of molecular markers in oral squamous cell carcinoma: A multivariate analysis. *Head Neck* **2009**, *31*, 1544–1556. [CrossRef]
56. Gröbe, A.; Hanken, H.; Kluwe, L.; Schöllchen, M.; Tribius, S.; Pohlenz, P.; Clauditz, T.; Grob, T.; Simon, R.; Sauter, G.; et al. Immunohistochemical analysis of p16 expression, HPV infection and its prognostic utility in oral squamous cell carcinoma. *J. Oral Pathol. Med. Off. Publ. Int. Assoc. Oral Pathol. Am. Acad. Oral Pathol.* **2013**, *42*, 676–681. [CrossRef]
57. Wang, Y.; Wang, X.; Zhang, Y.; Yu, L.; Zhu, B.; Wu, S.; Wang, D. Vasculogenic mimicry and expression of ALDH1, Beclin1, and p16 correlate with metastasis and prognosis in oral squamous cell carcinoma. *Int. J. Clin. Exp. Pathol.* **2018**, *11*, 1599.
58. Dong, Y.; Wang, J.; Dong, F.; Wang, X.; Zhang, Y. The correlations between alteration of p16 gene and clinicopathological factors and prognosis in squamous cell carcinomas of the buccal mucosa. *J. Oral Pathol. Med.* **2012**, *41*, 463–469. [CrossRef]
59. Yakushiji, T.; Noma, H.; Shibahara, T.; Arai, K.; Yamamoto, N.; Tanaka, C.; Uzawa, K.; Tanzawa, H. Analysis of a role for p16/CDKN2 expression and methylation patterns in human oral squamous cell carcinoma. *Bull. Tokyo Dent. Coll.* **2001**, *42*, 159–168. [CrossRef]
60. Ignatov, A.; Bischoff, J.; Schwarzenau, C.; Krebs, T.; Kuester, D.; Herrmann, K.; Costa, S.D.; Roessner, A.; Semczuk, A.; Schneider-Stock, R. P16 alterations increase the metastatic potential of endometrial carcinoma. *Gynecol. Oncol.* **2008**, *111*, 365–371. [CrossRef]
61. Goto, T.; Mizukami, H.; Shirahata, A.; Yokomizo, K.; Kitamura, Y.H.; Sakuraba, K.; Saito, M.; Ishibashi, K.; Kigawa, G.; Nemoto, H.; et al. Methylation of the p16 gene is frequently detected in lymphatic-invasive gastric cancer. *Anticancer Res.* **2010**, *30*, 2701–2703.
62. Barekati, Z.; Radpour, R.; Lu, Q.; Bitzer, J.; Zheng, H.; Toniolo, P.; Lenner, P.; Zhong, X.Y. Methylation signature of lymph node metastases in breast cancer patients. *BMC Cancer* **2012**, *12*, 244. [CrossRef] [PubMed]
63. Krupar, R.; Hartl, M.; Wirsching, K.; Dietmaier, W.; Strutz, J.; Hofstaedter, F. Comparison of HPV prevalence in HNSCC patients with regard to regional and socioeconomic factors. *Eur. Arch. Oto-Rhino-Laryngol. Off. J. Eur. Fed. Oto-Rhino-Laryngol. Soc. (EUFOS) Affil. Ger. Soc. Oto-Rhino-Laryngol. Head Neck Surg.* **2014**, *271*, 1737–1745. [CrossRef] [PubMed]
64. Shyamsundar, V.; Thangaraj, S.V.; Krishnamurthy, A.; Vimal, S.; Kesavan, P.; Babu, A.; Kmk, M.; Ramshankar, V. Exome Sequencing with Validations and Expression of p16/CDKN2A Shows no Association with HPV in Oral Cancers. *Asian Pac. J. Cancer Prev.* **2022**, *23*, 191–200. [CrossRef] [PubMed]
65. Wamsley, N.T.; Wilkerson, E.M.; Guan, L.; LaPak, K.M.; Schrank, T.P.; Holmes, B.J.; Sprung, R.W.; Gilmore, P.E.; Gerndt, S.P.; Jackson, R.S.; et al. Targeted Proteomic Quantitation of NRF2 Signaling and Predictive Biomarkers in HNSCC. *Mol. Cell. Proteom. MCP* **2023**, *22*, 100647. [CrossRef]

66. Rajagopalan, P.; Nanjappa, V.; Patel, K.; Jain, A.P.; Mangalaparthi, K.K.; Patil, A.H.; Nair, B.; Mathur, P.P.; Keshava Prasad, T.S.; Califano, J.A.; et al. Role of protein kinase N2 (PKN2) in cigarette smoke-mediated oncogenic transformation of oral cells. *J. Cell Commun. Signal.* **2018**, *12*, 709–721. [CrossRef]
67. Wahyuningsih, L.; Dwianingsih, E.K.; Risanti, E.D.; Tirtoprodjo, P.; Rinonce, H.T.; Hakim, F.A.; Herdini, C.; Fachiroh, J. Tissue P16 is Associated with Smoking Status Among Indonesian Nasopharyngeal Carcinoma Subjects. *Asian Pac. J. Cancer Prev. APJCP* **2019**, *20*, 2125–2130. [CrossRef]
68. Zhu, X.; Li, J.; Deng, S.; Yu, K.; Liu, X.; Deng, Q.; Sun, H.; Zhang, X.; He, M.; Guo, H.; et al. Genome-Wide Analysis of DNA Methylation and Cigarette Smoking in a Chinese Population. *Environ. Health Perspect.* **2016**, *124*, 966–973. [CrossRef]
69. Shao, Y.; Jiang, H.; Wu, X.; Luo, Y.; Tang, W. p16 promoter hypermethylation is associated with increased risk of nasopharyngeal carcinoma. *Mol. Clin. Oncol.* **2014**, *2*, 1121–1124. [CrossRef]
70. Lee, K.W.K.; Pausova, Z. Cigarette smoking and DNA methylation. *Front. Genet.* **2013**, *4*, 132. [CrossRef]
71. Rungraungrayabkul, D.; Panpradit, N.; Lapthanasupkul, P.; Kitkumthorn, N.; Klanrit, P.; Subarnbhesaj, A.; Sresumatchai, V.; Klongnoi, B.; Khovidhunkit, S.P. Detection of Human Papillomavirus and p16INK4a Expression in Thai Patients with Oral Squamous Cell Carcinoma. *Head Neck Pathol.* **2022**, *16*, 444–452. [CrossRef] [PubMed]
72. Machiels, J.P.; René Leemans, C.; Golusinski, W.; Grau, C.; Licitra, L.; Gregoire, V. Squamous cell carcinoma of the oral cavity, larynx, oropharynx and hypopharynx: EHNS–ESMO–ESTRO Clinical Practice Guidelines for diagnosis, treatment and follow-up. *Ann. Oncol.* **2020**, *31*, 1462–1475. [CrossRef] [PubMed]
73. Vrinceanu, D.; Dumitru, M.; Bratiloveanu, M.; Marinescu, A.; Serboiu, C.; Manole, F.; Palade, D.O.; Costache, A.; Costache, M.; Patrascu, O. Parotid Gland Tumors: Molecular Diagnostic Approaches. *Int. J. Mol. Sci.* **2024**, *25*, 7350. [CrossRef] [PubMed]
74. Riese, U.; Dahse, R.; Fiedler, W.; Theuer, C.; Koscielny, S.; Ernst, G.; Beleites, E.; Claussen, U.; von Eggeling, F. Tumor suppressor gene p16 (CDKN2A) mutation status and promoter inactivation in head and neck cancer. *Int. J. Mol. Med.* **1999**, *4*, 61–65. [CrossRef]
75. Calmon, M.F.; Colombo, J.; Carvalho, F.; Souza, F.; Filho, J.F.; Fukuyama, E.E.; Camargo, A.A.; Caballero, O.L.; Tajara, E.H.; Cordeiro, J.A.; et al. Methylation profile of genes CDKN2A (p14 and p16), DAPK1, CDH1, and ADAM23 in head and neck cancer. *Cancer Genet. Cytogenet.* **2007**, *173*, 31–37. [CrossRef]
76. Kuroki, M.; Iinuma, R.; Okuda, H.; Terazawa, K.; Shibata, H.; Mori, K.I.; Ohashi, T.; Makiyama, A.; Futamura, M.; Miyazaki, T.; et al. Comprehensive Genome profile testing in head and neck cancer. *Auris Nasus Larynx* **2023**, *50*, 952–959. [CrossRef]
77. Shen, A.; Ye, Y.; Chen, F.; Xu, Y.; Zhang, Z.; Zhao, Q.; Zeng, Z.L. Integrated multi-omics analysis identifies CD73 as a prognostic biomarker and immunotherapy response predictor in head and neck squamous cell carcinoma. *Front. Immunol.* **2022**, *13*, 969034. [CrossRef]
78. Deneka, A.Y.; Baca, Y.; Serebriiskii, I.G.; Nicolas, E.; Parker, M.I.; Nguyen, T.T.; Xiu, J.; Korn, W.M.; Demeure, M.J.; Wise-Draper, T.; et al. Association of TP53 and CDKN2A Mutation Profile with Tumor Mutation Burden in Head and Neck Cancer. *Clin. Cancer Res. Off. J. Am. Assoc. Cancer Res.* **2022**, *28*, 1925–1937. [CrossRef]
79. Dragomir, L.P.; Simionescu, C.; Mărgăritescu, C.; Stepan, A.; Dragomir, I.M.; Popescu, M.R. P53, p16 and Ki67 immunoexpression in oral squamous carcinomas. *Rom. J. Morphol. Embryol. = Rev. Roum. Morphol. Embryol.* **2012**, *53*, 89–93.

MDPI AG
Grosspeteranlage 5
4052 Basel
Switzerland
Tel.: +41 61 683 77 34

Current Issues in Molecular Biology Editorial Office
E-mail: cimb@mdpi.com
www.mdpi.com/journal/cimb